Virtual Futures for Design, Construction & Procurement

Virtual Futures for Design, Construction & Procurement

Edited by

Peter Brandon
Director of Salford University Think Lab and
Centre for Virtual Environment and Director Strategic Programmes
School of the Built Environment
University of Salford, UK

Tuba Kocatürk
Coordinator of Digital Design Programmes in
Architecture Co-chair FFD International (IASS)
School of the Built Environment
University of Salford, UK

Foreword by

William J. Mitchell
Massachusetts Institute of Technology

Blackwell
Publishing

© 2008 by Blackwell Publishing Ltd

Blackwell Publishing editorial offices:
Blackwell Science Ltd, 9600 Garsington Road, Oxford OX4 2DQ, UK
   Tel: +44 (0) 1865 776868
Blackwell Publishing Inc., 350 Main Street, Malden, MA 02148-5020, USA
   Tel: +1 781 388 8250
Blackwell Science Asia Pty Ltd, 550 Swanston Street, Carlton, Victoria 3053, Australia
   Tel: +61 (0)3 8359 1011

First published 2008

ISBN: 978-1-4051-7024-6

Library of Congress Cataloging-in-Publication Data

Virtual futures for design, construction & procurement / edited by Peter Brandon, Tuba Kocatürk.
     p. cm.
  Includes bibliographical references and index.
  ISBN-13: 978-1-4051-7024-6 (hardback : alk. paper)
  ISBN-10: 1-4051-7024-7 (hardback : alk. paper)
  1. Building–Data processing.   I. Brandon, P. S. (Peter S.)   II. Kocatürk, Tuba.

TH153.V55   2008
690.0285–dc22
2007039586

A catalogue record for this title is available from the British Library

Set in 9.5/12 pt Palatino by Aptara Inc., New Delhi, India
Printed and bound in Singapore by C.O.S Printers Pte Ltd

For further information on Blackwell Publishing, visit our website:
www.blackwellpublishing.com

Contents

Note on editors

Professor Peter Brandon, DSc, DEng, MSc (Arch), FRICS, ASAQS Director of Salford University Think Lab and Director Strategic Programmes, School of the Built Environment, University of Salford, UK.

Professor Brandon is a former Pro-Vice-Chancellor for research at the University of Salford, the only UK University to be awarded a 6-star rating in the Built Environment within the independent UK research assessment exercise. His research interests range across construction economics and management, information- and knowledge-based systems for construction and, more recently, sustainable development. He has published widely, including 17 books as author, co-author and editor plus over 250 papers worldwide. Several of the outputs of his research have resulted in commercial projects.

He has played a significant role in UK Construction Research Policy, including serving as Chairman of the UK Science and Engineering Research Council Panel for Construction and as Chairman of the UK Research Assessment Exercise Panel for the Built Environment (1996 and 2001).

Dr Tuba Kocatürk, (PhD, MSc, BArch) is an architect, researcher and a lecturer in Architectural Design and Technology. She is the Programme Coordinator, Digital Design Programmes in Architecture, School of the Built Environment, University of Salford, UK.

She co-founded and is current Co-chair of the Free-Form Design Sub-working Group of the IASS (International Association for Shell and Spatial Structures). Her main research interests are digital design tools and processes, CAAD/CAE/CAM process integration, cognitive models in digital design education and design knowledge modelling. She has given various lectures and conducted workshops at international level.

Contributors

Professor Mustafa Alshawi
Director, Research Institute for the Built and Human Environment
University of Salford, UK

Professor Chimay J. Anumba
Professor of Construction Engineering and Informatics, and Director, Centre for
Innovative and Collaborative Engineering
Loughborough University, UK

Professor Ghassan Aouad
Dean, Faculty of Business, Law and the Built Environment
University of Salford, UK

Dr Zeeshan Aziz
Research Associate
Loughborough University, UK

Professor Matthew Bacon, FRSA
Chief Technology Officer, Integrated FM Limited
Warwick, UK

Professor Andrew Baldwin
Professor of Construction Management
Loughborough University, UK

Ms Simona Barresi
Academic Fellow, Informatics Research Institute
University of Salford, UK

Professor Richard J. Boland, Jr
Professor of Information Systems, Weatherhead School of Management
Case Western Reserve University, USA; and
Judge Business School
University of Cambridge, UK

Professor Peter Brandon
Director of Salford University Think Lab and Director Strategic Programmes,
School of the Built Environment
University of Salford, UK

Dr Richard A. Buswell
Lecturer, Department of Civil and Building Engineering
Loughborough University, UK

Professor Fred Collopy
Professor of Information Systems, Weatherhead School of Management,
Case Western Reserve University, USA

Mr David Conover
Senior Advisor
International Code Council, USA

Professor Grahame S. Cooper
Professor of Applied Information Systems Engineering, School of Computing,
Science and Engineering, Informatics Research Institute
The University of Salford, UK

Professor Nashwan Dawood
Professor of Construction Management & IT, and Director of the Centre for
Construction Research and Innovation (CCIR), School of Science and Technology
University of Teesside, UK

Mr John Dean
Research Fellow, Centre for Construction Innovation & Research
University of Teesside, UK

Professor Robin Drogemuller
Professor of Digital Design
Queensland University of Technology, Australia

Professor Terrence Fernando
Director of Future Workspaces Research Centre
University of Salford, UK

Professor Martin Fischer
Professor, Civil & Environmental Engineering, and Director of the Center for
Integrated Facility Engineering (CIFE)
Stanford University, USA

Professor Alistair G. F. Gibb
Professor of Construction Engineering Management, Department of Civil and
Building Engineering
Loughborough University, UK

Professor Julia Grant
Associate Professor of Accountancy, Weatherhead School of Management
Case Western Reserve University, USA

Professor Dipl-Ing Manfred Grohmann
Principal, B+G Ingenieure Bollinger und Grohmann GmbH; and
Professor, University of Kassel, Germany

Mr H. L. Guo
PhD Student, Department of Building and Real Estate
The Hong Kong Polytechnic University, Hong Kong

Mr T. Huang
PhD Student, Department of Building and Real Estate
The Hong Kong Polytechnic University, Hong Kong

Dr Tuba Kocatürk
Programme Co-ordinator, Digital Architectural Design and Digital Design Initiatives
School of the Built Environment
University of Salford, UK

Mr C. W. Kong
Tutor, Department of Building and Real Estate
The Hong Kong Polytechnic University, Hong Kong

Dr Angela Lee
Programme Director, Architectural Design & Technology, School of the
Built Environment
University of Salford, UK

Professor Heng Li
Professor, Department of Building and Real Estate
The Hong Kong Polytechnic University, Hong Kong

Dr Ramesh Marasini
Research Fellow, Centre for Construction Innovation & Research
University of Teesside, UK

Mr Nicholas Nisbet
Consultant
AEC3 UK Ltd, UK

Professor Rivka Oxman
Professor of Digital Architectural Design and Director of Digital Design Initiatives
School of the Built Environment
University of Salford, UK

Professor Yacine Rezgui
Professor of Applied Informatics, Informatics Research Institute
University of Salford, UK

Mr Martin Riese, BArch OAA AIA
Director
GT Asia Limited, Hong Kong

Mr Martin Simpson
Associate Director
Arup, UK

Dr Rupert C. Soar
Senior Lecturer, Wolfson School of Mechanical and Manufacturing Engineering
Loughborough University, UK

Dr Souheil Soubra
Head of Division
CSTB (Centre Scientifique et Technique du Bâtiment), France

Professor Joseph H. M. Tah
Professor in Project Management, Oxford Institute for Sustainable Development
Oxford Brookes University, UK

Dipl-Ing Oliver Tessmann
Research Associate
University of Kassel, Germany

Professor Anthony Thorpe
Professor of Construction Information Technology, Department of Civil and
Building Engineering
Loughborough University, UK

Mr Jeffrey Wix
Consultant
AEC3 UK Ltd, UK

Dr K. D. Wong
Associate Professor, Department of Building and Real Estate
The Hong Kong Polytechnic University, Hong Kong

Dr Song Wu
Lecturer, School of the Built Environment
University of Salford, UK

Professor Lin Zhao
Assistant Professor of Computer and Information Science
Arkansas Tech University, USA

Note on Think Lab

THINKlab

This book arises from debate within the internationally leading University of Salford 'Think Lab'. This state-of-the-art facility has been developed for research into Information and Communication Technologies in many fields, including design and construction. It provides a forum for leading figures across the world to participate, both in person and through virtual collaborative technologies, to discuss topics relating to future developments in ICTs applied to various topic areas. For further information visit www.thinklab.salford.ac.uk

Acknowledgements

The editors of this book wish to thank all who have contributed and who, in innumerable ways, have made this book possible. Firstly, to all the authors, for sharing their knowledge and insight, and without whom this book could not have been produced. We would also like to sincerely thank William J. Mitchell for his positive response to our invitation to write the Foreword.

We received great encouragement and also financial support from the RICS[1] Foundation for the organization of the Virtual Prototyping workshop (which initiated and then provided the motivation for the creation of this book), and we would like to acknowledge their prominent role in making this book happen. We would also like to thank Hanneke van Dijk for her invaluable contribution and hard work at every stage of the book, from its conception to its delivery. The editors and publisher gratefully acknowledge those who have granted permission to reproduce material in this book.

[1] The RICS Foundation is a charitable body established by the Royal Institution of Chartered Surveyors to seek to improve the quality of the built and natural environment.

Foreword – Virtual worlds, virtual prototypes and design

William J. Mitchell

Architects and other designers inhabit a curious borderland between the virtual and the physical. They have always been concerned with conjuring up things that don't exist but might, imagining them in detail, and eventually finding ways to translate these visions into physical reality. Over the last half-century, computer-generated virtual worlds have played an increasingly crucial role in this process.

The first virtual world I ever saw was called Spacewar. It ran on early DEC computers in the 1960s and 1970s, and was later reincarnated as one of the first arcade games – more accurately, a bar game, since there weren't yet any arcades. A Spacewar 'world' had two inhabitants. You controlled one spaceship moving in the gravitational field of a star, your opponent controlled another and you tried to shoot each other down with missiles. It was very simple – limited, of course, by the available processing power and graphic display capabilities – but it had the essentials: a simulated spatial environment, simulated physics, control of something that represented you and multiplayer interaction.

Four decades later, we have entered the era of Second Life – which, when I last logged on to check, claimed to have 4 523 218 residents. Second Life describes itself as 'a 3D online digital world imagined, created and owned by its residents'. It runs over the internet, and its operators just keep adding servers as it grows. It presents itself in coloured, shaded perspective, with real-time motion. Players control avatars that can interact via text messages, buy and sell virtual artefacts, acquire virtual real estate and design and build virtual things. Numerous organizations, such as the *Reuters* news agency, have built sites in Second Life to transact business there.

Furthermore, Second Life and its competitors are beginning to blur familiar categories. As virtual worlds rapidly expand their capabilities, they increasingly overlap the established and hitherto fairly distinct domains of videogames, computer animation, computer-aided design systems, geographic information systems, engineering simulation systems as well as remote interaction and collaboration systems. As a result, architects and others can now design for the physical world, for virtual worlds and for hybrids of the two. And they must begin to take careful account of the varied and complex relationships that may exist between the physical and the virtual.

Connections of the virtual and the physical

A virtual world may, for example, function as a *mirror* of a fragment of the physical – perhaps by means of movement sensors on dancers that drive the motions of avatars, or thermal sensors on buildings that drive 3D digital models showing energy flows.

In principle, this is much like live television, but it allows much greater flexibility in defining the functions that map from physical to virtual.

Alternatively, a virtual world may serve as a *utopia* – that is, an 'improved' version of the physical. This seems to be key to the appeal of Second Life. Tremendous effort has been devoted to making it look like the physical world, and to function like it in some carefully selected respects, but avatars are generally richer, thinner and better looking than their physical counterparts. It never gets cold or rains. And you don't get sick and die.

A virtual world may, of course, simply present a *fiction* – showing things that do not exist in the physical world, never did and never will. Second Life has a great many wholly fictional environments that function much like stage or screen sets, within which avatars act out roles.

Physical and virtual worlds may also operate in parallel, but with *cross-linked* dynamics. The flourishing economy of Second Life, for example, is cross-linked to the economy of the physical world through mechanisms for the exchange of Second Life dollars for US dollars. Taxation departments are beginning to take note of this.

Looking to the past, a virtual world may present a *reconstruction* of something physical that no longer exists – of Rome at the time of Julius Caesar, for example. Or it might present something that was once proposed but never made physical, such as Tatlin's tower for Saint Petersburg. In this case, it is presenting a *counterfactual conditional*: *if* the tower had been built, *then* this is what Saint Petersburg would be like.

Looking to the future instead, a virtual world might present a *prediction* or *conditional prediction*. Thus an architect may build a virtual version of a project and say to his/her client, in effect: '*If* you invest in making this physical, *then* this is what you will get.' More strongly, it could function as a *promise* to the client, and even establish a contract.

Construction contract documents for buildings once took the form of drawings and specifications, and served as *prescriptions* of what was to be done. More recently, 3D digital models created by CAD (computer aided design) and BIM (building information modelling) systems have played this role. As the distinctions between CAD systems and virtual worlds become less rigid, it is not hard to imagine virtual worlds becoming repositories of construction documents. Architects would deposit them there, and contractors wanting to bid on projects would need to visit them.

Of particular interest to designers are *virtual prototypes* of products, buildings and urban settings. These are digital representations of design proposals – ideas put forward for exploration, analysis and evaluation, with the intention that they might one day be physically realized if they stand up to scrutiny. The emerging practice of virtual prototyping is beginning to connect virtual worlds, in a powerful way, to the long tradition of prototyping in design and construction.

Traditions of prototyping

A prototype is usually thought of as a partial, approximate, or abstracted realization of a component or system that is constructed before the real thing in order to advance the design process. It differs, by virtue of its particular, antecedent, functional relationship to the real thing, from superficially similar artefacts such as movie-set fictional buildings, historical reconstructions and art works. Over the centuries, prototypes have taken diverse forms and have served a wide variety of specialized purposes.

A full-scale clay model of an automobile, for example, allows designers to sculpt surfaces, check form and appearance and, eventually, seek approval from management before proceeding to tooling and production. An experimental prototype version of a nuclear weapon can be tested, at some convenient desert location, to see if it works – and perhaps to announce a threat to potential targets. An example of a key structural component of a building can be set up and loaded to destruction to demonstrate that its counterparts will be safe in use. A pre-production motorcycle engine can be run in an instrumented acoustic chamber to determine exactly what sort of noise the proposed product will make and whether it is acceptable. A lightweight, non-functional mockup of a building façade, erected on site, can serve the political purposes of informing nearby residents about what is proposed and convincing them that it will be good for the neighbourhood. A running prototype of a new software product can be demonstrated to potential investors to convince them that it has a market.

The economics of prototyping

The goals of prototyping are, in some mix: to provide information needed for further refinement and development of a design; to identify any design errors or potential failure points; to provide a basis for choosing among options or deciding whether to proceed to the next stage; and (in a less scientific spirit) to persuade decision makers. Achieving these goals comes at a cost, and the art of prototyping is to produce the required information with minimum expenditure of time and resources. A prototype fails if it is too incomplete or inaccurate to yield the information required in the relevant context. On the other hand, it slows down the design process and wastes precious resources if it is unnecessarily elaborate and costly for its purpose.

The economic imperatives that frame prototyping practices vary with stage in the design process and type of end product. In particular, there is generally a progression from quick, approximate, inexpensive prototypes that provide rough guidance at early stages to more carefully crafted, detailed and costly prototypes at later stages. A product design might, for example, initially take the form of rapidly constructed foam or paper prototypes, proceed to carefully crafted wood or plaster and then – as confidence in the design and commitment to it grow – eventually develop to precisely fabricated, fully functional metal and plastic.

The length of the expected production run also makes a difference. Where many physical instances will be produced from a single design, as with automobiles, the cost of a prototype may be multiples of the cost of a finished product. But, where just one finished instance will be produced, as with a building, the cost of prototyping usually must be limited to a small fraction of the cost of the singular product.

Some of the information that can be gained from prototypes is of particularly high value. Where design failure would be catastrophic – perhaps resulting in injury or death – it makes sense to invest heavily in prototyping. But, where the consequences of failure would be less severe, this may not be so necessary.

In general, a designer contemplating the production of a prototype confronts a set of cost–benefit questions. What sort of information or rhetorical effect do I need? What is the quickest and cheapest way to get it? What effects will this have on schedule and budget? Will the benefits justify the costs?

The rise of the virtual

In the past, the only option was to create, test and evaluate physical prototypes. During the nineteenth and twentieth centuries, before they were largely displaced by computer methods, careful model testing techniques played an indispensable role in many design domains (Cowan, 1968). Where designers needed to economize in this sort of process, they resorted to scaling down from full size (for obvious reasons, a particularly common practice in architecture and urban design) and to substituting inexpensive, easily worked materials for the intended actual materials.

Over the last couple of decades, though, the ongoing development of information technology has made the alternative of virtual prototyping increasingly feasible and attractive. In the manufacturing industry, the virtual prototyping of the Boeing 777 in the early 1990s was a particularly significant milestone in the application of this new approach (Sabbagh, 1996). In this case, digital models of the aircraft were used extensively throughout the design process, and the first physical prototype was a fully functional one that flew. In architecture and construction, the steadily increasing adoption of three-dimensional, computer-aided design systems has generated growing interest in the virtual prototyping of buildings.

Sometimes a virtual prototype will be less expensive to create than its physical equivalent, but this is not necessarily the case; production of careful, detailed digital models can be time-consuming and extremely costly. But part of the attraction of virtual prototyping (at least in principle) is that digital models typically need to be created *anyway*, to serve as design documentation. So, why not build digital models that can support documentation, visualization, engineering analysis and various forms of performance simulation? Then, the cost of modelling will be justified by numerous, varied benefits. This is the old dream of integrated computer-aided design, which has been around since the 1970s (Mitchell, 1977), and is finally approaching fruition in current building information modelling (BIM) systems (Goldberg, 2006).

A second attraction of virtual prototypes is that they are (when properly structured) relatively easy to edit and change, and thus provide designers and their clients with an enhanced ability to explore ranges of variants on design concepts. Part of this capability comes from standard CAD editing tools, which are typically easier to apply to a digital model than, say, saws and chisels to a wooden model. Increasingly now, parametric models – which are built from the beginning with variation in mind – support this kind of exploration even more effectively. Parametric modelling is actually a very old idea – it was a key feature of Ivan Sutherland's pioneering *Sketchpad* CAD system in the early 1960s (Sutherland, 1963) – but it had little impact on design practice until, much more recently, commercial software vendors began to integrate useful parametric tools with CAD products, and to market them heavily.

A third attraction is that virtual prototypes enable the substitution of virtual experiments – conducted by applying analysis or simulation software to digital models – for the slow, painstaking, costly, sometimes dangerous process of instrumenting physical prototypes and running actual experiments to generate data. Instead of subjecting a physical model of a structure to load testing, you can now run a finite element analysis; instead of carefully instrumenting a physical model of an auditorium for acoustic testing, you can run an acoustic simulation; instead of placing a scale model of a building in an artificial sky, you can run a lighting simulation. The development of useful analysis or simulation software may be a slow and costly process, but once the software exists the marginal cost of each application is typically very low compared to that of physical modelling, instrumentation and experimentation. Furthermore, while physical prototyping and test facilities tend unavoidably to be expensive to create and maintain, and generally are not widely accessible, analysis and simulation software is, in principle, available to anyone who can download it. Hence, as our stock of analysis and simulation software grows, the advantages of virtual prototyping become increasingly overwhelming.

Overall, then, the effect of virtual prototyping has been both to expand the range of benefits to design processes that can be achieved through prototyping and to transform the economics of prototyping. Until now, this has been accomplished through the use of computer-aided design and engineering simulation software. But this may change as more general virtual worlds, such as Second Life, continue to grow, expand their modelling and simulation capabilities and attract investment. My guess is that designers will increasingly see virtual worlds as attractive outsourcing sites – places where the local conditions and economies make it quick, cheap and easy to build prototypes.

Translation, abstraction and elaboration

It is tempting to think of virtual prototyping as a replacement for physical prototyping, but in practice the two often coexist and are inextricably interconnected. Automobile design, for example, still entails the use of both clay and curved surface CAD models, and there is little sign of this changing. The reason is that the affordances of clay and CAD overlap to some extent, but do not mutually substitute. Clay models of automobile bodies are not precisely mathematically defined, but they are tactile – which means a lot to craftsmen who are used to receiving information through their hands, and they offer more to the experienced and sophisticated eye than the most advanced raytraced rendering. Conversely, CAD models *are* mathematically defined, and they support a wide variety of analyses, simulations and visualizations, but they abstract away from tactility. Clay and CAD, it turns out, are complementary.

This complementarity has been reinforced, in recent years, by the emergence of techniques for efficiently translating between physical and virtual prototypes. Laser scanners and three-dimensional digitizers can be used to convert physical models into corresponding digital ones, while computer-controlled, rapid prototyping devices – laser cutters, 3D printers, multi-axis milling machines and the like – render digital models into material, tactile form. In architecture, the office of Frank Gehry has been an important pioneer of design processes grounded in both physical and digital modelling of form, with the use of sophisticated techniques to translate back and forth (Mitchell, 2001).

These physical/virtual translations are special cases of the translations among representations that occur continually in most serious design processes – from two-dimensional drawings to physical scale models, from building plans to circulation network diagrams, from detailed drawings of buildings to simplified urban massing models, from point clouds' output by scanners to structured CAD models, from non-functional foam and cardboard mockups to functional metal prototypes, and so on. These translations are rarely mechanistic or neutral, but are occasions for exercising design judgement. In some cases the task is, for a particular design purpose, to abstract away from detail that is unnecessary and distracting in the current context. In other cases, it is to carry the design to a higher level of elaboration by adding detail and structure.

Design processes benefit from the availability of diverse prototyping techniques, together with tools and strategies for translating among prototypes by abstracting and elaborating. As a result, virtual prototyping techniques have had a complex and somewhat paradoxical effect on design practice. They have partially displaced more traditional, physical techniques whilst – simultaneously, and probably more importantly in the end – they have also complemented them and thus enhanced their value.

The future

At this point, the future of virtual prototyping seems very bright. The technologies that support it continue to develop rapidly, to migrate from research laboratories to commercial software products and to move from there to practice. There are promising efforts to reduce impediments to adoption and effective application by establishing useful standards and by moving away from closed, proprietary software systems towards more open-source frameworks for software development and dissemination. Increasingly stringent demands to assure life safety and sustainability will motivate architects and other designers to invest more time and resources in careful prototyping.

Concurrently, virtual worlds are rapidly growing in scale and sophistication, attracting investment to support research and development, and increasingly becoming sites for everything that can benefit, in some way, from taking place in online virtual environments – mirroring, role playing, reconstruction of the past and many useful forms of prototyping. This is shifting much of the burden of software tool development from the relatively small and specialized domains of computer-aided design and engineering simulation to a much larger and better-funded community. Furthermore, it is producing effective combinations of modelling and simulation with social interaction and exchange capabilities. It seems likely, then, that virtual prototyping will become a specialized interest of sub-communities within the larger communities supported by virtual worlds.

The moment has come. Following some important pioneering work in the 1990s, the 2000s will be the decade of the virtual prototype – perhaps to be found in a larger virtual world. This book presents a comprehensive, up-to-date overview of this crucial topic in design and construction, with contributions by some of the leading theorists, researchers and practitioners.

William J. Mitchell, Professor of Architecture and Media Arts and Sciences at Massachusetts Institute of Technology, holds the Alexander W. Dreyfoos, Jr (1954) Professorship and directs the Media Lab's Smart Cities research group and the MIT Design Laboratory. He was formerly Dean of the School of Architecture and Planning and Head of the Program in Media Arts and Sciences, both at MIT.

References

Cowan, H. J. (1968) *Models in Architecture*. Amsterdam, Elsevier Science.
Goldberg, H. E. (2006) AEC From the Ground Up – BIM Update 2006. *Cadalyst* **1**, November.
Mitchell, W. J. (1977) *Computer-Aided Architectural Design*. New York, Van Nostrand Reinhold.

Mitchell, W. J. (2001) Roll over Euclid: How Frank Gehry designs and builds. In: *Frank Gehry, Architect* (ed. J. F. Ragheb). New York, Guggenheim Museum Publications, pp. 352–363.

Sabbagh, K. (1996) *Twenty-First-Century Jet: The Making and Marketing of the Boeing 777*. New York, Scribner.

Sutherland, I. E. (1963) Sketchpad: A man-machine graphical communication system, *Proceedings of the AFIPS Spring Joint Computer Conference*, May 1963, Detroit, Michigan, pp. 329–346.

Introduction – Virtually there…?

Peter Brandon

1. Context

Most people close to retirement in construction have spent a working lifetime which has broadly followed the advent and development of commercial and electronic computers. Looking back for them, it has been a fascinating journey. It started with computer bureaus which took basic information, punched it on to cards and then processed it before sending it back to you to undertake the corrections! The bureaus charged about one farthing (this is about one-tenth of a current new UK penny) for every character and this just about paid the firm to have Bills of Quantity, for example, prepared in this way. Unit quantities followed, whereby standard units, such as a door, could have all its components and decoration computed by the machine automatically, specified and costed. Computer-aided drafting developed alongside, but it was not until the microcomputer, as it was often called (now the PC), was developed that CAD and other programs under the control of the user began to make serious inroads into the market-place.

Computing then moved from the domain of the enthusiast to that of commerce. The rapid rise in the power of such machines and the parallel reduction in cost enabled the machine to come within the scope of the general public, and with the advent of the internet the machine was transformed from being just another tool to an essential aspect of daily life for most people in the developed world.

The process has been one of exponential growth in both the power and spread of the machine, which has seen the behaviour patterns of people changed out of all recognition through its influence. The dependency on the computer has increased in the developed world to the point where our manufacturing plants, our homes, our energy supplies and our mechanical devices could not work without the support of microprocessors. This has been extended still further to the point where the primary source of our knowledge and the way it is managed and presented is now found through the internet, and a massive explosion of knowledge has occurred. During the Clinton Presidency in the USA the number of internet sites rose from less than a hundred to over 350 million and is now well over one billion. This has created what Malcolm Gladwell (Gladwell, 2001) identified as a tipping point, a place where an epidemic of change occurs in the social, technical or cultural aspects of life similar to that of an epidemic in the health domain.

These developments are well known and do not need amplification, but where do we go from here?

2. Prediction and foresight

For many years people have been engaged in the process of predicting what would happen in the foreseeable future. As the speed of change increases then people want to know where they should place their investments, what they need to learn and where they can direct their efforts for maximum advantage. Increasingly, the predictions about the kind of life people will lead have been dominated by the power of technology and in particular the power of the computer. Many of these predictions have come true and there is now considerable dependence on the machine in nearly all aspects of daily living. Consequently, issues such as security of information have become of major importance (see Grahame Cooper, Chapter 16), and this, coupled with governance and intrusion into people's lives (often triggered by malign use of the technology or significant external events, e.g. terrorism), has opened up the likelihood of Orwellian possibilities for the future (Orwell, 1949) – Orwell just got the year of 1984 wrong!

At a more mundane level the construction industry has tried to predict what could, and in some cases should, happen to its structure and performance in the light of the technological revolution that surrounds it (see Nashwan Dawood *et al.*, Chapter 19). It would be true to say that construction has been one of the last major industries to fully embrace the technology. Whereas the large scale engineering industries such as aerospace and automobile manufacture have raced ahead with the adoption of computing technology, construction has remained stubbornly resistant. The technological developments have been progressing fast in other industries but construction has for some time fought off the challenge, as exemplified by the cartoon prepared by the author in 1998 (see Figure 1).

Nevertheless, various studies have identified the potential for harnessing the technological developments for construction (CICA Report, 1992; Hannus *et al.*, 2003; Rezgui and Zarli, 2006). Some of these are reflected in this volume. The industry itself is aware of the need to respond and, in a large scale study undertaken by the author in 2004, the construction industry in Australia placed developments in IT as the second most important factor to influence the industry over the next 20 years, after sustainable development (see Figure 2).

However, the industry seems hidebound by its present structure, the educational standard of much of its workforce and the lack of investment required to transform its working practices. These problems are often described as 'cultural' and there is some merit in looking at them in this way. They have arisen after centuries of adopting a particular kind of behaviour, and, in an industry that is so diverse and so geographically fragmented, it is going to take many years for this to change.

One of the problems might be that construction is often considered to be one large monolithic and homogeneous industry. But if the methods adopted for, say, free form structures together with the personnel employed were compared with the average house builder, then a major gulf would appear, certainly in the UK.. The former uses three dimensional modelling, laser scanning, high level mathematics and CAD/CAM systems for manufacture with a highly trained and informed personnel. The latter uses the craft technology of another era, admittedly now with many preformed components, but with a workforce that is under-educated and which has little knowledge of the potential of computer-based methods. Both sectors may have similar aims but their ability to use new technologies could not be further apart.

Other industries   **Construction**

Figure 1   Construction's response to the transfer of Information Technology? Brandon 1988).

Global Trends Affecting the Industry

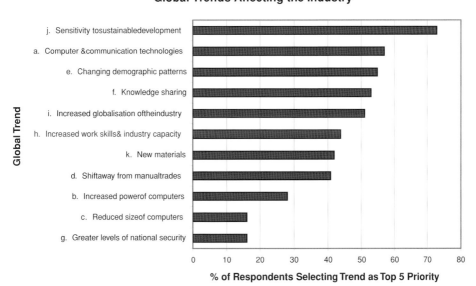

Figure 2   Global trends affecting the Construction Industry. 'Construction 2020 – A Vision for Australia's Property and Construction Industry'. Report CRC Construction Innovation, Australia 2004.

Despite this recognition of the influence of the information and communication technologies (ICTs) to change the industry, when the members of the industry had to choose the highest priority for the industry to address, changing the business environment came top. By this, the industry means the cut throat competition and procurement. The two are linked, of course, but it is not obvious to most individuals in the industry.

There is a difference between prediction and foresight. With prediction the event will happen and the recipient has to decide what his or her reaction might be. With foresight a desirable scenario for the future is developed and the person has to decide how they might make this scenario become achievable. In one the person is a passive receiver, whereas in the other the person can attempt to create and shape the future.

Flanagan and Jewell (2003), in a comparison of foresight studies in construction, identified similarities between the predictions of ten countries and found that four countries (France, UK, Singapore and USA) had not included information technology as a major issue. This is slightly misleading as the assumption of these countries was that information technology permeated everything and that progress in any area could not develop without it. They did not see a need to identify it as a separate issue worthy of addressing. This in itself is quite interesting as it may mean that the assumption of the industry in these countries is that the industry would continue in the way it always has but would be merely enhanced by ICT developments. If this is true then the mind set of the industry is not one of revolution (as has been seen in other manufacturing industries) but of incremental change within the existing paradigm. This is not a recipe for major advancement. The power of ICTs to revolutionize what is done in the industry is so strong that it cannot surely be left to a supporting role?

To some extent this may be the result of the timescale over which the study was expected to look. It is difficult to see a vast industry which is fragmented into sectors with a complex supply chain being able to change radically within 5–10 years. Consequently, incremental change becomes the horizon of the day. If, however, we go beyond the 15-year horizon then much of the immediate focus on current problems begins to fade away. The political and structural issues and the current barriers are perceived to no longer be as relevant.

3. The changing scene

This book and its authors have chosen to look at this longer horizon and beyond, to examine what might benefit the industry of the future. Some of the chapters provide case study material of advanced applications, existing now, since it is unlikely that there will be a 'big bang' change across the whole industry. There will be pioneers, and the case study material provides an indication of what will eventually become common practice.

There are clear signs that things are changing and this provides the context for this book. At the moment the revolution seems to focus around the use of 3D modelling of the design and process, and the use to which the model can be enhanced to be utilized throughout the construction and operation phases of development. This could, of course, span several decades. If the information can be used several times and a legacy of knowledge and intent produced for future users then the model becomes an organic repository for the history and development of the project, able to be mined at will for those who wish to adapt and develop the building in the future. This has been put forward as a major motivator for many years. Some have taken this concept much further on real-life projects towards 4D and nD modelling (see Ghassan Aouad *et al.*, Chapter 12 and Martin Fischer, Chapter 8), and the result is an integrated database of knowledge built around the 3D design model to which all the participants can contribute.

The potential of this approach is at last being recognized. Some major clients (see Martin Riese, Chapter 5) are beginning to demand that 3D modelling is used on their projects, and if the design teams and the contractors are not able to operate this technology then they will not be employed by that client. It is a strong incentive! This level of commitment usually means there is a 'champion' in the directorate of the company who has a vision for the future that encompasses computer approaches at the highest level. Some clients are already claiming that such technology is saving them more than 10% of the total cost of the project in design/construction and substantially more within the operational cycle. These are significant sums of money and as such provide significant competitive advantage to the firms concerned. An interesting project would be to see how these champions arise and what level of knowledge they require to provide the level of commitment needed to instigate change and provide the investment needed.

In other instances the nature of the building demands a much higher level of technology support. These buildings simply could not be built without the use of 3D modelling and the supporting knowledge structures. Conventional processes using 2D drawings will not allow the design forms to be represented for construction, let alone set out on site or checked for regulatory requirements. However, one of the problems that these buildings pose is that regulatory authorities have not yet adapted to the new approaches and are trying to interpret designs for conformance using tools from another era. 2D drawings, for example, do not allow a full understanding of, say, fire regulation requirements for a building which is free form with multiple floor levels within an irregular structure (see Nicholas Nisbet *et al.*, Chapter 17). Glymph has expressed this succinctly with regard to the buildings of Frank Gehry in which he was the senior partner responsible for the management and technological support (see Gehry, 2002).

It would appear that at last we have reached a watershed where at least some sectors of the industry will move forward, driven by client demand, and the knowledge that interesting forms and systems cannot be implemented without the use of advanced CAD/CAM models. The next step is to get the business-to-business models developed so that the interface with the supply chain is improved. One follows from the other in an integrated system, and the full benefits will not be felt until such time as the integration is complete. It is a case of 'the whole is greater than the sum of the parts', and this includes the regulatory systems and techniques which are used for evaluation.

4. Market pull or technology push?

But where is the motivation for change to come from? Is it to be driven by the inventiveness of the IT research community, or is it to be driven by the demand from the practitioners for assistance to solve their current problems? In some countries, particularly the UK, the prevailing view of government research funding agencies has been that it should be industry which pulls through any new technology into the market-place since the main aim of these funding agencies is to increase economic competitiveness. 'Relevance' is the order of the day. This approach is understandable as industry and commerce provide much of the *raison d'être* for the research and these sectors want to see quick results. However, this emphasis tends to skew the research towards current problem solving and quick wins at the expense of attempting to shape the future and preparing an applied sector for the technological future – hence the lack of preparedness for the expected virtual futures. Consequently, those countries which are less enslaved by long traditions may find themselves at the vanguard. It is no co-incidence that Gehry, for example, finds it easier to work in less developed countries than in those with a strong professional and regulatory environment.

It must be wrong to polarize the situation in this way, i.e. market pull or technology push. If the industry does not understand the potential of the technology both now and in the future then it is likely to invest poorly and shut its eyes to future innovation which can transform its performance. This requires education throughout the industry, particularly of those in positions of power, and it requires experimentation to ensure that the appropriate technologies are adopted and adapted for the sector concerned. A long term view is helpful in providing continuity of development and to avoid too many blind alleys. However, the need to reflect and adjust to changing business conditions is equally important because predictability is notoriously difficult in the area of information and communication technologies. This may change as the subject becomes more mature and stable, but for the foreseeable future this is unlikely to happen.

5. The changing technology

The past three decades have seen significant movement in the focus of the applications that have been attempted with electronic computers. In the early days the software was developed to either enhance or replicate human activities. Most of these activities were developed to deal with the inadequacies of the human mind and human limbs. These were summarized by Broadbent as long ago as 1973. Humans were not too good at calculating in a reliable manner, got tired quickly and were slow at writing things down. The machines were targeted at improvements in these areas, without changing the manner of the activity, to provide a more effective solution. As time went on, the power to hold data and manipulate them provided new forms of analysis and insight, which allowed a more informed decision to be made. Visualization of data then developed to aid understanding and communication, and the infrastructure for communication changed radically with the internet and mobile devices. Indeed the power of visualization has transformed how we present and interpret data (see Richard Boland *et al.*, Chapter 18) in the context of aspects of management and it is now an expectation for much of our routine decision-making. In many ways the technology is in advance of the human issues, and it is they which hold back the advancement (see Mustafa Alshawi, Chapter 21).

At various times different groups of researchers have explored the use of machine intelligence as a way to aid human decision-making. In the 1980s, in response to the Japanese 5th Generation Project, many countries invested in what were then called 'expert systems' and methods of knowledge representation. An example was the ELSIE expert system for construction pre-design planning, which in several real case studies outperformed human experts (Brandon *et al.*, 1988). Again, this exploration followed the pattern of most computer developments and largely focused on replicating human behaviour – in this case 'intelligence'. This would, of course, be a superb achievement, but it does also raise questions as to whether this is appropriate or whether a different focus might provide better service for future human use. These are not trivial issues and may well dominate our thinking for the next working lifetime.

6. The disappearing computer

The next step is probably the use of ambient technologies, where the computer is possibly not even visible to the user but nevertheless is communicating, monitoring and analysing as and when required in a more natural way than we expect today from our keyboards and other devices. Already, firms such as Phillips have designed computers which could, and do, reside in the wallpaper as a natural part of the building construction/finishing and through which communication with the outside world is facilitated. This facilitation does not have to be through conventional devices but through speech picked up from our key fob or through projection devices direct onto our retina or even, in the long term future, through our brain patterns.

Terrence Fernando and his colleagues across Europe (see Terrence Fernando, Chapter 20) are developing their work in the University of Salford 'Think Lab', investigating how ambient technologies might impact on future workspaces in a number of key industries. This work, sponsored by the European Union Framework Programs to the amount of 12 million Euros, has just started and includes construction and planning as industry sectors. It will be interesting to see how this work reveals the potential for the future.

This 'disappearance' of the machine has, of course, its downside. How will the user know it exists? What else is it picking up beside our messages and signals, and how is it manipulating what we provide for it? What is it doing to our privacy and our sense of self? Are we drifting into a position where we are integrated into a massive network of knowledge in which we are just a small (but dispensable?) part? How can we make our lines of communication secure?

Like all technologies ICTs can be used for good or ill, and do we really have the power in a globalized society, where the computers are linked together in a potential grid of knowledge, to protect the values of independence and freedom which we prize so highly? We use democratic mechanisms to adjust and balance our behaviour, but where are these within the confines of the integrated computer networks which we see developing today? This may seem a million miles from the operation of a CAD system or the preparation of a programme schedule, but they all contribute to the whole.

In the early 1980s the author was concerned that when writing FORTRAN programming for construction he would include statements which said that IF such an event occurred THEN the following action should be taken. These statements contained within them the value system of the program writer as he, first of all, had to determine which events should be considered before action was taken and then he had to decide what that action should be. These were often his viewpoint or the perceived wisdom of the day. The statements were encapsulated in the software routines and then those routines were embedded in others until it was difficult to find the original without much effort. Unlike normal human interaction there was no challenge to the assumptions because it was too difficult or even impossible to find the original. By its very nature computer software is designed to be used by many people (in general software it can be over a billion users) and therefore the values of one individual becomes an imposition on the many. It is extremely difficult to design systems which are transparent.

Grady Booch, the Chief Scientist for IBM, in the 2007 Turing Lecture at Manchester University, UK, suggested that the next three decades could be labelled as follows as far as software development is concerned:

- The first decade would be one of greater transparency in software development (although it is not clear upon what basis this statement could be made, but it sounds promising!)
- The second decade would reflect even greater dependence on the machine (so that we could not exist without its support)
- The third decade would be the decade of the rise of the machine (and presumably the decline of the independent human mind)

By themselves none of the developing technologies are harmful, but without control and devices for ensuring equitable performance attuned to the needs of the human race (something we cannot achieve yet even without machines) we are in danger of creating a potential disaster. Can we have the bright positive future without releasing the monster lurking beneath the surface?

For example, can you imagine trying to devise a system for evaluating sustainable development upon which all the world's nations could agree and which they and their machines will comply and from which they will exercise control? At the moment we cannot agree on a common set of measures. It is unlikely we could agree on how to change these measures as more knowledge becomes available and our mistakes become visible, and it is unlikely that we could agree as to how our machine would or should interact with that of another country with its own set of assumptions even if the concept of 'country' and national boundaries remain intact.

7. Technology supporting construction

In the light of the above, the problems of construction seem almost insignificant, but it is worth noting that accommodation is one of the most important aspects of how we judge our quality of life alongside food and health. If we get it wrong it has enormous impact. So what can technology do for us?

Two of the major problems often quoted as faced by construction are firstly, the separation of design from manufacture, and secondly, the inability to test and simulate the design and the processes of construction and occupation in advance of the building assembly. The size, complexity, expense and one-off nature of large building and civil engineering structures has made it economically impossible to provide physical prototypes which allow the examination and testing of design and processes in a way that some other high technology industries can achieve.

This is now beginning to change. With the advent of commercial 3D modelling of buildings now beginning to permeate the market-place it is now quite commonplace to provide 'fly-throughs' in real time with fairly realistic visual imaging of the building. At the moment the design approach has not changed dramatically with the technology, and it does appear that some traditional techniques employed by humans for centuries are still the most able way of being creative in the visual fields.

However, a new form of 'digital craftsmanship' is emerging which may allow new methods of designing that have not yet been fully defined by the design and construction community. The use of distributed objects, which can be gathered together from around the world, laid out for assembly in a virtual environment and moulded at will, may be one approach. Another may be the use of 'agents' within the virtual worlds (see Joe Tah, Chapter 15) that exist, which will seek out information, as and when required, acting as a kind of 'knowledge slave' for the design and construction professional. Indeed Lawson (2004) has suggested that this kind of support might be the way forward to enable true computer-aided design – at the moment it is a misnomer.

The communication of ideas through new ambient technologies and the monitoring of performance as design and construction develop will bring new thinking to the methods adopted for design. It will also transform the design process itself, and the digital environment has yet to be explored and examined as a new paradigm for design (see Rivka Oxman, Chapter 1). The new media may change the way designers think and develop their creative acts.

These technologies will also provide a rich environment of knowledge where the traditional boundaries between the professions will become almost meaningless. The interrogation of models for examination and investigation is one of the real assets of the computer-based approach. It enables designers (in the broadest sense) to test and examine what they are proposing before the commitment to build. It not also allows clients and other stakeholders to examine in a similar way, but also to take the model into the operation phase of the development cycle in a continuous and seamless transition. Knowledge capital is now a key ingredient in the search for improved solutions (see Chimay Anumba and Zeeshan Aziz, Chapter 11; Yacine Rezgui and Simona Barresi, Chapter 13; and Matthew Bacon, Chapter 14) and its power is only just emerging. This is changing the way we perceive design and the manner in which support for design is constructed, and it is by no means clear what formalisms may yet appear for this purpose (see Tuba Kocatürk, Chapter 4)

One of the most significant aspirations in construction in recent years has been the desire to build a building in a virtual environment before starting the procurement process, thus enabling the buildability of projects to be tested. In the past two years this has become a reality, and Heng Li and his team at Hong Kong Polytechnic University (see Andrew Baldwin *et al.*, Chapter 7), working with CATIA, have developed a visual simulation of the construction process which has enabled significant savings to be made by contractors and enabled design solutions to be tested for their viability and efficiency. This is a major step forward and challenges the existing models of evaluation that are often locked into a historical precedent which is no longer applicable.

Designers (see Manfred Grohmann and Oliver Tessmann, Chapter 2; Martin Simpson, Chapter 6) have experimented with the new visualization approaches which can give a much improved understanding of the aesthetic of the construction and also its structural integrity and its method of construction. The result is some stunning architecture which would just not be possible without the engagement of the machine. The direct link to computer-aided manufacture (also pioneered by Gehry and Glymph) is changing the relationships between members of the design team. On the Experience Music Project in Seattle, Glymph has stated publicly that the Project Manager had to stand aside and allow the negotiation of the structure to be held directly between the fabricator and the designer through the machine.

Others (see Robin Drogemuller, Chapter 9; Souheil Soubra, Chapter 10) have pushed the boundaries of evaluation and the integration of new priorities for assessment into the traditional 3D models to engage the agendas of sustainable development and whole-life cycle evaluation. Gradually the complete development life cycle will be represented in the model and the virtual development will have been complete. Of course, it will never be complete as models are always inferior representations of the real world and therefore they are always capable of refinement.

A big question that hangs across the whole of this development of technology is 'what are we prepared to allow the machine to decide upon?' To what level are we prepared to allow the machine to replicate the processes of our mind, and at what point will we recognize that the machine is better than us at undertaking creative acts? The automation of the mind, encroaching on our role as decision makers will be a key issue for many years to come. Already some routine decisions, largely involving calculation, are already trusted to the machine, but at what point are we prepared to allow our judgement, particularly those related to human behaviour, to be delegated?

This relationship between human and machine is being taken to what must be the ultimate link between the two. The 'jacking in' of the computer directly into the brain to enhance performance is being suggested. It is already underway for dealing with certain human physical disabilities such as hearing and visual dysfunctions, but the forecast is that it will eventually be used to stimulate the brain or enhance it for lifestyle requirements. There may come a point when we may have to decide the characteristics of the brain that we can place on a chip that will improve the performance of an architect or engineer! This may be a long way off but we should not allow this kind of technology to creep up on us unawares. The next generation will have a whole host of ethical issues to deal with as the technology begins to develop machine intelligence which may compete with human intelligence and be available to only a small sector of the population.

8. Learning from others

It is unlikely that construction will lead the way in exploring these technologies. It is already some distance behind other large manufacturing sectors such as the aerospace and automobile industries. The supply chain for construction is not yet sufficiently computer literate or developed, except for some very special cases, to engage in a technology-centred procurement and manufacturing process. Some CAD/CAM examples exist, largely associated with fabrication of the steel frame and cladding, but generally traditional craft methods prevail. These are, of course, broad generalizations and many professional firms in the more advanced building sectors are incrementally introducing elements of a seamless design and construction system.

One of the most interesting innovations being considered at the moment is the possibility of 'printing' buildings direct from a design (see Tony Thorpe *et al.*, Chapter 3). In this case the principles of 3D printing, where the machine lays down a succession of solid layers of material, has been transported to a large concrete pour creating a structure direct from the design model input. This concept is still in its infancy and others are being considered, such as direct extrusion of buildings with lasers creating voids and openings. It is possible that these experimental programs may transform the way in which we think of the act of construction.

Some major clients realize that something has to be done to improve industry performance, not only in embracing the technology but in changing the underlying structure of the industry so that the technology can be harnessed to improve efficiency and effectiveness. British Airports Authority are a good example where the construction teams for the Heathrow Terminal 5 have been formed under just a few headings, with a supply chain dedicated to each team enabling the development of systems linked by computer models to be introduced. It is no coincidence that when this type of innovation is introduced it is often driven by senior personnel from other industries taking a leading role. In the case of British Airports Authority it was Sir John Egan who came from the car manufacturer, Jaguar. Perhaps it is the role of enlightened large clients to provide the uniting vision and driving force to change the nature of the industry.

There is a sense that much of what we are able to achieve will be dependent on the technology infrastructures that are being developed and are available for any group to adopt and use. The internet has been such a vehicle which has changed the way we communicate and study and undertake our business. All industrial sectors, and construction is no exception, will have to take cognisance of what is happening in the wider world and test each development's suitability for its own use. However, in addition, the industry may attempt to shape the direction of these developments to suit its own ends and to secure its own future. It then becomes more than a technology watch – it is a technology driver.

In any case, the boundaries between industries/activities will become fuzzy and there will be migration of knowledge from one group to another. The introduction of knowledge grids, where several computers act together and share each others' knowledge (similar to an energy grid), is an interesting concept which may well impact on the way construction is perceived. Nobody quite knows what will happen when knowledge sharing takes on a new dimension beyond what we have seen even with the internet. It opens enormous possibilities for collaborative working across all participants in the development process.

9. The future and this book

This book brings together some of the leading international thinkers and practitioners in a multi-disciplinary discussion about what the future might be for both information and communication technologies impacting on the design and construction industries, looking forward into the medium-term future some 15 years or more. We know that we cannot predict with any certainty what will happen, but we can suggest the priorities and shape of the technological infrastructure that will allow us all to perform our roles within the design and construction process in a more informed and effective manner. However, those roles are changing because of the technology, and this in turn may impact on the structure of the industry itself. Do we need the same professional infrastructure? Do we need the separation of design and manufacture? Do we continue with on-site manufacture? Does the new media mean that we think in a different way? These are important questions which arise out of this exploration. The authors vary, not only in their professional and academic backgrounds but also in their experience of the new technologies. They each provide a different perspective on the problem, and from the editors' point of view this has been fascinating and we are indebted to them for their willingness to put their views down on paper for us all to share.

We will all have a view of where the future might lead us and we hope this volume will play its part in stimulating discussion of not only what we should be doing as an industry to harness the power of the technology but also what we should not be doing. It encourages us not to sit back and be mere passive recipients of new technologies but to become active participants in the way these transforming technologies develop.

Some of the questions we might ask ourselves are:

- How do we build on what we have now?
- What can we expect from technological advances in the future?
- What impact are the new technologies going to have on the future of construction?
- What are the social and managerial impacts of such changes?
- What are the barriers to be overcome?
- What changes are needed in the industry?
- What changes can we expect in procurement and contract arrangements?
- What actions need to be taken now?
- What are the social impacts of such changes?
- What are the dangers of allowing machines to take more of a decision-making role?

Clues to the answers to these questions can be found in the book. There is probably no other significant development that is likely to change our industry so dramatically as the information and communication technologies. This collection of chapters provides new insights and directions for the future and prepares the wider community for the virtual futures which are to come.

It is a wonderful challenge but also an awesome responsibility for all of us engaged with design and construction.

References

Brandon, P. S., Basden, A., Hamilton, I. W. and Stockley, J. E. (1988) *Expert Systems. The Strategic Planning of Construction Projects*. The RICS/Alvey Research for Chartered Surveyors, RICS in Collaboration with The University of Salford, UK.

Broadbent, G. (1973) *Design in Architecture:* Architecture & the Human Science. London, John Wiley & Sons, pp. 314–316.

CICA (Construction Industry Computing Association) Report (1992) "*Building IT 2000*". Sutherland Lyall. Manchester, CICA. [http://www.cica.org.uk]

Flanagan, R. and Jewell, C. (2003) *A Review of Recent Work on Construction Futures. CRISP Commission 02/06*. London, Construction Research and Strategy Panel.

Gehry, F. O. (2002) Commentary in: *Gehry Talks: Architecture + Process* (ed. M. Friedman). New York, Rizzoli Publishing.

Hannus, M., Blasco, M., Bourdeau, M., Bohms, M., Cooper, G., Garas, F., Hassan, T., Kazi, A. S., Leinoinen, J., Rezgui, Y., Soubra, S. and Zarli, A. (2003) *Construction ICT Roadmap*, ROADCON project deliverable report D52, IST-2001-37278. Available at http://www.roadcon.org [Accessed 2 December 2006].

Gladwell, M. (2001) *The Tipping Point*. London, Abacus.

Lawson, B. (2004) Oracles, draughtsmen and agents: The nature of knowledge and creativity in design and the role of IT. *Proceedings of the INCITE 2004 Conference: Designing, Managing and Supporting Construction Projects through Innovation and IT Solutions*, 18–21 February, Langkawi, Malaysia. Construction Industry Development Board.

Orwell, G. (1949) *Nineteen Eighty Four*. London, Secker &Warburg.

Rezgui, Y. and Zarli, A. (2006) Paving the way to the vision of digital construction: a strategic roadmap. *ASCE Construction & Engineering Management Journal*, **June**.

Turing Lecture (2007). Available at:

http://www.bcs.org/server.php?show=ConWebDoc.10367 [Accessed April 2007].

Part 1
Design, engineering and manufacturing challenges

1 Emerging paradigms and models in digital design – Performance-based architectural design

Rivka Oxman

In the field of design and the built environment, survey and analysis of the applications in virtual prototyping identify major issues and define the unique requirements of architectural modelling. In view of current developments in the theory and technology of digital architectural design, certain potential new directions for architectural modelling in virtual prototyping are beginning to emerge.

This chapter presents an approach to architectural modelling in virtual prototyping termed 'performative prototyping'. The term 'performative' combines both the concepts of performance and formation, or 'form generation'. Performative prototyping is postulated as an approach to modelling of the architectural object in virtual prototyping environments, in a manner that supports integrated performance-based generation in the design process.

The limitations of traditional architectural modelling techniques for virtual prototyping are presented. A distinction is made between typological modelling and topological modelling as a basis for modelling in virtual design environments. The potential of performance-based models to overcome these limitations is proposed. Current performance-based architectural design models, methods and techniques are presented as being particularly relevant to virtual prototyping for architectural design. In the proposed approach, performative factors that are incorporated within simulations can potentially be integrated with generative procedures in the modelling of the design prototype. Experiments in digital architectural design are presented as examples of this approach to performative generation in architectural modelling.

1.1. Introduction: Virtual prototyping for architectural design

Virtual prototyping is the exploitation of simulation processes for the test, evaluation and modification of prototypes in virtual design environments. This chapter considers how ongoing developments in digital architectural design will have potential effects upon the theory, concepts and technologies of virtual prototyping. The underlying motivation is that in digital architectural design we abandon the conventional concept of the prototype as a formal construct necessary for design, and we are, therefore, entering a new period of virtual prototyping.

In considering this transition to a new prototype model in virtual prototyping for architectural design, we have identified two basic problems associated with the conventional models:

(1) One problem results from the way in which the prototype is conceived and modelled in architectural design. The traditional conception of the prototype is via 'typology', which operates on the basis of typological sets of relationships with instantiations of the modification of sub-type variables. This process of typological prototype modelling is characteristic of many design disciplines and constitutes a limitation to innovative and creative design in virtual prototyping. Therefore, while virtual prototyping may function well for analytical and evaluative procedures, can it equally be advanced to support generative procedures? If so, does the way in which the prototype is modelled as a typological design constitute a limiting factor?

(2) The second problem is the logic behind the formulation of simulation processes in virtual prototyping. Virtual prototyping remains within the boundaries of a conventional formulation of its objectives. In spite of having a wide range of digital tools for the analysis and evaluation of performance aspects, none of them currently provides additional integrated generative capabilities. In current virtual environments, performance simulation is still coupled with methods of traditional design. That is, after reviewing the results of the simulation/evaluation procedure, required or desirable modifications are determined by the designer, the model is adapted and the process repeated.

Current theories in digital technologies, and particularly current applications in digital architectural design, contradict certain of these root concepts of this traditional typology-based usage of VP (virtual prototyping) technologies. Digital architecture as a design discipline has become rich in concepts that are transforming the modelling content of the architectural artefact and, perhaps more significantly, transforming design methodologies (Oxman, 2006). Furthermore, this approach to new forms of digital models is creating options for continuities in the modelling process – from design through fabrication and construction. As current digital modelling approaches and technologies advance in architectural design, they demonstrate new potentials for generative paradigms. Assuming the successful development of generative models and approaches in architectural design that can transcend the limitations of the typological approach, architectural design can potentially contribute new definitions of the design prototype for VP.

These developments are based upon new performative and parametric design technologies that are moving architecture from traditional typological design towards topology-based design – from form-based processes towards process-based design (Oxman, 2006) – and are developing new roles for the human designer (Aish, 2003) in interaction with digital technologies and resources. This fundamental reorientation of design roles and relationships between the human designer and the digital medium can potentially contribute to a redefinition of the function of virtual prototyping. As new roles for the human designer as 'tool maker' emerge, and as the digital model becomes a multi-stage continuous model, we expect that there will be a concomitant evolution of the theory and technology of VP.

We are currently engaged in looking at the influence of the representation of the architectural model upon VP, of its ability to support generative procedures beyond typology, and of the ability of such models to be integrated with synthesis and evaluation procedures. From a design methodological point of view, we are seeking to integrate design synthesis processes with existing design simulation/evaluation processes and thus expand the architectural function of VP.

This chapter addresses the following issues:

(1) *Typology and its limitation*: How can we overcome the typological limitation of conventional architectural prototype models?
(2) *Beyond performance-based simulation*: Is there a new role for virtual prototyping? Can we promote an expanded functionality of VP environments beyond analytical and testing simulation to simulation that can drive synthesis and generation?

In the following sections we present certain ongoing developments in digital architectural design and analyse their potential effects upon VP concepts and future technologies. Currently, external performative influences are the basis for simulation/evaluation affected upon the model of the design prototype. Can such simulations also become an integral part of generation processes for design modifications? Here the distinction between the nature of the prototype model as a high-level typological format supporting design modification and that of the digital prototype as a parametric generative formalism is an important one. Following this logic of remodelling the architectural design prototype as a morphogenetic model, a new role for virtual prototyping in architecture is possible. This model integrates simulation/evaluation/and generative modification in a continuous process. We refer to this expanded design cycle of modelling, performance simulation, evaluation and generative remodelling as 'performative prototyping'.

1.2. Virtual prototyping

In a survey of current approaches to virtual prototyping in the fields of engineering design and construction management in the built environment, it became apparent that research is needed on the development of virtual prototyping as a design environment. In the face of new digital technologies and conceptual changes in digital architectural design today, there is an even stronger need to formulate potential directions for architectural design in VP. We assume that any such changes for an expanded VP design functionality will eventually have a general impact on virtual prototyping itself. This section begins with a review and analysis of current VP approaches and technologies.

Current applications in virtual prototyping are centred on product modelling and the modelling and management of building information (Kiviniemi, 2005). The ICON project was an interesting attempt to build a theoretical basis for the creation of a generic information structure for the construction industry. A central feature of the information model was to allow information to be viewed from different 'perspectives' and at different levels of abstraction, according to the requirements of the user (Brown *et al.*, 1995). The ICON project has become the basis for the development of the nD-modelling approach (Aouad *et al.*, 2005). This approach became a basis for the creation of dynamic virtual models to be visually simulated along the whole life cycle of a construction project. Today, other applications focus on developing VR environments for modelling and simulating construction processes, and are particularly useful for planning and scheduling. An interesting application presents an implementation of virtual prototyping in a residential project in Hong Kong. This application can dynamically simulate a real construction process and demonstrates its future VP potential for other applications in the field of construction management (see Chapters 5 and 7). Most of these applications are based on the integration of a 3D CAD (computer-aided design) model with application software for functions such as scheduling. Other functionality expands the processes of visual simulations by integrating the time dimension, 4D, with a 3D CAD model that evolves in time. Baldwin expands and discusses how different VP technologies have been developed to enable both clients and construction organizations to model and visualize complex construction processes (see Chapter 7). Recent work has introduced the simulation of a physical environment, including the behaviour of its inhabitants (Kalay, 2004). Encompassing this range of simulation and evaluation processes for building evaluation and construction, a report for a future VR roadmap for the built environment presents the current state of the art and explores new directions for 2030 (see Chapter 19).

The review of virtual prototyping in the field of the built environment and the construction industry revealed that many current works are related to the development of VP for simulating construction and construction-related processes. There currently appears to be fewer works exploring conceptual directions related to ways in which design processes can be simulated and supported in virtual environments.

New developments in digital media for architectural design and related design fields have now enabled the formulation of novel approaches to design modelling and simulation that can potentially support manipulation and generation. The relationship between new digital technologies and their potential impact on virtual prototyping in architectural design have not yet been fully explored by the research community. These changes are affecting our very understanding of the term, and thus also presenting new perspectives for VP technologies relevant to other fields.

It is the definition of such developments and the research implications for VP that is the objective of the current chapter. In the following section, key concepts of major technologies in virtual prototyping for design are presented. Following this, ongoing changes in the field of digital architectural design are considered, and their potential effects upon current VP concepts and technologies presented and discussed.

1.2.1. Introduction: the conceptual and methodological foundations of virtual prototyping

In order to understand basic principles in developing future virtual prototyping environments for design, the conventional role of the physical prototype in design needs to be briefly stated. Physical prototypes have always been the traditional way to investigate design issues such as visual and spatial quality. However, there are many limitations to the physical prototype. They are difficult to build and difficult to modify. Early CAD prototypes replaced the physical prototype by introducing digital modelling techniques and supported the manipulation of the graphical representations of digital objects. Various geometrical modelling/rendering software has been developed to support modelling needs (Kalay, 2004).

Virtual prototyping originated in the manufacturing industry where industrial products such as cars and airplanes were tested before manufacturing. Today, virtual prototyping exploits simulation processes for the testing, evaluation and modification of

CAD prototypes in virtual environments. Virtual prototyping is carried out on the CAD model, by performing the same tests in computational environments for virtual proto- typing as those formerly undertaken on the physical prototypes. The role of simulations here is to obtain quick design feedback and to help make expedient and well-founded judgements in order to support efficient design modification processes.

Virtual prototyping and current simulation technologies are used to evaluate the performance of the design prototype, mainly for product development, manufacturing and construction purposes. Such simulations are based on quantitative and analytical results, and can vary from visual analysis to various engineering analysis simulations testing temperature, acoustics, behaviour under physical conditions, etc.

1.2.1.1. *Prototyping technologies*

Several techniques have been developed in the field of virtual prototyping to support efficiency and accuracy. According to Chua there are various technologies associated with the terms 'prototype' and 'prototyping in design'. The two most accepted tech- nologies are virtual prototyping and rapid prototyping. The two have different roles in design and require further definition (Chua *et al.*, 1999). Virtual prototyping is the replacing of the physical model and the requirement for physical manipulation and the need to rebuild a physical model for further analysis and simulation; while rapid prototyping is the various systems that rely on CAD systems to control machines that generate a physical model.

Virtual prototyping (VP)
The role of VP is to test, evaluate and modify design prototypes in virtual environ- ments with the intention of obviating the need for physical models. Virtual prototyp- ing typically involves analysing CAD models for different end applications, such as manufacturability or assemblability, and applying these results to later phases. Virtual prototyping exploits simulation processes for the testing, evaluation and modification of design prototypes from conceptual design to the final product. For example, ana- lytical tests and evaluations can accelerate production, so providing information for the different applications involved in the life cycle of a design. For example, it can help designers visualize the interactive results of design modifications, allow tests to be performed during each phase of product development, help programme managers identify programme risks, enhance economic competitiveness, etc.

Rapid prototyping (RP)
RP is the production of a physical model from a computer model. In the last decade, a number of RP techniques have been developed. These technologies have also been referred to in the literature as 'layer manufacturing', 'material deposit manufactur- ing', 'material addition manufacturing', 'solid free-form manufacturing' and 'three- dimensional printing' (Chua *et al.*, 1999). A recent review has surveyed and analysed current technologies (Sass and Oxman, 2006). Prototyping technologies have differ- ent suitability and effectiveness for various stages of the design cycle with respect to their relevance in design and manufacture. In fact, this is the main issue currently un- derlying the distinction between virtual and physical reality (see Chapter 3). The RP technologies of use different materials and techniques in order to enhance testing and evaluation that are based on physical accuracy.

1.2.2. Network environments

These concepts can also be applied in shared network environments. A virtual prototyping environment can be a multi-disciplinary collection of models, simulations and simulators focused on guiding product design from idea to prototype. Synthetic environments are internet-based simulations that can reach a high level of realism. These environments are fundamentally different from traditional simulations and models. They are created by wide area networks and augmented by realistic special effects and accurate behavioural models that allow visualization of, and total immersion in, the environment being simulated.

Analysis by digital experts on the internet is another application for analysing CAD models for testing different aspects and characteristics of a virtual model. Sharing a virtual prototype with various team members in different locations can potentially make VP a very powerful interactive analysis tool. Analysing CAD models for different end applications may be performed with the aid of a digital expert agent/module via the internet.

1.2.3. Summary

New virtual technologies are driving developments that are contributing to greater efficiency and accuracy. They can even be integrated in a single process. For example, virtual prototyping and rapid prototyping can be integrated and used interactively. Both the physical model and the virtual model can be integrated in a single process. The designer can benefit from employing the two modes: a virtual model can be converted into a physical model, and a physical model can be converted into a virtual one. For example, a virtual prototype can be digitally made, tested and evaluated to be printed by a rapid prototyping system. While performance aspects can be better tested on the digital model, the physical model may be a better medium for testing physical accuracy.

However, current technologies, advanced as they may be technologically, are generally driven by a traditional design methodological basis. Therefore, they currently have little conceptual impact or qualitative effect on the design process. They are still based on traditional design processes and human centric design approaches. For example, the human designer/engineer must evaluate the simulation results and determine the required modifications. In the following section we discuss certain of these conceptual problems in current approaches and raise issues that may lead to a meaningful shift in developing a new role for virtual prototyping in architectural design.

1.3. What's in a prototype? Problems and issues in virtual prototyping for architectural design

1.3.1. The typological prototype and its limitations in VP

Design is conventionally viewed as a process of repetitive cycles of generation/evaluation/modification until convergence is reached with the design objectives. The need to modify and re-design the model by modification and manipulation on a CAD model are key issues in developing applications in VP design environments.

One problem is the way CAD models are conceived in design. Most CAD prototypes are still based on a traditional conception of the prototype. They support a typological view of design.

In most design engineering domains the accepted paradigmatic approach to the modification of a CAD model is based on the acceptance of traditional design models: designers generate design changes by specifying variations in the basic parameters of the essential variable of the type being designed. For example, in architectural design, procedures of refinement and adaptation are based on typological knowledge (Oxman and Oxman, 1992).

This paradigmatic approach to modification in the CAD model in engineering design domains accepts the existing variables and parameters of a given prototype. Instances, which vary typologically by modifying the essential variables of the prototype, produce variations or sub-types. From a cognitive perspective, the conventional design prototype is a model that exhibits the essential features of the type. Thus modifications of the prototype model are generally based upon variations within the typological family.

An interesting direction that attempts to transcend the limitations of typological models derives from the development of virtual prototyping tools to support digital modifications that are similar to 'hand-craft design'. Human-centric design devices in VP environments are replacing the physical prototype. In the field of mechanical design, a virtual environment, within which the mechanical assemblies can be constructed by direct manipulation, was being developed almost ten years ago (Thompson *et al.*, 1998) – and it is still a major and valid concept that is employed today with newer software. Operations such as assembly and surface manipulation are supported by these environments. The research literature reports on the development of techniques such as haptic interfaces to allow forces of contact to support surface tracing, assembly forces and grasping are being simulated. Another prototyping tool that supports digital modifications is based on a motion tracking system used for tracking hand movements and for recognizing hand gestures as input into a VR environment (Murray and Fernando, 2003; Murray *et al.*, 2003). Construction of these typical assemblies is assisted by the provision of geometric constraint-based environments for the assembly and disassembly of engineering components. These developments support hand-driven modifications to replace conventional hand-driven manipulation of a physical prototype. However, this type of haptic, interactive design is not necessarily suitable for architectural design.

Digital technologies are promoting new methods for architectural design that are highly suited to the demanding physical performance of building design and which also furnish a modelling medium to transcend the typological limitations. The emergence of parametric and performative systems for design is affecting our very understanding of the term 'prototype', and thus also presenting new perspectives for VP technologies.

1.3.2. Performance-based simulation and its limitations

The main goal for virtual prototyping in design is to test, evaluate and modify prototypes in virtual design environments. Currently, simulation processes in virtual prototyping remain within the boundaries of the conventional formulation of the design prototype. A wide range of digital tools is available for analysing and evaluating aspects of performance. However, none can provide generative capabilities. Current theories and digital technologies suggest a shift from analytical simulation to simulation for synthesis and generation. This approach identifies performance with generative processes (Kolarevic, 2003). This distinction is very significant. Instead of analysing the performance of a design prototype, and modifying it accordingly, it is proposed that performance-based simulations be used to directly generate and modify the digital prototype. In such an approach the desired performance can be selected and activated as a performative mechanism that can generate and modify designs.

For example, a typical simulation that characterizes testing and evaluation in a conventional prototyping process is the Finite Elements Method (FEM). This type of simulation typically analyses structural stability. We propose that, on the basis of such analytical procedures, we require simulations that may also generate the structural modifications required by the analytical findings.

Evaluations provide feedback for design modifications. Performance here is defined as the ability to manipulate directly the properties of a digital prototype. Its implications can be broadened and include simulations of both quantitative and qualitative aspects such as spatial, social and cultural factors, in addition to technical simulations such as structural and acoustical performance. This direction requires the formulation of new types of simulations that can be oriented towards the generation of designs.

In the following section we present ongoing developments in performative digital architectural design and analyse their potential effects upon the theory, concepts and technologies of virtual prototyping. Current directions in performative architectural design appear to provide a future promise to advance the architectural design relevance of VP environments. Performative systems for architectural design:

- Are consistent with parametric and non-typological models of the architectural artefact
- May be capable of developing integrated generative as well as evaluative models
- Require a generative parametric representation of the design artefact that is both generic and generative and can support significant physical simulations such as structural performance and spatial organization
- Should be consistent with concepts of 'core project models' that continuously run through all phases of the design and construction processes

1.4. Performative design

Architecture as a design discipline has become rich in new ideas that are transforming the content of the architectural artefact, of design methods and methodologies (Oxman, 2006). In the following section we define some relevant key concepts in the field of digital architecture related to developments in performative design and identify their future relevance for architectural design in virtual prototyping. Potential approaches that will contribute to digital design in VP are defined, presented and discussed. Emphasis is placed upon current performance-based models, methods and techniques as being particularly relevant to these changes. In the following sections key concepts in digital design that underlie performance-based design are reviewed and discussed. Three types of performance-based design are classified and presented. Finally, we discuss how performative considerations may affect both the design prototype and the type of simulations that can become an integral part of the design generation/modification in virtual environments.

1.4.1. Key concepts in digital design

1.4.1.1. *Generation: Process driven versus form driven*

Design representation, so central a concept in traditional models of design, is no longer a valid concept for understanding digital design processes. Designers are moving away from the idea of design generation being based upon the manipulation of formal representations. The idea of the designer 'thinking through his drawing' was central to the way we conceived of design for centuries. Particularly in architecture, design conceived as a process of manipulation of formal representations, through drawing or modelling, has been central to our understanding of the design process.

In digital design, design is conceived in terms of processes' models. Furthermore, in certain generative processes of digital design the formal implications of the concept of representation are negative and unproductive. The usage of formal concepts such as: typology, generic knowledge, concept formation, conceptual design, *partis*, etc. have been rejected by Lynn and other theoreticians (Lynn, 1999). Topological representation of form and the use of animation and morphing as digital processes in design are replacing the conventional theoretical basis underlying architectural design (Lynn, 1999).

Design generation without categories of form requires a new definition of the concept of form. Here the distinction between form and process, form as a static description and formation as a dynamic process, becomes significant. Digitally generated forms are not designed or drawn in the conventional way, 'Instead the designer constructs a generative system of formal production, controls its behaviour over time and selects forms that emerge from its operation' (Rahim, 2005). This principle of generation is a process that leads to the 'emergence of form' and is currently driving the shift from the 'form making' to 'form finding'. Again, ideas of typology, generic design, precedent, or case-based design are completely antithetical to a view of design as an emergent process. Emergence is a process of form finding based upon performance rather than preconceived formal content. Processes of emergent form finding are not only antithetical to the logic of form-based generation, including the centrality of representation and formal languages, they also require new modelling processes. So such models must be conceived completely differently in virtual environments.

1.4.1.2. *Formal basis: Topology versus typology*

The reconsideration of topology and non-Euclidean geometry as a methodological basis for digital design has contributed to the exploration of new geometrical possibilities. Topology is the study of the relational structure of objects. It is the study of those properties of objects that do not change when homeomorphic transformations are applied. Therefore, topological structure can be defined in a variety of geometrically complex forms. Emmer (2005) has investigated the role of topology as a new formation process in digital design.

These design theoretical tendencies have been supported by new digital techniques and software technologies that have opened up a universe of interactive topologically based geometric manipulative possibilities. In such digital design media the static coordinates of shapes and forms of conventional digital media are replaced by computational dynamic constructs, including topological surfaces or hypersurfaces. Furthermore, interactions with digital modifiers (NURBS: non-uniform rational b-splines) and modelling operations such as 'lofting' are opening up new techniques for the creation and interactive manipulation of complex geometrical shapes in design.

Given the formal characteristics that have became associated with these developments, generically referred to as 'topological design', these design phenomena were significant in advancing the theoretical and practical significance of digital design in the second half of the 1990s. New terminology also emerged, including: 'free-form design'; 'hyper-surface design', 'blob architecture', 'hyper-body architecture', etc. In addition to this descriptive terminology, topological design is also associated with a body of theoretical concepts related to the morphology of complexity such as hyper-continuity and hyper-connectivity. References are also made to biological conditions of networked connectivity and rhizome-like complexity. Among designers whose work demonstrates this approach are van Berkel, Caroline Bos, Greg Lynn and Kas Oosterhuis.

Topological design may be seen as characterizing the departure from the static and typological logic and design methodologies that supported previous CAD generations.

1.4.1.3. Associative geometry

There is a difference in philosophy between explicit and associative geometry due to the topological effect of digital environments enabling the reconfiguration of parameters of a geometrical structure (Burry and Murray, 1997; Burry, 1999). Associative design is based on parametric design techniques that exploit associative geometry. In parametric design, relationships between objects are explicitly described, establishing interdependencies between the various objects. Variations, once generated, can be easily transformed and manipulated by activating these attributes. Different value assignments can generate multiple variations while maintaining conditions of the topological relationship.

Today, formal techniques in associative parametric technologies provide design environments in which the designer can define the generic properties of a geometrical structure within a user-defined framework. In current parametric design techniques, complex non-standard geometry can be generated and manipulated. Currently, Gehry Technologies is offering 'Digital Project', based on CATIA (a high-end modeller originally developed as a parametric, automotive and aerospace modeller). Another known digital technology is Bentley Systems' 'Generative Components'.

This combination of interactivity, transformability and parametrically controlled perturbations that generate discrete structural variations within design formation processes is an emerging characteristic phenomenon of digital design. Parametric systems are becoming cornerstones in the more complex performative digital environments described below. Within the framework of these behavioural characteristics, the body of theoretical concepts related to parametric formations includes adaptability and change, continuity, proximity and connectivity.

1.4.1.4. Geometrical complexity: Versioning versus variation

'Versioning suggests that design is an evolving and dissolving differential data-design that no longer simply "exist" but rather becomes as it becomes informed' (Rahim, 2005). 'Versioning' is an operative term meant to describe designs that become informed. Focusing on prototyping, the employment of digital processes of versioning is a manifesto of a new way to design.

1.4.1.5. New role for the designer

Another fundamental aspect that should be reconsidered is the new role currently recognized as a new skill that characterizes the digital designer. The traditional role of 'the designer as a user' is extended to 'the designer as a tool builder'. Tool builders can define their own generative components and define their transformational behaviour (Aish, 2003). Designers in this type of formation model are provided with interaction facilities, both to the formal modelling as well as to a set of digital techniques. Furthermore, users can interact with 'integrated inner tools' using scripting methods and tools that enhance design freedom and control. This way users can control their type of and level of interaction with the representational medium, manipulate and refine complex geometries while working in such problem areas as the design of structural shape.

1.4.2. Performance-based design

Performance has long been recognized as an important issue in design. Performance-based design has always been significant in architectural design. Methods and techniques known as 'appraisal aids' were developed by Tom Maver during the early days of CAD. Figure 1.1 presents two distinct cycles related to digital design methodologies (Oxman, 2006). The conventional cyclic process (see Figure 1.1(a)) is based on a repetitive cyclic design process typically known as an iterative 'analysis–synthesis' process. The second type of cyclic model in design is based on performance-based design. In the conventional process, analysis and simulation are performed on a geometric model. On the basis of the results, the design is modified and changed geometrically in order to improve its performance. Analytical feedback and visualization techniques provide performance evaluations and guide changes and modifications to achieve a better performance. Current analytical computational techniques such as FEM are based on the same conceptual approach. Today, these types of methods use formal representations that interconnect mesh elements and make it possible to test and make dynamic analyses such as structural, energy and fluid analyses. These can deal with geometries of free form design or any other geometrical complexity.

In contemporary performance-based design, or performative design, building performance can be regarded as a guiding design principle (Kolarevic and Malkawi, 2005). Well known examples of performance-based design are the Greater London Authority Headquarters building (2002) and the Swiss RE building (2004) designed by Foster & Partners. The optimization of energy and acoustical performance was achieved while

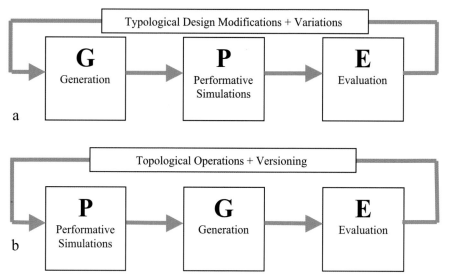

Figure 1.1 Models of performance-based design: (a) conventional CAD model of performance-based design; (b) digital model of performance-based design. (Figure supplied by Neri Oxman.)

the surface of the curvilinear façade of each building was modified by analyses made by performance-based techniques and simulation programmes. The objective was to maximize the amount of natural lighting and ventilation while reducing each building's energy consumption, which was achieved by employing environmental performance techniques. In the case of these buildings, the complex geometry was the result of the performative requirements rather than predetermined formal preferences. Form was 'found' on the basis of performative requirements, rather than complex form being a specific objective and starting point of the design.

Thus not all performative design results in free form. Performative design is a design technique for form finding based upon performative techniques. However, performative techniques are generally analytical rather than generative.

Performance-based models of design (see Figure 1.1(b)) can be regarded as topological models where digital simulations of external forces are applied in driving geometrical formation processes (Oxman, 2006). As defined above, in such a model, 'versioning' is an operative term used to describe evolving design modifications that become informed and responsive to differentiated performative effects.

External forces may include structural loads, acoustics, transportation, site, programme, etc. Information itself can also be considered as an external 'force' that can manipulate and activate digital design processes. In design, the external influence of forces can also be applied to inform the complex behaviour of a model that can be deformed and transformed. This may be relevant to dynamic objects where a dynamic simulation can be computed considering environmental influences as the driving forces. Lynn's New York Bus Terminal (Lynn, 1999) is an example in which a particle system is used to visualize the flow patterns presented on the site. These were created by simulating force fields associated with the movement and flow of pedestrians, cars and buses across the site.

1.4.3. Performance-based generative design

Performance-based generative design may be considered as an extension of the performance-based model. It is based on generative processes driven by performance. In a performance-based generation model, the data of performance simulations drives generation processes that generate the form. An example that suggests the important future potential for integrating a performance-based model with generative design tools is described by Shea *et al.* (2003). They describe a structural generative method, integrating a generative design tool, 'eifForm', and an associative modelling system, 'Custom Objects', through the use of XML models. The system integrates grammatical shape generation, performance evaluation, behavioural analysis and stochastic optimization. The eifForm software developed by Shea is perhaps the best example to illustrate this approach: although it is limited to structural roof design, it demonstrates the generative principles. The generative mechanism in this work is built upon the shape–grammar formalism. In this approach the initial design is repeatedly modified and improved until its performance is satisfactory. The difference between this approach and the former one is that in this case the form and structure are automatically generated and integrated with the evaluation of structural performance.

1.4.4. Conclusions

The performative prototype is fundamentally different from conventional CAD prototypes. Existing analytical CAD prototypes are not programmed to integrate the dynamic processes of formation based on specific performance. In CAD models the tools are based on performance evaluation of the object itself. In the performative approach there is no need to separately evaluate the design. Performance is considered to be a shaping force rather than evaluative criteria. The prototype is changed and modified automatically by selected performative simulations. Design performance may include various classes of performance requirements, including environmental performance, cost performance, spatial, ecological and technological perspectives. Performance-based design employs analytical simulation techniques that produce detailed parametric expressions of performance. These, in turn, can produce formation responses to the various classes of performance requirements.

In performative design a formal system may be considered as a digital prototype that is adaptive to appropriate kinds of digital performance simulation. The emphasis shifts to the process of form generation responsive to the data input of performative simulations. Developing generative mechanisms for solution representations that are responsive to the data input of a performative digital simulation is an important methodological challenge for the advancement of performative design. Recent work in the field of morphogenesis may provide relevant examples.

A new formulation to support this approach to performative design for virtual prototyping should go beyond the formal image-based representation of the prototype that currently characterizes CAD models. This will require the following:

- A geometrical representation based on topological relationships
- A performative mechanism that informs a generative mechanism

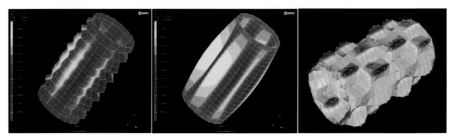

Figure 1.2   Simulation for test and evaluation. (Figure supplied by Neri Oxman.)

- An adaptive mechanism that activates geometrical modifications
- A mapping and integrating mechanism that keeps the performative integrity in response to multiple performative aspects

1.5. Performative prototyping in digital architectural design

Design may be the result of a performative act and a form finding process. Desired conditions such as structure and physical context, of solar or acoustical control can be formulated as performative simulations that may directly inform generative and formative processes in the VP model. 'These effects are seamless between environment, user, formation, structure and material distribution' (Rahim, 2005).

For example, a typical simulation that characterizes testing and evaluation in conventional engineering design prototyping processes is the Finite Elements (FE) analysis. This type of simulation analyses structural stability. In such a case, we propose that, in place of FE analysis we would integrate a simulation model that would also generate the structural form in response to evaluative findings. Two illustrative examples are presented in Figures 1.2 and 1.3. Figure 1.2 illustrates a conventional FE analysis. Figure 1.3 illustrates modifications of the model informed by the structural simulation (Oxman, 2005).

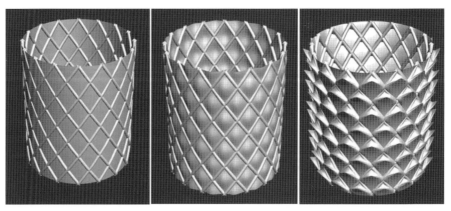

Figure 1.3   Performance-based simulation for generation and modification. (Figure supplied by Neri Oxman.)

Most engineering design disciplines today employ simulations in VP that are oriented towards efficiency and optimization. However, in creative domains, such as architecture, product design, art and design, there currently exists great potential to employ simulations as a means for creative design. Simulations that translate qualitative considerations as well as quantitative ones may instigate novel methods for design. Innovative design can be created by selecting and formulizing abstract ideas and goals as an input for the selection of certain simulations. Designs do not have to be based exclusively on efficiency and optimization.

The following section presents experiments in digital architectural design to illustrate some principles of this approach. The study was part of an experimental programme in the digital design studio exploiting 3D modelling, simulation and animation software in experiments related to performative design that integrates performance and formation.

1.6. Performative design

The performative approach is fundamentally different from conventional CAD simulation processes. Existing CAD prototypes are not programmed to integrate formation processes informed by performance-based simulations. Traditional CAD tools are based on performance evaluation of the object itself. In a performative model the prototype is dynamically modified. For example, desired physical effects, such as the effects of light or behaviour, can be formulated as performative systems, which include formative transformations as well as evaluative simulations.

We will now discuss and demonstrate how performative considerations can be formulated as simulations and become an integral part of design models. Two experimental projects that illustrate the design method are presented below.

The first of these illustrates the incorporation of light qualities as morphogenetic factors. Light has long been a primary component in architectural design. The system, called 'Tex-Light', integrates structural principles based on textile patterns with the lighting simulations. Different types of textile principles and weaving techniques were exploited to define geometrical and structural relations, parametric transformations, and variations.

The structural pattern is adaptable and designed to respond to changing light conditions. Light filtering elements are modulated and transformed according to desired qualitative and quantitative measures through a protocol of predefined transformations and manipulations. The control of light effects was a design objective, and light simulations were conducted to modify and activate structural transformations in order to achieve the desired effects. Visual simulations were performed to study the lighting behaviour of various modulations of the structural system. The performative simulations were intended to determine how light effects inform the parametric transformation of the structural pattern (see Figure 1.4).

The second experiment in performative modelling involved the design of responsive building skins that might protect a building from excessive wind loads, solar penetration and acoustical contextual problems such as urban noise. The responsive wall is a project that integrates a constructive skeleton which supports a dynamic surface structure. The skeleton has built-in sensors that can inform and simulate the dynamic motion of the skin, which is designed as a system of scales.

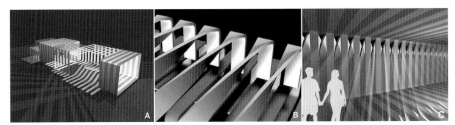

Figure 1.4   Tex-Light architecture: (A) spatial organization; (B) twisting manipulations of structural components; (C) performative light simulations. (Figure supplied by Tal Kesten and Alex Eitan.)

The skin design itself has been generated by a performative simulation of wind force and light penetration. These forces produce dynamic effects on the skin. Animated techniques were employed to produce these dynamic simulations and performative effects. The objective in this model is that the formation of the skin/structure assembly will be generated by multiple performative conditions, all of which are dynamic. This condition of design models, that support multiple performative analyses in their componentized assembly, is highly representative of complex wall assemblies. In this case, the design objective is to support the dynamic behaviour in the various components in response to the dynamic nature of the wind and light loadings of the building surface. Animation studies were employed to model the formation process in which form generation was controlled according to types of simulations (see Figure 1.5).

1.7. Summary and conclusions

Performative design introduces new directions in architectural digital design and has important implications for future developments in virtual prototyping for architecture. In per-formative design, new approaches to simulation models are employed that integrate form modelling, performance evaluation and generative procedures into a virtual environment. Instead of using simulations for testing and evaluation, design generation and design evaluation should be integrated in digital technologies for modelling the physical qualities of design systems, the simulations of their performance and for redesigning their form according to the findings of performance evaluations.

New developments in virtual prototyping will be based upon such integration possibilities of generation and evaluation in performative design. One of the outcomes

Figure 1.5   Responsive wall: (A) skeleton; (B) animated skin; (C) responsive wall. (Figure supplied by Shoham Ben Ari and Roey Hamer.)

of these studies is the need to research the formulation and design of such integrated performance/generation systems. Future directions for development are new digital techniques that couple principles of performance with principles of generation. However, much of the problem of future research in the design of performative architectural systems for virtual prototyping lies in the need for a new conceptualization of the architectural prototype. What is in a prototype? What is the content that will make both generation and evaluation possible?

In both projects illustrated, simulations were employed as 'design tools' in the hands of the designer. This is a novel direction that will promote the development of user interfaces for the new digital designer. If the conceptualization of the new architectural prototype is the first prerequisite for research and development, then the new model of the 'designer as a tool maker' must find expression in that design space that we call 'virtual prototyping'.

We have only considered the question of the architectural model and how it will evolve in order to support a performative/generative process in VP. What then might be the limits of creativity of the virtual prototype? If performative design takes us beyond the formal into a territory in which form and formal preferences are not significant considerations, is the formlessness of performance about to generate a new world of 'natural' forms that are predetermined only by their response to physical context? Where then is the dividing line between the architect as a form giver and the tool maker as a form receiver? When and how does one intervene with physical laws?

Acknowledgements

This chapter presents the work of the DDGR experimental group at the Faculty of Architecture and T.P. at the Technion, Israel. My students Tal Kesten, Alex Eitan, Shoham Ben Ari and Roey Hamer are acknowledged for their contributions. Neri Oxman of MIT, USA, is acknowledged for her theoretical input as well as for permitting publication of excerpts from her final diploma project at the Architectural Association, London (Oxman, 2005).

References

Aish, R. (2003) Extensive computational design tools for exploratory architecture. In: *Architecture in the Digital Age* (ed. B. Kolarevic). New York, Spon Press, pp. 245–252.

Aouad, G., Lee, A. and Wu, S. (2005) From 3D to nD modelling. *ITcon* **10**: 15–16.

Brown, F. E., Cooper, G. S., Ford, S., Aouad, G., Brandon, P., Child, T., Kirkham, J. A., Oxman, R. and Young, B. (1995) An integrated approach to CAD: modelling concepts in building design and construction. *Design Studies* **16** (3): 327–347.

Burry, M. (1999) Paramorph: Anti-accident methodology. *Architectural Design* [Special issue: *Hyperspace 2*] **September**: 78–83.

Burry, M. and Murray, Z. (1997) Computer aided architectural design using parametric variation and associative geometry. *Challenges of the Future, ECAADE Conference Proceedings*, 17–20 September, Vienna.

Chua C. K., The, S. H. and Gay, R. K. L. (1999) Rapid prototyping versus virtual prototyping in product design and manufacturing. *International Journal of Advanced Manufacturing Technology* **15**: 597–603.

Emmer, M. (2005) Mathland: The role of mathematics in virtual architecture. *Nexus Network Journal* **7** (2): 73–88. [Basel, Birkhauser.]

Kalay, Y. E. (2004) *Architecture's New Media*. Cambridge, MA, MIT Press.

Kiviniemi, A. (2005) Requirements management interface to building product models. *PhD Dissertation*, Civil and Environmental Engineering, Stanford University, pp. 92–144.

Kolarevic, B. (2003) Digital morphogenesis. In: *Architecture in the Digital Age* (ed. B. Kolarevic). New York, Spon Press, pp. 7–19.

Kolarevic, B. and Malkawi, A. M. (ed.) (2005) *Performative Architecture: Beyond Instrumentality*. New York, Spon Press.

Lynn, G. (1999) *Animate Form*. New York, Princeton Architectural Press.

Murray, N. and Fernando, T. (2003) Development of an immersive assembly and maintenance environment. *Proceedings of the International Conference on Advanced Research in Virtual and Rapid Prototyping*, 21–23 October, University of Salford, UK.

Murray, N., Goulermas J. Y. and Fernando T. (2003) Visual tracking for a virtual environment. *Proceedings of HCI International*, 15–17 September, University of Salford, UK.

Oxman, N. (2005) *Final Diploma Project*. London, Architectural Association.

Oxman R. (2006) Theory and design in the first digital age. *Design Studies* **27** (3): 229–266. [Special issue on Digital Design (ed. R. Oxman).]

Oxman R. and Oxman, R. (1992) Refinement and adaptation in design cognition. *Design Studies* **13** (2): 117–134.

Rahim, A. (2005) Performativity: beyond efficiency and optimization in architecture. In: *Performative Architecture: Beyond Instrumentality* (ed. B. Kolarevic and A. M. Malkawi). New York, Spon Press, pp. 201–216.

Sass, L. and Oxman, R. (2006) Materializing design. *Design Studies* **27** (3): 325–335. [Special issue on Digital Design (ed. R. Oxman).]

Shea, K., Aish, R. and Gourtovaia, M. (2003) Towards integrated performance-based generative design tools. *Proceedings of the ECAADE, 21st Conference on Education in Computer Aided Architectural Design in Europe,* 17–20 September,*Graz University of Technology, Austria*, pp. 553–560.

Thompson, M. R., Maxfield, J. H. and Dew, P. M. (1998) Interactive virtual prototyping. *Proceedings of Eurographics*, 25–27 March, Leeds, UK.

2 Algorithmic design optimization

Manfred Grohmann and Oliver Tessmann

Computer simulation is an essential tool in engineering complex structural systems. Over the last few years we have made several attempts in our research to establish and improve the link between structural simulation, evaluation and architectural design with a focus on optimization.

Optimization is not perceived as a mere structural issue but as a negotiation process between shapes envisioned by the architect and their structural needs. Topology and geometry are not fixed but open to modification in areas that significantly exceed defined stress values. This process can be operated by simulating and evaluating structural performance, iterative optimization and the use of evolutionary algorithms that yield solutions not achievable with conventional techniques. This research is exemplified by a range of projects developed in the office of Bollinger + Grohmann.

Beyond this research, on a geometrical level, which still has a lot of potential for further investigation, we see a need for the extensive use of simulation in the field of material research in combination with rapid prototyping and rapid production technologies to overcome the modernist notion of standardized materials. Differentiation on the level of material properties could have a great impact on structural and architectural design.

2.1. The notion of optimization

Virtual prototyping is used in the realm of structural engineering to simulate the stress response of structures under imposed loads, the impact of temperature and material properties. Due to sophisticated analysis software, engineers are able to evaluate rather complex design proposals increasingly emerging in the field of architecture.

A project has not necessarily to be decomposed to a basic structural system, but can be analysed as an entire model with a multitude of structural members and their interactivity.

Both architects and engineers use digital tools, but the data input requirements of engineers differ from the ones provided by the architects. Instead of volumes and offset surfaces expressing an architectural concept, centre-lines and axis-surfaces are required to evaluate structural performance. To improve the cooperative design approach of architects and engineers, either an integrative digital model, incorporating all information, is needed or the process of deriving data from an architectural model and feeding them into analysis software has to be streamlined. The second approach is the one we aim for in our office. The following projects and studies show different

Figure 2.1   'Take-off' object suspended from the terminal ceiling. (Figure supplied by Reger Studios GmbH.)

approaches and developments for optimization over the last few years. Bridging the interface gaps is also part of our research undertaken at the University of Kassel, (Germany), Faculty of Architecture, Department of Structural Design.

2.2. Project 1: 'Take off' – an object of visual–kinetic interactivity

'Take off' is an installation in Munich airport (Terminal II) by Franken Architekten that aims to anticipate the acceleration, speed and dislocation of travelling (Figure 2.1). The structure consists of an array of 363 lamellas mapped with images on both sides and two load-bearing offset tubes. A visual–kinetic interactivity between the object and passengers is generated simply by the movement of the observer with no physical movement of the object itself. The impact of the media is like a short film or an interactive narrative referencing the trajectory of travelling (Franken, 2004).

The entire object is suspended from the ceiling of the terminal building. The cable suspension system is constrained by a multitude of parameters that define cable directions. The structural system of the building provided specific nodes that were able to bear the additional loads of the 'Take off' object. At the other end of the cables, the offset tubes had to be suspended at specific points to achieve equilibrium.

Part of the space between the ceiling and the object is occupied by a mobile maintenance device, which meant that the cables had to change orientation: starting from the ceiling the cables were redirected and connected to specific nodes at the offset tubes. The redirection is solved by a node that was generated by the cooperative approach of the architect and the engineers, both using different kinds of simulation software.

Figure 2.2 Five different cable directions in different planes. (Figure supplied by Bollinger + Grohmann.)

The architect aimed for a curved continuous change of direction: a rubber-like morphology (Figure 2.2) that reflects the internal tension of the node. The node was generated with the help of animation software capable of simulating fabric behaviour under the influence of external forces. The applied fabric properties and tension forces generated a form that was afterwards exposed to the actual forces of the node.

This structural simulation was carried out with a Finite-Element software package (Figure 2.3). In a manual iterative process the nodes were generated, analysed and

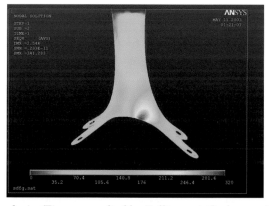

Figure 2.3 FEM analysis. (Figure supplied by Bollinger + Grohmann.)

altered again in the animation software. This alternating process went on until the morphology and the required structural performance converged.

This generation/analysis loop was carried out with manual data exchange and iterative slight adjustments made by hand in the animation software. Optimization was achieved by trial and error without using computational power to generate different versions of the node. Thus the study exemplifies the cooperative design approach that seeks for convergence of formal and structural needs. Two different simulations of forces and material properties finally matched in one model. The optimization of fabrication was not incorporated into this process. The node was milled out of a solid block of steel.

2.3. Project 2: The Mariinsky Theatre – differentiate structure

The new Mariinsky Theatre by Dominique Perrault is an extension to the existing historical building in Saint Petersburg, Russia. It is wrapped into a crystalline golden shell that creates an impressive public space between the theatre and the city, which is used as a lobby, restaurant, foyer and reception (Figure 2.4).

The shell is composed of 56 triangular facets that create the crystalline appearance. The facets act as a surface-active folded plate system. Activating their in-plane stiffness allows a remarkable reduction of columns in the interior space. Also to optimize the structural system, each 'facet' is elaborated into an inward-orientated tetrahedron with four non-uniform faces to increase maximum spans. Each face of each tetrahedron is subdivided into a series of primary and secondary structural beams (Figure 2.5). In one plane they collectively act as surface-active systems that form a common structural element: the tetrahedron.

A kinetic system formed by the group of tetrahedrons is supported by pinned columns at roof level and at ground level, transferring the vertical and lateral forces.

Figure 2.4   The golden shell in its urban context. (Figure supplied by © Dominique Perrault Architecture.)

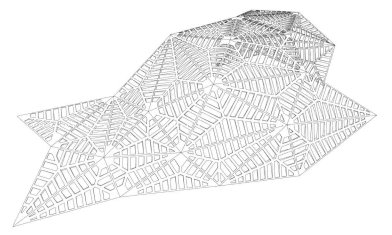

Figure 2.5   Volume model of the golden shell. (Figure supplied by Oliver Tessmann.)

Both the topology and structure of the distorted tetrahedron are described by a set of precise design rules that can be applied to any scalene triangle that acts as the base of the tetrahedron (Figure 2.6). These design rules are translated into an algorithm that generates a complex structural system represented by a 3D centre-line model. The goal of this study was to derive a maximum of output information by giving a minimum of input information to reduce digital geometric modelling, in numbers: 420 poles are generated on the basis of five input parameters.

The implementation of the design rules was carried out in two ways:

- *The scripting approach*: First a script was developed in Rhinoceros, a 3D modelling and CAD (computer-aided design) application that provides a programming language based on VBScript. As input parameters the three vertices of the base triangle, the sloping angle of the surfaces and two beam-distances are needed to generate a total of 420 structural elements. This approach results in a mere static representation of a possible solution. Geometric relationships, defined and used

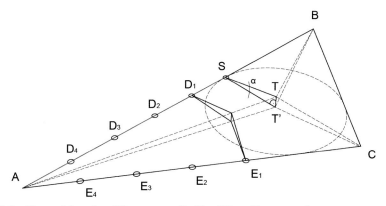

Figure 2.6   Geometric rules. (Figure supplied by Oliver Tessmann.)

during the generation process, are not present in the finished model. Altering input parameters means deleting and regenerating a new structure. Even so, the digital craftsmanship is significantly reduced due to automation; an investigation of the solution space is still time-consuming and based on the repetitive work in creating 56 tetrahedrons.

- *The associative parametric model*: In a second approach, generative components (GC) were used to implement the tetrahedron structure. GC are a 'model-oriented design and programming environment which combines direct interactive manipulation design methods based on feature modelling and constraints, with visual and traditional programming techniques' (Aish, 2004). The environment is based on the CAD application Microstation XM.

The design rules proved to be algorithmically expressible but remained static in the first approach. A further exploration of a solution space was expected by the use of GC and its ability to tackle complex geometric relationships within a 3D model. The underlying design rules could be captured by predefined 'features' in GC. The associative geometry system remained active in the finished model and is represented via a symbolic dependency graph (Figure 2.7).

The model then serves as a generic prototype with defined input and output parameters. Instances of this prototype were then placed along the predefined shell geometry of the theatre. Dynamic representation could be enhanced by defining global variables that control properties of the tetrahedrons like side-beam distance and sloping angles of the faces. Changes of these variables affect all instances of the tetrahedrons, which made dynamic changes and adaptations possible.

Generating multiple solutions makes it increasingly important to define precise evaluation criteria and streamline the evaluation process itself. In this case, the evaluation method is structural and quantitative, in terms of material use. The geometry generated in a CAD application has to be transferred to RSTAB, a structural analysis software (Figure 2.8).

To keep the digital automation continuous during data exchange, the common procedure of sharing a DXF file was replaced by a Visual Basic (VB) application that controls the generation of structural nodes and assigned poles, based on data provided by a script or readout from a CAD file. The Drawing Exchange Format became unnecessary since it interrupted the automation and was exclusively used to reduce information inscribed in the existing model. In contrast, the VB application can save information added to every node during analysis. Beyond pure x, y, z coordinates, a node number and its assigned poles are stored. A change of coordinates is possible without losing the structure's topological information. The evaluation navigates a range between structural performance and necessary material. The results are presented as a spreadsheet.

The golden shell of the Mariinsky Theatre can be perceived as 'a system of differential repetition' (Reiser and Umemoto, 2005). The geometry of a tetrahedron is not fixed but associative; design rules define the structural topology. The actual geometry is different in every tetrahedron. This new perception of seriality in architecture has to be reflected in the engineering practice as well. Traditional evaluation techniques would be a bottleneck within the design development. The improved interface between CAD

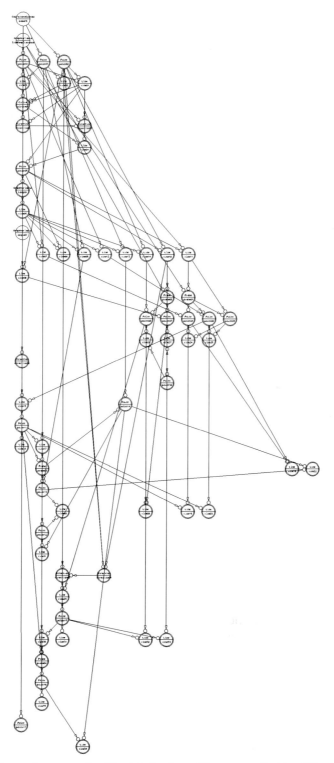

Figure 2.7 Dependency graph of the tetrahedron. (Figure supplied by Oliver Tessmann.)

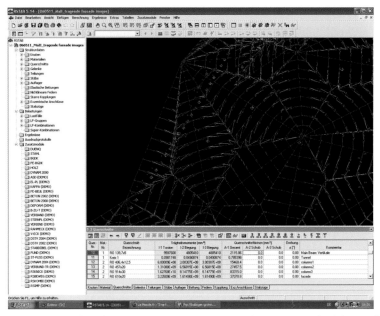

Figure 2.8 Nodes and poles represented in RSTAB. (Figure supplied by Bollinger + Grohmann.)

application and analysis software accelerated the generation/analysis loop, indispensable for the development of the structure.

2.4. Project 3: Naples – evolving roof

For the Piazza Garibaldi in Naples, Italy, Dominique Perrault designed a large roof based on a system of triangulated faces. In cooperation with Fabian Scheurer from the Chair of CAAD at the ETH Zurich, Switzerland, we conducted a study to optimize the folded roof structure with the help of genetic algorithms.

The roof structure can be described topologically as a two-dimensional plane based on a system of self-similar triangles folded in their third dimension. Each of the 289 nodes is assigned a random z-coordinate within defined thresholds. A tube-like column folded out of the roof reaches the ground and acts as a support structure.

To achieve cantilevering capacity and a minimum of node displacement, just by folding the triangulated plane, the behaviour of the entire structure was simulated in RSTAB (Figure 2.9). A programming interface provided by the software made it possible to generate the structure, change its morphology and analyse it afterwards.

By encoding the z-coordinates of all nodes into a genome and using a genetic algorithm, which allowed for crossover and mutation, the performance of the structure could be significantly improved. As a fitness criteria, the displacement of the nodes under self-weight was calculated by the analysis software, the worst node defining the inverse fitness for each individual. After 200 generations with 40 individuals each, the displacement repeatedly reached a minimum of 129mm – at a cantilever of

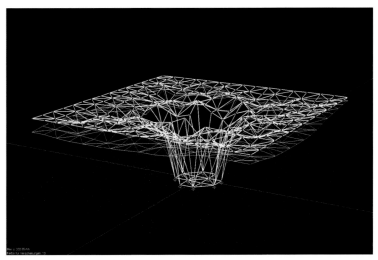

Figure 2.9   RSTAB model showing deflection of nodes. (Figure supplied by Fabian Scheurer.)

25 meters and with a diameter of 193mm for the members [Figure 2.10] an impressive value, according to the engineers. (Scheurer, 2005.)

The study shows a very selective use of computational power to generate and analyse a multitude of solutions for the roof structure. Architectural design shifts from pure modelling to the definition of organizational principles and systems with a specific

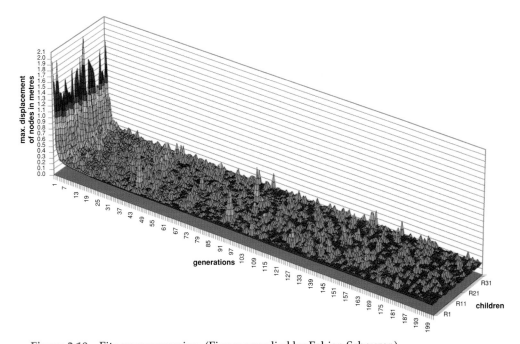

Figure 2.10   Fitness progression. (Figure supplied by Fabian Scheurer.)

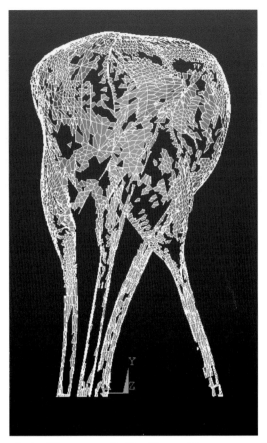

Figure 2.11   FEM analysis. (Figure supplied by Gregor Zimmermann.)

behaviour. The actual form emerges from a process seeking for optimal performance. The solutions derived from this process do not necessarily match conventional structural systems but gain performance by the self-organization of its members (Sasaki, 2005).

2.5. Project 4: D-Tower – differentiate material systems

A different approach to match structural performance with architectural design is achievable by the differentiation of material. Anisotropic materials, like reinforced concrete or glass-fibre reinforced plastics, can be adjusted to locally changing stress values in complex shapes. For the realization of the D-Tower, designed by NOX, FEM analysis was operated by Gregor Zimmermann (University of Kassel, Department of Structural Design) (Figure 2.11). The 11-metre high, glass-fibre reinforced sculpture coalesces architectural form, skin and structure into one single element (Figures 2.12 and 2.13). Simulation of form and material behaviour under vertical and lateral loads led to a subtle differentiation in surface thickness between 4.5 mm and 18 mm and fibre orientation. The resulting objects needed no further structural components.

Figure 2.12   D-Tower by day. (Figure supplied by Bollinger + Grohmann.)

But virtual prototyping and simulation can also become effective on microscopic scales, though having repercussions on the level of structural systems.

2.6. Project 5: Ultra high performance concrete

The Department of Structural Engineering/Werkstoffe des Bauwesens (construction materials) at the University of Kassel is currently developing 'Ultra high performance concrete' (UHPC). An extremely dense microstructure generated by powder-like silica sand among others offers a compression strength that is almost comparable to steel. The self-compressing material can contain steel fibres as integrated reinforcement that provides high tensile strength. This composite features amazing structural and aesthetic properties that go far beyond known concrete materials.

Figure 2.13   D-Tower illuminated by night. (Figure supplied by Bollinger + Grohmann.)

Figure 2.14   UHPC sample panel (University of Kassel). (Figure supplied by Oliver Tessmann.)

The idea of replacing conventional reinforcement through fibres integrated in a matrix seems intriguing against the background of non-standard architecture. Complex shapes are possible without integrating reinforcement meshes in a curved mould (Figure 2.14). One reason among others that holds back the use of UHPC in current constructions is the problem of proper calculation and dimensioning of the material. Extensive testing is being carried out at the University of Kassel to gain data and experience, but one major unknown factor is the fluid dynamic behaviour of the steel fibres while pouring the material into the mould. Calculations are hardly possible without knowing the direction and alignment of the fibres. The self-organizing behaviour of fibre alignment is a complex process and has to be further investigated by computational fluid dynamics software.

2.7. Outlook

Architecture and engineering will experience a further transfer and incorporation of simulation technology from different disciplines. Environmental simulation will be enriched by fluid dynamic simulations from medical and aeronautic research. Structural analysis can be refined by the precise simulation of material behaviour.

The analysis of micro-scale properties has repercussions on larger scales and can already be observed in other disciplines. Computational fluid dynamics (CFD), for example, is used to optimize the skin of swimsuits. By analysing and mimicking the structure and microgrooves of shark skin, the speed of swimmers could be increased through the reduction of passive drag (www.fluent.com/about/news/pr/pr69.htm). Speedo, a company producing swimsuits, transferred this technology from race-car design to the hydrodynamics of swimming and swimsuits. Why not transfer these technologies to architecture and construction?

Simulation can help to overcome today's standardization of material properties. Instead of invariably uniform products, material properties can be adjusted to actual forces. We see a great potential in the differentiation of material systems like glass-fibre reinforced plastics or steel-fibre reinforced concrete. In combination with strategies developed for rapid prototyping, even porous sponge-like materials with changing densities are adaptable to structural needs. Cellular materials are used in the field of tissue engineering. Sponge-like carrier materials approximate the properties of bone tissue for the fixation of fractured bone or after the removal of bone tumours. At the same time it acts as a scaffold for healthy cells. Cell proliferation not only takes place on the surface but also inside the porous material. The morphology of the scaffold defines the overall shape (Manjubala, 2005).

Beyond the technological possibilities it will be exciting to see how these aspects will diffuse into the design practice of architects and engineers. Simulation and virtual prototyping have to be conceptualized in architectural and structural design instead of being a pure checking procedure of the finished designs. This can only be achieved by collaborative working and 'active interaction' rather than 'intense exchange' of all involved parties on the project and research levels.

References

Aish, R. (2004) Extensible computational design tools for exploratory architecture. In: *Architecture in the Digital Age: Design and Manufacturing* (ed. Branko Kolarvic). New York, Taylor & Francis Group, pp. 243–252.

Franken, B. (2004) Take-off-spatial installàtion in Terminal II at Munich Airport. In: *Diversifying Digital Architecture* (ed. Yu-tung Liu). Basel, Birkhäuser, pp. 84–88.

Manjubala, I. (2005) Biomimetic mineral–organic composite scaffolds with controlled internal architecture. *Journal of Materials Science: Materials in Medicine* **16**: 1111–1119.

Reiser, J. and Umemoto, N. (2005) *Atlas of Novel Tectonics.*. New York, Princeton Architectural Press.

Sasaki, M. (2005) *Flux Structure*. Tokyo, Toto Shuppan.

Scheurer, F. (2005) http://www.designtoproduction.com/

3   VR or PR: Virtual or physical reality?

Anthony Thorpe, Richard A. Buswell, Rupert C. Soar and Alistair G. F. Gibb

The visualization of ideas is critical to the design process. A designer will iterate a design until the desired functionality or aesthetics emerge, a design team must communicate thoughts between individuals in order to establish consensus. Further into the design, coordination, interface, manufacturing and assembly issues need to be resolved. Designs that impact on the built environment may require public feedback. All designs interact with end-users, and the need for communication of design information to a wider audience underpins success.

The following sections consider the issues associated with the medium of that communication; the modelling of ideas through virtual reality and through the creation of physical models using rapid prototyping (RP) are contrasted. These digitally driven technologies offer unprecedented ways to present information, and most buildings already contain components that have been created using digital fabrication techniques. So, if we have the technology to produce physical models of buildings and building components directly from the computer and we can model it all virtually, can we go one stage further? Shouldn't it now be possible to design and visualize construction digitally and then click 'print' for your building?[1]

3.1. Design visualization

Visualization is an important part of design communication. The information contained in computer-aided design (CAD) modelling software and/or the design documents contains all the information about how the geometry of the various components fit together and how they should look; information that can now be coordinated through a single 3D building information model (BIM) (see Chapter 5). One of the key problems with buildings is how to convey the environment that will be created. Products such as cars and buildings are similar in that humans interact with them in an environment defined by the product. VR can be a useful tool in evaluating how the product functions, because it can simulate real scale, in real time, give sensory feedback and hence generate the look and feel of the environment, albeit with limitations. Rapid prototyping is very good at conveying the geometry of an object and is used widely in manufacturing for creating visual and functional items for evaluation during the design cycle (Dickin, 2001). Figure 3.1 depicts the relationship between the information conveyed through

---

[1] 'Click print for your building' is a phrase coined by Sweet (2003).

Figure 3.1 Information conveyed by RP and VR.

a physical model manufactured directly from data from the CAD modeller by RP and the environment simulated by virtual reality (VR).

The use of virtual reality and rapid prototyping in designs for the built environment are depicted in Figure 3.2. RP models are finding their way into the conceptual design process. Foster and Partners (Foster and Partners, 2006) have two Z Corp 510 Spectrum 3D-printers (Z Corp, 2006) in-house that build 'throw away sketch models'. The models are created using scripts to generate solid geometries on the Bentley Microstation platform (Bentley Systems, 2006). Colour can be added to these models as part of the printing process, although evidence suggests that it adds little value to architectural evaluation of form. Colour is useful, however, for visualizing the output of analysis, such as solar radiation falling on a façade or for identifying/differentiating functional parts. This use, however, is not widespread. The size of Foster and Partners allows the incorporation of the Specialist Modelling Group: a group of mathematically minded architects who specialize in form rationalization and who oversee the use of the RP machines. Their work complements a 'traditional' in-house model making team.

VR tends not to be used in the early stages of the design as a direct design tool, except for media-rich environments, or where the flow of humans and operational logistics are important (Whyte, 2002). Visualization is more commonly found once the design has been established, or for considering coordination and scheduling issues (see Chapter 7).

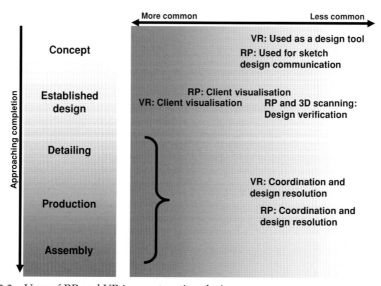

Figure 3.2 Uses of RP and VR in construction design.

It is also worth considering the extent to which CAD software and digital fabrication fit into Figure 3.2. The CAD[2] model is the linking factor in the generation of the information required to drive a VR simulation, an RP machine, meshing and geometry data for analysis such as CFD/FEA and CNC machine code generation for digital fabrication. The manufacturing sector has fully embraced the advantages of solid modelling and system automation (Bedworth *et al.*, 1991). The benefits of modelling construction in 3D solids are beginning to be realized (see Chapter 5).

Some of the advances being made by Bentley are to bridge the gap between creating the conceptual design and detailing with their Generative Components software. The Autodesk platform (Autodesk, 2006) is still the construction industry 'standard' (by market share) and there are numerous packages that can be bolted onto the standard system. These include visualization tools for creating rendered images, fly-throughs and VR simulations. Although the software is not a true solid modelling tool, solid objects can be created and exported. Companies such as Fulcro (2006) create building services' systems' models in 3D and use the data to generate a fabrication code directly for manufacturing ventilation ductwork. Unfortunately, rapid prototyping equipment is less forgiving than CNC machinery and a good solid model is required to generate successful builds. There are considerable difficulties, prohibitive in most cases, in generating RP models from non-solid CAD modelling tools. Although it is possible to generate solids from the Autodesk platform, it is time consuming and may slow down the uptake of such technologies. Possibly one of the most advanced construction related software packages is Tekla Structures (Tekla, 2005), a fully parametric solid modelling tool for the design and manufacture of structural steel and precast concrete. It allows multiple materials to be modelled down to the last nut and bolt. The CNC fabrication code is generated automatically. Models created using this and similar software (see Chapter 5) contain all the information for the design and construction of VR simulations, RP models, fabrication of the end components and modelling of assembly.

3.2. Virtual reality

Unlike rendered CAD images, VR gives a dynamic, real-time environment that can be explored. Full-emersion VR is less common, the majority of businesses and home entertainment systems view through a 2D screen. Amongst the applications of VR in the built environment are (Whyte, 2002) (and see Chapter 1):

- people-flow modelling
- as a design tool in media-rich environments
- architect to client communications
- clashes and coordination of site works
- supermarket and factory layout logistics
- safety applications (visualizing train control signals)

There are barriers and drivers to the widespread take up of VR. One of the disadvantages is that to make environments interactive, data has to be left out to reduce

---

[2] 'CAD' refers here to 3D solid modelling software and can be thought of as a modeller, or an 'aid to design'.

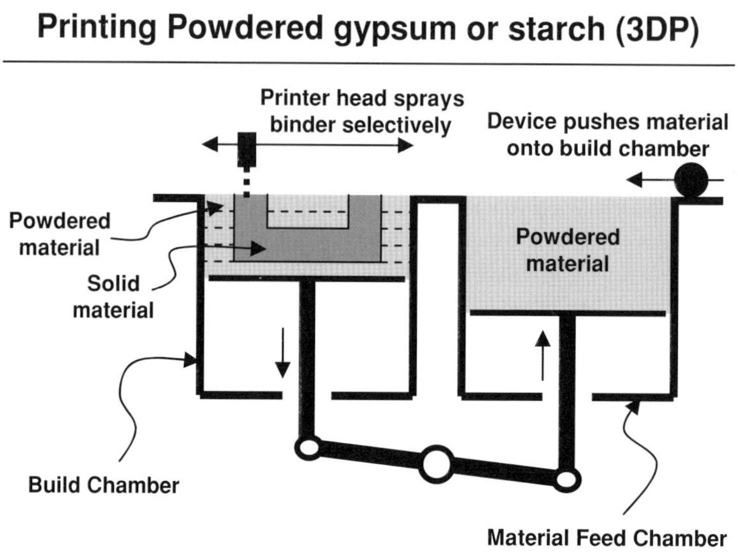

Printing Powdered gypsum or starch (3DP)

Printer head sprays
binder selectively

Device pushes material
onto build chamber

Powdered
material

Solid
material

Powdered
material

Build Chamber

Material Feed Chamber

Figure 3.3   Diagram of the principal features of a rapid prototyping process.

the computing power needed to run the simulation. This is called 'optimization' and, although increasing computer power is reducing this problem, buildings remain large structures to simulate in real time. One of the downsides to this optimization is that certain assumptions need to be made about what is seen. These choices are subjective and can bias another person's opinion of the simulation. The principal drivers for the use of VR in the built environment, however, do not seem to be in architectural presentations to clients. This is possibly because architects tend to work on small unique projects in which the cost–benefit of VR is not realized (Whyte, 2003). The real value appears to be in coordination tasks and the simulation of dynamic operation.

3.3. Rapid prototyping

A family of names is used to describe essentially the same type of fabrication technology: additive manufacturing; rapid manufacturing; rapid prototyping; and solid freeform fabrication. This method of making physical components is delivered by many types of process, examples of which are: thermo jet; selective laser sintering; stereolithography; fuse deposition modelling; and 3D printing (depicted in Figure 3.3) (Wohlers, 2004). Typically, each of the processes can use a range of materials and all have advantages and disadvantages suited to particular tasks. They all 'print' 3D structures typically up to 500 mm in the x, y and z directions. A design is usually created using 3D-CAD solid modelling. A model is first tessellated, in much the same way as a Finite Element Analysis mesh is generated, then sliced into layers according to the specific machine parameters. The machine then builds the component by sequentially creating and bonding each layer to its predecessor to reproduce the 3D artefact. The process depicted in Figure 3.3 pushes a layer of powdered gypsum over the print area. The machine then deposits a binder through standard inkjet printer heads in just the areas that need to be solid. The next layer is added, and so on. The final part is then removed from the cake of unbound powder.

3D printing devices are virtually desktop items, currently about the size of a free-standing office photocopier. Models produced using 3D printing are in the £10–£30 sterling region and are therefore relatively inexpensive, although costs vary considerably depending on the RP process and materials used. The key difficulty is creating CAD models that can be translated into an error-free STL file, which is required to ensure successful production of the part. STL stands for 'standard triangulation' or 'stereolithography' language and is the standard RP machine interface data format. RP machines are sensitive, and creating STL files from surface/wire frame models is almost impossible. A solid CAD model is required and the construction industry predominantly uses non-solid modellers. Bureau services for the industry, such as Solinova (2006), will remodel the object in solids from the design drawings in order to build the 3D model. Resolving this technical problem would undoubtedly accelerate the uptake of this technology.

3.4. Comparison of VR and PR

There are some differences and similarities between VR and RP. The VR model allows the exploration of a virtual environment in real time, with feedback. A physical model is static, but where (as is often the case) the virtual model is viewed though a 2D screen, the physical model is viewed 'naturally' and only needs to be adjusted for scale rather than mentally translating two dimensions into three. Viewing ports are limitless in VR but are limited with a physical model. VR can give a 'correct' impression of the full-scale, real-world view to the viewer. RP modelling can only hope to generate scale models of reality. Surface texture is important in architecture. VR, however, can only simulate the look of a surface that has a particular texture. Depending on scale, RP can model precise texture.

Technical issues can blight both technologies. This is principally in the translation of data from CAD to STL/VRML, which is never fully automatic. Checking the quality of the model, making decisions about what should appear in the VR model, or regarding the minimum feature size and scale with the physical model all require human input and decision-making. For the application of both technologies, there has to be a clear need and desire to bring these tools into design practice. They do, however, require specialist knowledge to be used successfully, and only large projects or organizations are likely to invest in the technical personnel and equipment to use these technologies.

VR can simulate full-scale environments. If RP could produce full-scale models, a real environment could be created for evaluation, but if this can be built at full scale then why not print for real? The digital technology is available to achieve this, it is the materials and process that are lacking.

3.5. Physical reality: Printing buildings

In the manufacturing sector, automation using industrial robots and machines that used direct numerical control took hold in the 1960s. The development of microprocessors delivered computer numerical control in the 1970s, and the computer revolution in the 1980s brought CAD software. In the 1990s, with the advancement of CAD and the increasing power of low-end computer systems, virtual reality software products became viable. At the same time, advanced parametric modelling was introduced and the industry has since enjoyed the development of the integration of design and analysis tools and machine control. All these technologies can be found in construction (Howe, 2000; Whyte, 2002; Kolarevic, 2003; Schodek *et al.*, 2005).

The introduction of CAD/CAM (computer-aided design/manufacturing) for the creation of huge structural components for buildings is driving the development of digital manufacturing technologies for construction. Construction design software is available that can model parametric 3D solids and support design and analysis through to CNC manufacture. These technologies are proven, and the world's leading designers and engineers are identifying the need for new technologies. Cutting edge designs for buildings are becoming increasingly unrealizable using the current state-of-the-art methods and there are calls for new materials and processes (Egan, 1998; European Construction Technology Platform, 2005).

The control of material placement and a reduction in the number and quantity of materials will play a key role. The manufacturing sector is turning to rapid prototyping technologies for solutions, especially for the production of highly personalized products exemplified by the Invisalign tooth aligner product (Invisalign, 2006). The construction industry is beginning to explore the potential that automated additive technologies offer for solving these problems. It is likely that new processes will drive down the cost of existing construction, while raising the bar of achievable construction design solutions. Initially, new technologies are most likely to find niche applications, but will ultimately filter down to the domestic sector – exemplified by the Tunnelform system (The Concrete Centre, 2004) which has moved from civil engineering to residential applications.

Constructing building components in an automated layer-by-layer fashion, by selectively depositing materials in their final location, requires a move away from conventional construction practice. Pegna (1997) realized that automating construction by employing machines that reproduce human processes is prohibitively difficult. He argued that the key is to build using a process that automates a few simple elemental operations, the control of which allows structures of sufficient complexity to be useful. The rapid prototyping model is such a process. Pegna experimented with selective bonding of sand and cement to create freeform structures from traditional building materials. Rapid prototyping processes are capable of automating huge mould-making processes today (Anon, 2005). It is not yet possible to print buildings using a process controlled directly from a digital model of a building, but, worldwide, two institutions are currently developing full-scale, layer-based technologies with this capability.

The first demonstrated, large-scale freeform process for construction, Contour Crafting, has been developed in the laboratory at the University of Southern California (Khoshnevis, 2002). It has been used to produce large (more than 1 m) freeform wall structures that would replace the structural concrete block wall similar to that found in UK house construction. The process extrudes the internal and external 'skin' of the wall to form a permanent shutter that is then backfilled with a bulk compound similar to concrete. Using thixotropic materials with rapid curing properties and low shrinkage characteristics, consecutive layers of the wall can be built up rapidly. The wall material deposition process is a two-stage operation. In order to improve the finish of the visible surfaces, the shutter material is shaped by a secondary manipulator, or

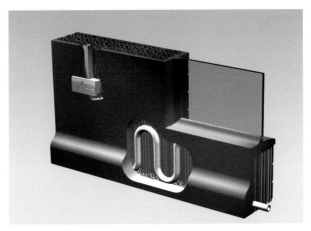

Figure 3.4   A concept wall constructed using a large scale, layer-by-layer approach.

trowel, as it is extruded. The combination of processes results in a system that can deposit (relatively) large quantities of material while maintaining a high quality surface finish.

Contour Crafting, however, cannot take full advantage of the extended functionality that can be embedded within the wall structure if the principles demonstrated by existing rapid manufacturing processes are applied. Precise control of very small volumes of build material (1–10 mm^3) could potentially allow the wall to be constructed, ground up, with all the internal pipe work, conduits and channels in place, removing structurally redundant material. The implications of such an approach could lead to: better design solutions using geometric freedom; a smaller number of build materials and a reduction in the material resource required for construction process; simplified on-site operations with a reduction in trade coordination; resolution of interface issues, part count reduction. The process would be capable of reproducing any geometry within the process parameters. Freeform construction processes would allow a designer to build to design rather than design to build. This would mean that designing for disassembly and recycling at the end of life would be more straightforward than with conventional construction systems. In addition, acoustic, permeability and thermal characteristics could also be modified by 'printing' appropriate, optimized geometry, so adding value to the structure (Buswell *et al.*, 2006).

Figure 3.4 depicts a concept domestic wall structure that might be designed using freeform construction principles. The key features are: optimized thermal and structural performance using a single material; appropriate voids left for service conduits and pipe work; internal and external surfaces finished; window frames, etc. moulded in the primary construction material, removing the requirement for second fixing.

Work is currently being undertaken at Loughborough University (UK) to develop a process capable of delivering the concept 'Wonderwall' at full scale[3]. Building components will be manufactured on a machine with a build bed measuring 3 m × 4 m × 5 m (see Figure 3.5). With any of these processes, the size of the build bed will restrict the size of the component that can be manufactured, with larger parts being made by

---

[3] IMCRC (EPSRC) project – 'Freeform Construction: Mega-scale rapid manufacturing for construction' – commenced in December 2006.

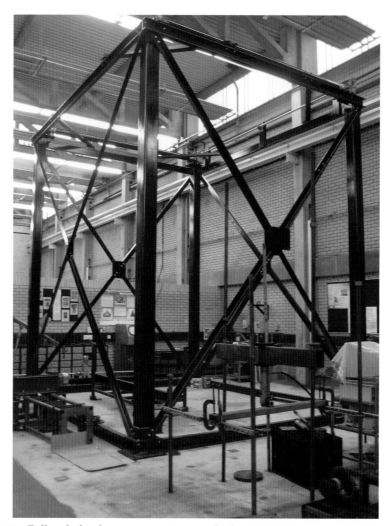

Figure 3.5   Full scale freeform construction test facility.

joining smaller components from several builds. To print a building, this deposition device could be mounted on some sort of mobile gantry system that would sit over the plot of the house; becoming an in-situ, mega-scale RP machine. A smaller machine could print sections or components of a building that are then assembled on site. The machine could either print components in-situ, or could be used to manufacture parts offsite.

3.6. Changes in design thinking

Conventional construction design involves exploration and iteration, where concepts are investigated in order to establish a suitable solution to the requirements. Once the concept design has been established, there is a clear definition of what is required.

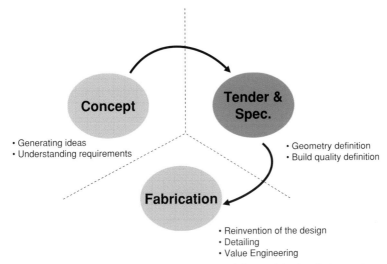

Figure 3.6   Conventional building procurement and flow of design information.

The task then is to rationalize this and describe these requirements so that others can detail the design and construct the project. The components of the building are then often analysed, value engineered, redesigned and coordinated with the other building systems, as depicted in Figure 3.6. Interface issues between components and between the contracts under which the project work has been let, are resolved. This redesign in effect serves to validate the original design.

These operations are facilitated if BIM is used as a virtual prototyping tool. The manufacturing sector has long been aware of the value of prototyping – from the creation of component assemblies using 3D solid modelling (digital prototypes) to the production of physical prototypes (Wiedemann and Jantzen, 1999). Resolving design issues computationally has led to shortened design times (Hilton and Jacobs, 2000). In construction, the application of the virtual prototype is being proved to be key in resolving many coordination issues as well as reducing cost and time on site (see Chapter 5). The ability to digitally prototype a building is also influencing the design and evaluation of performance (see Chapter 1).

A BIM is created using a 3D solid modelling tool, design software that has other advantages. Over the last two decades, solid modelling tools have developed alongside materials for rapid prototyping. End-use parts can now be produced using rapid prototyping processes and hence rapid manufacturing is possible (Hopkinson *et al.*, 2006). A key advantage of rapid manufacturing over manufacturing processes such as injection moulding, is the removal of the requirement for tooling. Within the parameters of the process, any geometry in the solid modeller can be reproduced by rapid manufacturing. The designer no longer has to worry about problems such as re-entrant shapes, draft angles and slit-line locations. The product can be manufactured to design rather than designed for manufacture (Hague *et al.*, 2003). The next step for layer-based manufacturing is to control material properties by mixing materials at deposition or by controlling the material micro structure at deposition. This is enabled by the ability to completely model a component digitally and use this information to directly control the build parameters of a manufacturing process.

Possibly the biggest challenge facing the construction sector is the move away from 2D design tools to 3D model representation. Designers would no longer have to convey intent through the description of surfaces, it can be done using solids. This will promote a shift in design thinking based on solutions that select sheets of material with different properties, forming and fixing them together. As with rapid prototyping, freeform construction design solutions can be thought of in terms of adding value through the increase of function by exploiting the control of the geometry of the final material. One example where the solid design thinking has changed and inspired innovation is in the design of acoustic absorbers (Godbold *et al.*, 2007). Virtual reality and digital prototyping will become more important for proving design before construction. The notion of the surface being an entity will disappear and the surface will simply become a property of the solid function objects such as the walls and floors of buildings.

3.7. Material and fabrication process requirements

The selection of construction materials is based on a balance of aesthetic appeal and functional requirement and many materials are available. It seems that the increasing variety and function of materials is giving more options for the designer, and consequently more materials are used to produce today's buildings. There are thousands of materials that could potentially be used in modern construction (Addington and Schodec, 2005).

Despite the increasing number of materials available to the designer, for specific tasks, core structural materials (such as timber, steel and concrete) have not changed greatly in the last 100 years, and are arguably still the most versatile construction materials available (Margolius, 2002).

Freeform construction technologies will require new materials, or at least the development of existing materials. The materials must be deposited selectively and precisely by computer controlled deposition systems. The materials used by these processes will have build specific parameters associated with them (Ashton, 1969; Austin and Robins, 1995):

- cure time
- rate of strength development
- final compressive and tensile strength
- ability to self-support (arch forming)
- form holding (slump)
- delivery rate
- mixing and delivery system
- shrinkage characteristics
- mix and cure thermal and moisture characteristics
- final material thermal and moisture performance characteristics
- finish characteristics
- weathering capabilities (water penetration, solar radiation, freeze/thaw, thermal cycling, thermal shock, airborne/ground carried chemical reactions)

These parameters will affect the design of structures. The design solution may not be 'I want that, how are we going to build it?' But more that the final form and function of the building could be influenced by the build parameters of the selected material and freeform construction process.

3.8. Conclusions

VR and RP are two technologies that will develop and become more commonly available tools for construction design, management, manufacturing and supply teams. The technical issues in converting information from one source (often modelling software) into data that can be used for either VR or RP are an obstacle. Human intervention is currently required as part of this process. RP has the additional disadvantage in that solid modelling tools are required to produce the information to control the RP machine.

VR gives the ability to digitally prototype a building, which allows interaction and real-time feedback through interface equipment such as audio, visual and haptic devices. Not only can the finished building be experienced, but the construction sequence can be explored. RP can generate physical objects directly from digital data. These are currently restricted to scale models of, or parts of, buildings. The concept is scalable, but the processes are not.

Mega-scale rapid prototyping is realizable using the RP concept to drive a purpose-built process that will use appropriate materials. Design optimization and integration of function and form will be possible. This requires a rethink of the traditional design paradigms that will require greater analysis and visualization. VR will play a key role in envisaging how the processes will operate and how they will be integrated into current construction practice.

References

Addington, M. and Schodek, D. (2005) *Smart Materials and Technologies for the Architecture and Design Professions*, 1st edn. Oxford, Architectural Press.

Anon (2005) RCT takes layer-by-layer approach to mold, core production. *Modern Casting* February 1. Available at: http://www.allbusiness.com/manufacturing/ fabricated-metal-product-manufacturing/350621-1.html [Accessed 10th October 2007].

Ashton, H. E. (1969) Weathering of organic building materials. *Canadian Building Digest* **CBD-117**. Available at: http://irc.nrc-cnrc.gc.ca/pubs/cbd/cbd117_e.html [Accessed 10th October 2007).

Austin, S. A. and Robins, P. (1995) *Sprayed Concrete, Properties, Design and Application*, 1st edn. Latheronwheel, Caithness, Scotland, Whittles Publishing.

Autodesk (2006) *Autodesk Home Page (UK)*. Available at: www.autodesk.co.uk [Accessed 10th October 2007].

Bedworth, D. B., Henderson, M. R. and Wolfe, P. M. (1991) *Computer-Integrated Design and Manufacturing*, 1st edn. New York, McGraw-Hill.

Bentley Systems (2006) *Bentley Systems Home Page*. Available at: www.bentley.com [Accessed 10th October 2007].

Buswell, R. A., Soar, R. C., Thorpe, A. and Gibb, A. (2006). Freeform construction: Mega-scale rapid manufacturing for construction. *Automation in Construction* **16** (2): 222–229.

Dickin, P. (2001) Computer aided design and rapid prototyping. In: *Rapid Prototyping Casebook*, 1st edn (ed. J. A. McDonald, C. J. Ryall and D. I. Wimpenny). Wiltshire, The Cromwell Press, pp. 5–12.

Egan, J. (1998) *Rethinking Construction*. London, Department of the Environment.

European Construction Technology Platform (2005) *Strategic Research Agenda for the European Construction Sector: Achieving a Sustainable and Competitive Construction Sector by 2030.* European Construction Technology Platform, www.ectp.org

Foster and Partners (2006) *Foster and Partners Home Page*. Available at: www.fosterandpartners.com [Accessed 10th October 2007].

Fulcro (2006) *Fulcro Engineering Services Home Page*. Available at: www.fulcro.co.uk [Accessed 10th October 2007].

Godbold, O., Soar, R. C. and Buswell, R. A. (2007) Implications of solid freeform fabrication on acoustic absorbers. *Rapid Prototyping Journal* **13** (5): 298–303.

Hague, R., Campbell, I. and Dickens, P. (2003) Implications of design of rapid manufacturing. *Proceedings of the Institution of Mechanical Engineers, Part C: Journal of Mechanical Engineering Science* **217**: 25–30.

Hilton, P. D. and Jacobs, P. F. (eds) (2000) *Rapid Tooling: Technologies and Industrial Applications*, 1st edn. New York, Marcel Dekker.

Hopkinson, N., Hague, R. J. M. and Dickens, P. M. (eds) (2006) *Rapid Manufacturing: An Industrial Revolution for the Digital Age*, 1st edn. Chichester, John Wiley.

Howe, S. (2000) Designing for automated construction. *Automation in Construction* **9**: 259–276.

Invisalign (2006) *Invisalign Home Page*. Available at: www.invisalign.com [Accessed 10th October 2007].

Khoshnevis, B. (2002) Automated construction by contour crafting: related robotics and information sciences. *Automation in Construction* [Special Issue: *The Best of ISARC 2002*] **13** (1): 2–9.

Kolarevic, B. (2003) *Architecture in the Digital Age: Design and Manufacturing*. New York, Spon Press – Taylor & Francis Group.

Margolius, I. (2002) *Architects & Engineers = Structures*, 1st edn. Chichester, Wiley-Academy.

Pegna, J. (1997) Exploratory investigation of solid freeform construction. *Automation in Construction* **5** (5): 427–437.

Schodek, D., Bechthold, M., Griggs, K., Kao, K. and Steinberg, M. (2005) *Digital Design and Manufacturing: CAD/CAM Applications in Architecture and Design*, 1st edn. Hoboken, NJ, Wiley and Sons.

Solinova (2006) *Solinova Realising Innovation*. Available at: www.solinova.co.uk [Accessed 10th October 2007].

Sweet, R. (2003) Buildings that build themselves. *Construction Manager* **October**: 14–17.

Tekla (2005) *Tekla Corporation Home Page*. Available at: www.tekla.com [Accessed 10th October 2007].

The Concrete Centre (2004) *High Performance Buildings: Using Tunnel Form Concrete Construction*. Camberley, Surrey, The Concrete Centre.

Whyte, J. (2002) *Virtual Reality and the Built Environment*, 1st edn. Oxford, Architectural Press.

Whyte, J. (2003) Innovation and users: Virtual reality in the construction sector. *Construction Management and Economics* **September** (21): 565–572.

Wiedemann, B. and Jantzen, H. (1999) Strategies and applications for rapid product and process development in Daimler-Benz AG. *Computers in Industry* **39** (1): 11–25.

Wohlers, T. (2004) *Rapid Prototyping, Tooling & Manufacturing: State of the Industry*. Fort Collins, Colorado, Wohlers Associates.

Z Corp (2006) *Z Corp Home Page*. Available at: www.zcorp.com [Accessed 10th October 2007].

4 Digital affordances: Emerging knowledge and cognition in design

Tuba Kocatürk

This chapter stresses the need for a critical understanding of the structure and state of the knowledge that has emerged with the new digital approaches in architectural design and production. This chapter will review and assess the emerging technologies, tools and processes facilitating the design and realization of digital design, with a special focus on freeform architectural practice. Digital technologies not only assist designers in the creation and realization of complex architectural forms, but the different capabilities they provide also start to define new tasks, values and concepts which will shape the image of the emerging design practice. The semantic relationships and dependencies between the emerging properties of architectural form and the digital processes characterize the contextual and dynamic nature of the domain knowledge. The emerging design knowledge is identified in a technological and interdisciplinary context, with comparisons of the digital and predigital design culture in order to highlight the transformation of established understanding about design and its associated knowledge.

4.1. Introduction

Digital technologies have affected architectural design and production on three distinct planes. First is the extent to which these technologies are integrated into the design and production processes, and the emerging skills that architects are expected to exhibit, which pertain to the diverse set of digital tools that have been introduced into the profession. The second is the emerging formal and tectonic qualities and varieties in architectural form facilitated by the generative digital media, which have provided the means to generate and describe complex architectural forms. Apart from its representational role, the new CAD (computer-aided design) software has also attained a generative and evolutionary function as opposed to the traditional use of digital media. The new generative function allows designers to start programming information even before the creation of an object. In other words, we observe a slight shift from object to information that not only challenges the traditional role of architects but also their relationship with other disciplines. According to Chaszar (2003), such techniques and approaches to form generation have given rise to a different cognitive model of form as well as a different vocabulary of forms. Thirdly, as observed by Benne (2004), technology is not only added to the existing design processes, but it facilitates 'a combined techno-organizational change', where the respective roles and links among the participants change along with the technology and the knowledge content.

At the most generic level, a design process starts with the generation of a form, according to the formal, functional, tectonic, aesthetic and methodological intentions of the designer. This form needs to be physically and/or mathematically described and represented for design evaluation as well as for the subsequent engineering and production processes. In the meantime, the overall building form has to be divided into rational cladding components combined with a rational supporting structure, during which the fabrication methods, alternatives and economies have to be taken into consideration in relation to the formal and behavioural properties of the materials comprising the tectonic elements of the surface and the structural system. There is actually no definitive or linear order between these phases, rather a cyclical loop during which the design is continuously regenerated and reshaped. For conventional design and production processes, designers could manage these iterative processes intuitively, given their experience and familiarity with the standardized building elements and construction methods, which constitutes the general design knowledge of a designer. Nonetheless, the emerging digital processes extend and add to the existing design knowledge with the definition of new tasks, concepts, organizational structures and dependencies between the cross-disciplinary processes.

In order to identify the knowledge elements that have emerged in relation to the digital technologies in architecture, we need to focus on the characteristics of this knowledge, which are rarely generic but rather dependent on the domain to which each applies. Among the domain-dependent types of knowledge, we can identify knowledge related to:

- domain terminologies
- the formal characteristics of the artefact
- the representation of the artefact (geometrical and non-geometrical properties)
- processes (from design generation to production)
- the semantic relationships and dependencies between various processes

Domain knowledge is also related to particular tools, since tools have their own independent processes for particular classes of problems.

In the following sections, we will particularly focus on the three stages of digital design and production, namely: form generation, representation and physical production, as well as various forms of new interactions between these stages which contribute to the emerging collaborative knowledge in the domain.

4.2. Emerging concepts in form generation

4.2.1. Digital/virtual generation of architectural form

Today, architects are offered an immense set of generative design tools, each requiring new skills. These tools also have enormous influence on the users' creativity, which contributes to the emergence of new form generation strategies. Digital design tools and techniques allow various approaches to form generation, conception and the search for new design vocabularies to be explored by designers.

One approach focuses on the geometry, and is rooted mainly in the aesthetic and conceptual intentions of the architect. The most common approach is the creation of shapes by the direct use and manipulation of computational tools (e.g. lofting, sweeping, Boolean operations) found in most modelling environments.

In another approach, shapes are generated according to the predefined sets of rule structures that lead to controlled parametric shape variations. The data sets and algorithms driving these approaches can vary widely; they may have a performance (e.g. construction, structural) rationale behind them, or they may be driven purely by aesthetic and conceptual intents. The digital environments that are widely used in CAD/CAM applications (e.g. CATIA, SolidWorks, Unigraphics) provide performance-based capabilities of parametric design (Schodek *et al.*, 2005). Some of the animation software (like MAYA) also has parametric capabilities, allowing the users to define animation paths along which particular instances of a form can be created by freezing the form at certain instances. The designer may also set variables of dimensional, relational or operative dependencies between the parts of an architectural form.

Rule-based procedures, which rely heavily on scripting, allow the creation of complex models that could otherwise be difficult to generate only by dimensional variation. Some rule-based approaches comprise the generation of designs via various forms of growth and/or repetition algorithm, expressed as a set of generative scripts, defining the evolution of new forms. For example, 'genetic algorithm' is a known method of evolutionary form generation mimicking the rules of nature: such as mutation and reproduction.

Another approach carries the discussion of architectural space into a virtual plane. It claims that architecture should not only be concerned with designing analog space but also with digital space, and seek to dissolve the boundaries between the virtual and the physical (Andia, 2002).

4.2.2. Mixed approaches: Physical and virtual prototyping in form generation

The alternative approach to digital form generation is to sketch and sculpt forms using analog/physical media for the concept generation, and then to build a digital model by fitting the mathematically defined curves and surfaces to the original hand-shaped model (Mitchell, 2001). The advent of digital three-dimensional scanning techniques has facilitated and enhanced the transfer of the geometry of the physical model to the computer environment. From a physical model, a digital representation of its geometry can be created using a 3D digitizer, by capturing vertex, edge and surface coordinates. This digital data is then ready to generate NURBS (non-uniform rational B-splines) curves and surfaces, which is then used to build a digital model that is utilized as the basis for creating a new physical model. Rapid prototyping (RP) devices, such as 3D printers, can be used to build physical models for visual inspection and comparison with the original model.

4.2.3. The shift from analog to digital and the emerging cognitive processes

According to Oxman (2006), the main difference between digital-based design and paper-based design is the explication of the cognitive processes during the generation and evaluation of designs. While design problem structuring was earlier a very implicit process and mainly took place in the designer's head, advanced design media offer tools to explicitly represent these problem structures, to build and communicate alternative structures. In other words, digitally mediated design offers alternative ways pertaining to how to mentally construct designs (internal representations) and how to communicate information about these designs within the design team (external representations). A variety of these tools exist, which differ according to the extent to which they bridge the gap and/or provide alternative links between the digital constructs and their material representations. Consequently, we observe an abundance of the traditional views of design and the emergence of new cognitive models/constructs based on the emerging relationships between the designer, the design object (artefact) and information in an interdisciplinary context.

4.3. Multiple and cross-disciplinary modes of representation

The changing form of representation from analog to digital has also affected the nature of information that is required along with these representations. 'Representation' refers to the representation of information (geometrical or non-geometrical) that is embedded within the design object for design evaluation, collaboration and for the subsequent analysis and production processes. Earlier generations of CAD software (e.g. AutoCAD) only represented the geometrical properties of the architectural elements. These drawing entities only included geometrical aspects of the objects they represented (Ibrahim and Krawczyk, 2003). The advent of BIM (building information modelling) introduced the value of integrating information and graphics within the building model. Embedded information can describe the geometry, as well as materials, specifications, code requirements, assembly procedures, production and any other related data associated with how the object is actually going to be produced. This value is particularly crucial when the embedded information is spontaneously propagated in the geometry of the building model, which can be updated automatically in coordination with the overall configuration. Geometrical manipulation can also be integrated with the analysis and simulation methods and 3D modelling. Such advanced digital performance can also have a dramatic effect on the coordination of interdisciplinary design teams. Consequently, the digital representations raise the question of appropriate geometric representational formats, and the level of data development appropriate for each stakeholder's function in the design and production process. Accordingly, the quality and the quantity of the information to be exchanged vary depending on:

- the parties sharing this information according to their design task
- the media of exchange
- the stage of the design process

   As opposed to the conventional forms of 2D representations (plans, sections, elevations, etc.) the complex spatial forms are not defined separately in different plans, elevation and sections, but directly as a virtual, 3D model that is constructed on the computer (Ruby, 2001). The modelling environment of the 3D geometry is very influential for the subsequent engineering and production processes and for the digital continuum, as well as for post-processing. Whether a 3D model is represented using

NURBS or a solid modeller becomes crucial when the data is exchanged between the architect, engineer and the manufacturer.

While some environments are better for conceptual and preliminary design, others support the more structured design development phase. It is not the intention of this chapter to give a complete overview of various design software: however, the range of qualities that the design tools exhibit is useful for evaluating them in relation to the design tasks and the design phase they support.

Conceptual design is very different from the design development or design for construction and production. During the conceptual design phase, which rather focuses on visualization, designers require immediate feedback from the digital (or physical) 3D models. Hence, conceptual modelling environments are primarily used for conceptual design and rendering (e.g. MAYA, form-Z, 3D studio max), relying on both solid and surface representations. As described by Schodek *et al.* (2005), there is no built-in intelligence in these environments that would point out conflicting geometry as can be found in various design development environments. Many of these environments have little or no interface for constructability or structural efficiency analysis. Different from the conceptual design tools, the design development tools have evolved primarily to support the rational development phase of the design, which is more structured and involves the evaluation of the design intentions developed during the conceptual design phase. These tools and their environments consist of representation capabilities for parametric surface patches and solid representations (e.g. CATIA, SolidWorks, Pro/Engineer) based on BIM systems, in which embedded information can describe both geometric and non-geometric specifications.

The way in which models are built and manipulated is quite different in these various digital environments in regard to the support they provide at different phases of the design with different tasks. Similarly, data transfer from one application to another may require varying file exchange standards for each application. Moreover, although geometric data transfer is quite successful for most applications, non-geometrical information embedded in these representations may be lost due to the lack of support and extent of the design task each application provides. Therefore, the designer must initially have a clear understanding of the expected design outcome to use these environments efficiently. The power of these environments becomes more evident once the basic understanding of the intent and structure of these modelling environments is gained. This requires not only new skills for architects but also an understanding about the changing nature of tools, their potentials and limitations, and, above all, having a critical understanding of the theories behind these tools to be able to use them effectively.

4.4. File-to-factory processes: CAD/CAM and the transition from digital to physical prototypes

CAD/CAM (computer-aided design/manufacturing) technologies enable the direct and indirect manufacture of digitally generated architectural parts and components, usually by plotting them on a 3D output device (e.g. a computer numerically controlled (CNC)-cutter) or by having them carved out from a solid material (e.g. by a milling machine). CAD/CAM technologies can be distinguished from the industrial production processes in the sense that they allow the design and manufacturing of complex, non-orthogonal custom products with high speed and ease. However, these technologies do not dictate a certain type of design approach or a fixed formal language, but allow a new design vocabulary to be explored by designers (Schodek, 2000). This approach contradicts the modernist approach to technology, which was characterized by the search for a formal language to reflect it and employ it as its stylistic expression.

CNC manufacturing processes may be examined by categorizing them from a number of perspectives. Designers may be interested in the geometric varieties these machines could allow and/or the range of materials they are capable of processing. In current practice, they are mainly categorized according to their distinct process types – as subtractive (based on material removal), formative (by applying heat and force), additive (adding material layer by layer) – which are also referred to as direct processes eliminating the need for tooling. There are generally three application areas in which additive, formative and subtractive processes can be usefully applied in the design and production of complex double-curved forms. During early design phases, they can facilitate the production of scaled study models. Alternatively, during the form development process, CAD/CAM techniques can also be effectively used to create prototype components and assemblies for both the evaluation and verification of design decisions (Rotheroe, 2001). The most common application of CNC fabrication during the early stages of design is the 'additive process'. Also known as 'layered manufacturing' or 'solid freeform fabrication', rapid prototyping is a computer-controlled, additive fabrication process used to fabricate physical objects directly from CAD data sources. Rapid prototyping technologies employ unique methods of adding and bonding materials – polymers, paper or powdered metal – in successive layers to create objects that can be used as concept models or functional prototypes. However, they have a limited application area in the actual production of large scale building components due to the restriction in the size of objects and the high production costs associated with them. Among the methods currently in use are stereolithography, selective laser sintering (SLS), fused deposition modelling (FDM) and three-dimensional printing (3DP). Rapid prototyping is extensively used to enable project teams to visualize concepts, evaluate new designs and test functional models of complex curvilinear geometries before committing to expensive tooling.

4.5. Dimensions of emerging design knowledge

In current digital design practice, several differences have been observed in the ways and extent to which designers incorporate the digital and traditional tools and technologies into their working processes. Similarly, how the design and implementation processes are structured and conducted vary from one practice to the next, which have immediate consequences on the formal qualities of the final artefact. A comparative analysis of two projects[1], exemplifying how the design and implementation processes (e.g. design intentions, representation, rationalization, materialization and fabrication) have been structured and conducted in each, has contributed to the understanding of the different dimensions of the emerging design knowledge and the factors contributing to its complexity. Figures 4.1 and 4.2 both serve to formalize and exemplify the strategic and situational interrelatedness of some of the generic concepts that have evolved in response to the extensive use of digital tools and techniques in design and

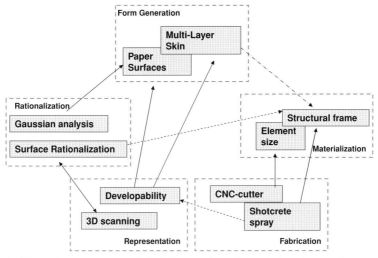

Figure 4.1 Emerging concepts and their interactions (knowledge elements) across various design stages. (Experience Music Project, Seattle, designed by Frank Gehry.)

production. These concepts and the emerging interactions among them are identified as the main sources of emerging design knowledge in the domain of digital freeform design (Kocatürk, 2006). For example, the emerging concepts under each category such as 'Gaussian analysis' (Figure 4.1) or 'Cold-forming' (Figure 4.2) provide the declarative and procedural elements of the emerging knowledge domain, which is referred to as 'process' and/or 'task' knowledge. While declarative knowledge comprises the facts that people know (what), procedural knowledge comprises the skills people know how to perform (Anderson, 1985). According to Jong and Ferguson-Hessler (1986), an effective knowledge repertoire for solving problems in semantically rich domains also contains strategic and situational knowledge elements (how and why) (Jong and Ferguson-Hessler, 1986).

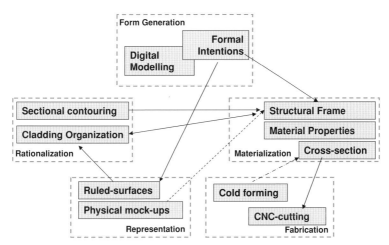

Figure 4.2 Emerging concepts and their interactions (knowledge elements) across various design stages. (Dynaform, Frankfurt, designed by Bernard Franken.)

In Figures 4.1 and 4.2, the links and associations created between the concepts (declarative and procedural knowledge elements) can be recognized as the strategic and situational dimensions of the knowledge. While situational knowledge is necessary to recognize problems for the selection of the relevant concepts, strategic knowledge refers to the conscious and tactful decisions made. These links may change both in meaning and form with regard to how they are associated according to the viewpoint of the project participant who creates an association between two or more concepts. An additional observation concerning the dependencies between concepts is that they can be linked at any phase of the design process even though they belong to different phases of design. For example, constructability criteria can influence the form generation process even at the very early stages of the design. Similarly, a specific production technology (e.g. CNC-cutting) will influence the maximum thickness and size of the cladding materials that can be processed using this specific technology, and will, in turn, inform the organization of the cladding components.

The comparative study of various cases has also contributed to an understanding of the differences between collaborative and individual creation of knowledge (Kocatürk, 2006). The highly non-linear process of knowledge exchange between stakeholders facilitated by digital design environments has initiated the creation of shared meanings between members of the design team. The resulting knowledge is defined as 'collaborative', which is constructed through the interaction of multiple actors, and embodies the dynamic elements of knowledge. This view contradicts the product oriented view of knowledge and propagates a rather process oriented view of knowledge, which is the key to understanding the emerging collaborative knowledge in the digital design realm.

4.6. Conclusions – The implications of the new technological context

This chapter has stressed the changes in design culture due to the extensive use of digital media facilitated by various virtual/physical prototyping technologies. The following have been recognized as the key areas facilitating the emergence of the new knowledge content of digital design/production in architecture, and the ongoing cognitive transformation:

- The extent to which digital technologies are integrated into the design process starting from generation through to the actual production of artefacts
- The emerging formal/tectonic qualities and varieties
- The changing socio-organizational roles and responsibilities of stakeholders

Several differences have been observed in the ways and extent to which designers incorporate digital tools and technologies into their working processes. Similarly, the ways in which the design and implementation processes are structured and conducted vary from one practice to the next. The emerging interactions between the design and production processes become highly non-linear and dynamic, leading to the emergence of a new, cross-disciplinary and collective body of design knowledge. In this respect, the emerging design knowledge becomes intrinsically interdisciplinary, which is constructed collaboratively between the various parties taking part in the design and implementation process. An interesting area of research, in this respect, would be the re-definition of the concept of 'design cognition'. The emerging interactions between design (mental creation) and production (material creation) processes and the unprecedented ability to construct various complex representations – facilitated by digital media – serving different cognitive purposes is the key to understanding the emergent cognition. At this point, the concept of cognition needs to be extended by including the 'interacting individuals'. This view is defined as 'distributed cognition', which moves the discussion away from one single discipline or individual to a group of multidisciplinary individuals situated within the context of complex environments. In the light of these developments, understanding the characteristics of the emerging design knowledge and design cognition is key to the future developments of virtual technologies to provide the necessary support for collaborative design. This should be achieved not only by facilitating 'data exchange' among disciplines, but also by providing the means for collaborative and creative generation and utilization of design knowledge.

References

Anderson, J. R. (1985) *Cognitive Psychology and its Implications*, 2nd edn. New York, W. H. Freeman.

Andia, A. (2002) Reconstructing the effects of computers on practice and education during the past three decades. *Journal of Architectural Education* **56** (2): 7–13.

Benne, B. (2004) *Techno-Organizational Models to Support Construction Industry Work Processes*. PhD thesis, University of California, Berkeley.

Chaszar, A. (2003) Blurring the lines: An exploration in current CAD/CAM techniques. *Architectural Design* **73** (1): 111–117.

Ibrahim, M. and Krawczyk, R. (2003) The level of knowledge of CAD objects within the building information model. *Proceedings of the 2003 Annual Conference of the Association for Computer Aided Design in Architecture* (ed. K. Klinger), Indianapolis, CA. [Accessed 20th August 2005, from the CUMINCAD Online database: www.cumincad.scix.net/ .]

Jong, T. D. and Ferguson-Hessler, M. G. M. (1986) Cognitive structures of good and poor novice problem solvers in physics. *Journal of Educational Psychology* **78** (2): 279–288.

Kocatürk, T. (2006) *Modelling Collaborative Knowledge in Digital Free-Form Design*. PhD thesis submitted to Delft University of Technology, Faculty of Architecture, Department of Building Technology.

Mitchell, W. J. (2001) Roll over Euclid: How Frank Gehry designs buildings. In: *Frank Gehry, Architect* (ed. J. F. Ragheb). New York, Guggenheim Museum Publications, pp. 352–363.

Oxman, R. E. (2006) Theory and design in the first digital age. *Design Studies* **27** (3): 229–265.

Rotheroe, K. C. (2001) Manufacturing freeform architecture. In: *New Technologies in Architecture I – Digital Design and Manufacturing Techniques* (ed. M. Bechthold, K. Griggs, D. Schodek and M. Steinberg). Cambridge, MA, Harvard Design School, Technology Report Series 2001-1.

Ruby, A. (2001) Architecture in the age of digital productability. In: *Digitalreal: Blobmeister – First Built Projects* (ed. P. Cachola Schmal). Basel, Birkhäuser Verlag, pp. 206–213.

Schodek, D. (2000) Closing comments. In: *New Technologies in Architecture, Digital Design and Manufacturing Techniques* (ed. M. Bechthold, K. Griggs, D. Schodek and M. Steinberg). Cambridge, MA, Harvard University Press, pp. 74–75.

Schodek, D. L., Bechthold, M., Griggs, J. K., Kao, K. and Steinberg, M. (2005) *Digital Design and Manufacturing: CAD/CAM Applications in Architecture and Design*, Hoboken, NJ, John Wiley & Sons..

Part 2
Challenges for implementation: From virtual through to construction

5 One Island East, Hong Kong: A case study in construction virtual prototyping

Martin Riese

To know where the construction industry will be in terms of innovation and design process collaboration in 15 years' time, one needs only to look at the aerospace and automobile industries today. Highly pervasive applications of digital tools enable the integration of geometric coordination, structural analysis, simulation, systems and manufacturing processes and life cycle and business management into a single, all-encompassing 'process dashboard', which delivers an overall improvement in value of at least 50% over all aspects of design and fabrication, across the entire supply chain.

5.1. Introduction

The One Island East project in Hong Kong is one of the most substantial implementations of electronic 3D building life-cycle management ever undertaken. This owner-driven process has combined the entire traditional design and construction team into one virtual 3D electronic building design and construction information process. The progress of the 3D building information modelling (BIM) has been kept ahead of the construction and has prevented literally thousands of clashes and design errors from reaching the construction site.

The technology has also enabled a number of design changes to be incorporated efficiently and later in the process. The result is that the owner is anticipating a 10% overall cost reduction and a significant schedule reduction on this $US500 million project. Taking into account the substantial use of energy and other resources by the construction industry as a whole, the savings brought about by new design and construction technologies and working methods are essential to preserving our environment. These powerful new tools also bring with them design possibilities that would otherwise be impossible.

5.2. One Island East, Hong Kong

One Island East is a new 70-storey office tower in Quarry Bay, Hong Kong. The project owner, Swire Properties Limited (SPL), established a Building Information Model project office adjacent to the construction site. The owner, the project design consultants, the project BIM consultant, Gehry Technologies (GT), the main contractor, Gammon

Construction Limited, together with a number of subcontractors, worked together in the same project space to create and maintain a single, 3D electronic construction building information model. The owner implemented BIM as part of the initiatives to help to bring about the following improvements in this project, and ultimately to the local construction industry as a whole:

- Comprehensive 3D geometric coordination of all building elements prior to tender
- Enhanced quantity take-off from the BIM to improve the speed and accuracy of the preparation of the bill of quantities in Hong Kong Institute of Surveyors format prior to tender
- Lower, more accurate, tender pricing resulting from the reduction in the contractors' unknowns and risks
- Automation and interoperability of 2D documents with 3D building information model data
- Creation of a reusable catalogue of intelligent parametric building parts (knowledge capture)
- Management of construction sequence and process modelling using the BIM elements
- Reduction of waste in the construction throughout the entire process
- Reduction of contractor requests for information (RFIs)
- Reduction of claims on site resulting from incomplete design coordination
- Quicker construction
- Lower construction cost
- Standardization of the construction supply chain and regulatory authorities
- Enhanced site safety
- Better build quality
- Facilities maintenance using the BIM elements

Project team members from the organizations of the owner, the architect, the structural engineer, the mechanical and electrical services design consultant, the quantity surveyor, potential main contractors and potential subcontractors were trained in the use of Digital Project – the software tool chosen to create the BIM for this project. The advantages of this particular software tool are:

- Built on CATIA, the software that helped to revolutionize the automobile and aerospace industries, Digital Project provides the best possible geometric modelling capabilities. Digital Project has been specifically developed to adapt sophisticated 3D tools from other industries to construction.
- Automated clash detection and management.
- Comprehensive mechanical and electrical systems' routing tools.
- Built-in scripting functionality to enable integration of the project's requirements for customization.
- Automated concurrent file versioning and file sharing management.
- Ability to manipulate and manage very large amounts of data.
- The same software used by other industries such as automobile and aerospace (transfer of knowledge between industries).
- Integration and interoperability with Microsoft Project and Primavera scheduling software.

- Designed to streamline and reuse process knowledge.
- File sharing and collaboration over the internet using emerging data compression technology.

In this project, the BIM implementation began after the scheme design phase. The BIM model was used as a vehicle to enhance detailed design phase coordination and cost estimation, and to enable the potential main contractors to provide more accurate bids for the construction. The owner instructed the successful main contractor and major subcontractors to use the BIM model to coordinate, record and communicate all aspects of the construction of the actual building.

Experience showed that the contractor brought an order of magnitude increase in granularity to the BIM. In general, the contractor greatly enhanced the BIM model that had been provided by the design team. In this process he introduced construction level detailing as necessary to create and vet the combined builder's work drawings (CBWDs), which he used on site to construct the building. (Site workers still use these 2D drawings to lay out the built works.) The culture of passing design responsibility to the contractor is still inherent to the procurement route selected for the project. It is well within the capability of the technology as it exists today to fully automate the production of 2D drawings – at least to the general arrangement drawing level. As a true virtual building prototype, the BIM model itself is the most detailed record of what has been constructed. The traditional combined services' drawings (CSDs,) which record the 'as built' condition, will be produced automatically from the BIM model.

On this project, the owner and the contractor share and maintain the master BIM in the project office. Changes and improvements in the design and construction process initiated by the contractor are communicated to the owner using Digital Project, and are then merged into the master BIM by the contractor (Figure 5.1). The owner is confident that by implementing BIM, the building will be completed quicker, cheaper and better.

Another central purpose of BIM is to improve the construction process (Figure 5.2). It is commonly known that the relative success of the construction process is largely influenced by the quality of the information provided to the constructors. Figure 5.3 shows the order of magnitude of information that needs to be coordinated in just one portion of a modern conventional building, prior to construction. All BIM data is placed strategically in a 'tree' structure, which makes it easy to locate elements, even in very large databases. This also helps to optimize performance in large databases (Figure 5.4).

Central to the success of a team of 25–50 designer-modellers, working together to produce a single BIM, is the ability to identify and manage clashes. All clashes (roughly 2000 in number) between the plumbing, HVAC (heating, ventilation and air conditioning) and electrical systems and the surrounding structural and architectural elements were identified and resolved before inviting potential contractors to tender for construction (Figure 5.5). Prior to using BIM, this kind of event may not have been detected until construction, potentially causing the project cost and time penalties.

Another significant advantage provided by the Digital Project is that it allows rule-based parametric object creation. In this project a number of simple rules for the assembly of a part of the building stability system (including structural engineering requirements) have been inputted via a graphically friendly template (Figure 5.6). Steel plates and stiffeners are no longer modelled one at a time, but rather are generated directly by the programme according to the rules and governing geometric framework. This saves

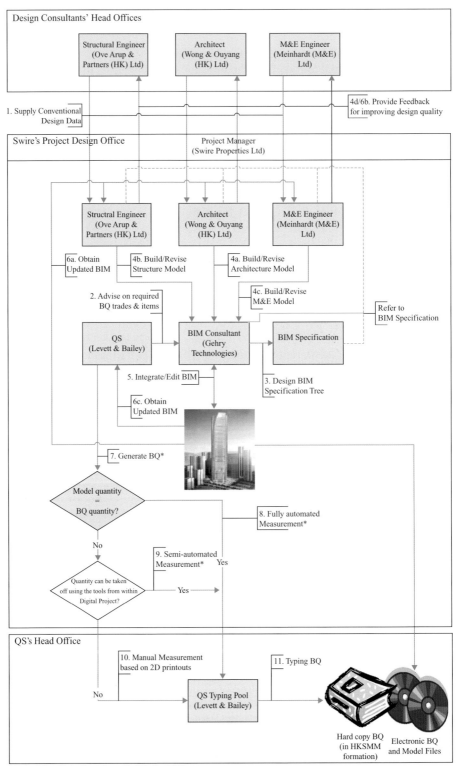

Figure 5.1 The (pre-tender) collaboration framework. (Figure supplied by K. D. Wong, T. C. Tse, Y. M. Leung and M. Riese, 2006.)

Figure 5.2   The construction site: OIE, Sept 26, 2006. (Figure supplied by the author Martin Riese.)

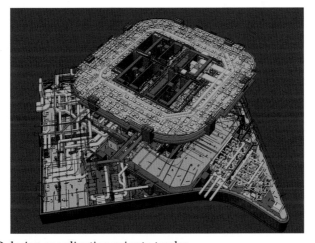

Figure 5.3   3D design coordination prior to tender.

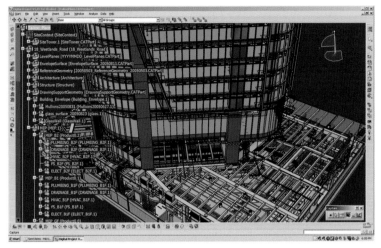

Figure 5.4   Partial BIM data (tree) structure.

time and reduces errors, and by enabling dynamic change management, enhances the collaborative design process.

All necessary information about building elements, such as size, material, weight, location and sequence, is organized and integrated into the BIM model. On this project, quantities taken from the BIM were formatted in the Hong Kong Institute of Surveyors format using automated scripting functions. As the design developed, the database of quantities was automatically updated. The quantity surveyors were able to track costs more quickly and accurately during the design process (Figure 5.7). Rather than spending time using scale rulers to try to measure quantities on different sets of large-scale 2D paper drawings, the quantity surveyors were able to spend more time researching the market to find where the best prices for the project could be obtained. This helped

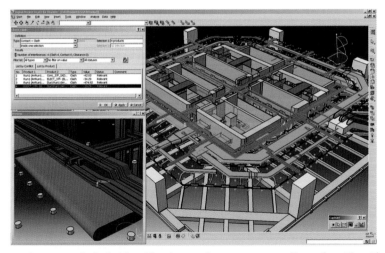

Figure 5.5   Automated clash identification and management allowed to identify a clash between an electrical cable tray and an air supply duct.

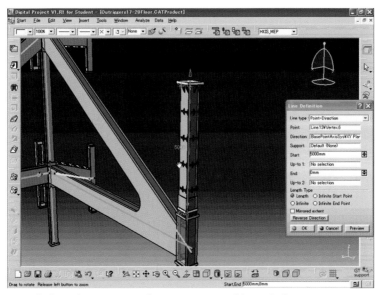

Figure 5.6   Rule-based parametric object creation and knowledge capture.

to save the project money and gave the owner and the design team quicker feedback on the development of the design.

5.3. File sharing and interoperability over the internet

New forms of 3D data compression, combined with emerging file sharing protocols, make it possible for construction project teams to develop and share large, complex BIM models over the internet. The highly effective project team collaboration that took

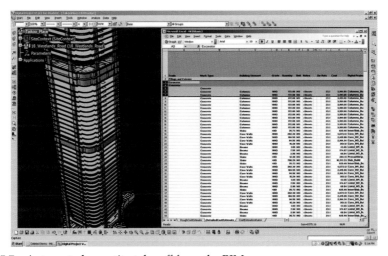

Figure 5.7   Automated quantity take-off from the BIM.

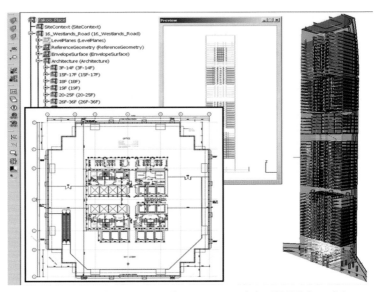

Figure 5.8   Interoperability between 2D documents and the 3D BIM model.

place in this project BIM office can also take place in the virtual project office of any construction team. Owners and project managers can have full, continuous and instant visibility into the current state of the BIM database without the need for drawing issues or special meetings. All elements of the BIM contain hyperlinks to individuals, manufacturers or design teams relating to those elements, thereby enabling instant collaboration at any level of the decision-making chain – over the internet.

5.4. Interoperability between 2D documents and a 3D BIM model

In the automobile and aerospace industries, the need for paper drawings has been greatly reduced. However, due to the nature of construction, paper drawings and tables are likely to be with us for the foreseeable future. Current BIM software technology provides interoperability between the 2D paper drawings, Excel spreadsheets, construction schedules and other 2D information and the BIM model (Figure 5.8). If the design is changed in the 3D model, the 2D information is automatically updated. Similarly, if the design is changed in the 2D document, the 3D model is also automatically updated. This interoperability saves time, reduces errors and can provide advantages to the designers, while ensuring consistency between all types of information documenting the project.

5.5. The importance of accurate MEP modelling

The One Island East project showed the importance of comprehensive MEP (mechanical, electrical and plumbing) modelling. Simple space reservation for MEP elements does not provide sufficient accuracy to coordinate the various elements of the project to the level required for cost certainty. It is the removal of these unknowns prior to tender

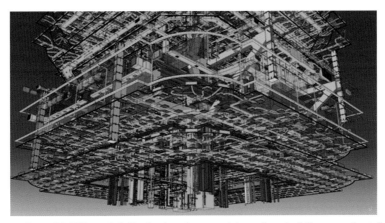

Figure 5.9   Partial MEP model displaying the transition of typical floor to plant floor.

that helps to bring the tender prices down. A large majority of the effort required to create the One Island East BIM model was related to MEP elements (Figure 5.9). Effective coordination of MEP elements with other MEP elements often has an effect on the ultimate coordination of the MEP elements with the rest of the project.

A number of MEP design options were investigated using the 3D process. Once the MEP elements have been created with the automated tools, changes are often easier to make. This process of iterative change management is the model for all future design. The quality of the design improves as the number of iterations is incorporated. In future, functional and engineering aspects of the MEP system will become part of the process. The BIM model will be a functioning virtual prototype of not just the geometry and the cost, but of performance aspects of the system. The tools already incorporate a high degree of automation, which approaches the level of an 'expert system' – and even the beginnings of 'machine intelligence'. Building upon this functionality, the overall building design will be informed by MEP system performance both before and after construction. The virtual construction prototype will go on to be the backbone of the building management system by processing real-time feedback from construction and as-built equipment.

Engineering analysis, simulation and business process management tools that are fully integrated into the BIM will ultimately provide an exhaustive combined simulation of the actual structural, MEP and life cycle uses. This will give the decision-makers greatly enhanced transparency into – and control over – the overall efficiency and effectiveness of the entire process. On the One Island East BIM project, the weekly team virtual building prototype walk-through became a primary tool for identifying coordination issues. Often problems were identified and solutions found within minutes of each other. This reduced the volume of interconsultant correspondence and reduced the need for e-mail chains.

5.6. Construction sequence modelling and primavera integration

The elements of the project created for the BIM model can be used by the constructor to explore, design and optimize the sequence of construction. Figure 5.10 shows how

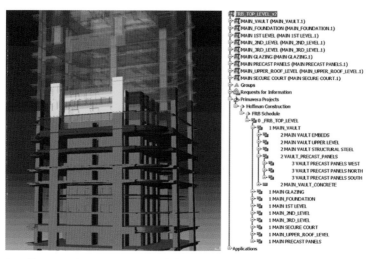

Figure 5.10   Construction sequence modelling: Primavera sequence 1 and 2.

the construction sequencing information is fully integrated with the BIM data 'tree' structure. By comparing different construction sequences and methodologies, the contractor can have a much clearer understanding of the time required – and the associated risks – before he submits his price. When he initiates changes on-site, he can determine in advance what effect these changes will have on the outcome of the entire sequence. It is anticipated that in the future this construction process modelling will be linked with other construction process dashboard functions to enable fully informed high level decision-makers to effectively drive the entire process from the beginning and from the top.

Primavera integration allows Digital Project users to link 3D BIM components to Primavera™ activities and to simulate construction activities in 3D (4D navigation). It integrates the project schedule into the BIM model, which enables:

- Import projects, work breakdown structures, and activities from Primavera™ into Digital Project
- Link imported activities to master model geometry
- Visualize in 4D: play a construction schedule over time
- Propagate any changes made in Primavera™ to the Digital Project environment

Integrating 3D design models and construction schedules reduces risk through:

- Greater transparency between trades
- Early identification of schedule conflicts (e.g. shared work zones, blocked access, etc.)
- Early identification of process-induced design clashes (e.g. permanent works that block crane or labour access, etc.)

The BIM elements can be used as part of the contractor's process of developing the construction sequence and methodology. The tools enable sequencing of the elements

Figure 5.11   One Island East pre-tender 4-day cycle construction process model. (Figure supplied by Gammon Construction Limited and Hong Kong Polytechnic University Construction Virtual Prototyping Laboratory, Professor Heng Li and Stephen Kong.)

and visualization of the construction methodology. When designing his construction sequence and methodology in 3D using the actual parts of the building, the constructor can precisely predefine most of the tasks that will be performed by his operatives on site. By encouraging the operatives not to deviate from preplanned construction sequences, site safety is improved (Figure 5.11). In future, the tools will be accessible from the construction process main dashboard, and will manage all aspects of construction productivity. Although not used on One Island East, the implementation of radiofrequency identification (RFID) of all elements of the project will integrate the material, human resource, business and overall process management functions with the design process, the life cycle management process and the process dashboard. The process dashboard will be the instrument of the ultimate decision-maker on the project. This could be the owner or the contractor, and its development and implementation may become the domain of specialized process engineering consultants.

BIM images are sometimes criticized for not being 'sexy' enough. It is commonplace in the industry today to produce seductive 3D visual images about how wonderful the future built environment will be. But in reality, to get to those images, buildings have to be built first. In order to reduce waste and become more efficient, we need to start by knowing where every duct, pipe, light and door hinge is – relative to each other. It is a huge process, which benefits greatly from the power of the computer to store, manipulate and organize very large amounts of information. Using BIM technology, the global construction industry is now beginning to implement the process of transformation. The experience of the One Island East BIM project shows that the construction industry is already beginning to benefit in the following ways:

- Fully coordinated design information prior to tender resulting in lower tender prices
- Semi-automated and more efficient quantity and cost control
- Industry Foundation Classes (IFC) interoperability and compatibility

- New design possibilities
- 3D construction sequence modelling to shorten build time and increase safety
- Reduced risk and reduced claims resulting from the construction process
- Buildings built sooner, cheaper and better
- The emergence of machine intelligence in the design process
- The eventual disappearance of paper drawings
- BIM lead-in to advanced facilities management tools

5.7. The future of the collaborative construction environment

The digital design director of Skidmore Owings and Merrill (New York Office, USA), Paul Seletsky (2005), defined the future of the collaborative construction environment as the following:

> New computationally-driven simulation methodologies being developed both within academia and commercially, can (and will) virtually simulate everything from basic lighting, energy, wind, and pedestrian circulation conditions to more advanced construction, fabrication, code, material, and security conditions. Easily misunderstood as supplemental engineering data (the mundane, statistical informa-tion, commonly applied after-the-fact to design projects) – the new Digital Design argues that digital building simulation will embody the future of architectural prac-tice; that those practitioners seeking a wider role beyond that of form-giver will be significantly empowered by the use of tools generating such analytical information as it is applied *before and after* the fact – from the project's conception, into its design and construction phases, and then well beyond, into its occupancy and life cycle management stages. Properly understood and utilized by the profession this entails a significant rise in the architect's stature, as the advantages of informed, rather than speculative, decision-making become self-evident. *Properly ignored*, the results may very well promote Construction Managers into a lead decision-making role, whereby architectural design is subsumed as a service within the construction firm – and in instances where more recognizable architectural talent is desired, can then be readily *'licensed'*.

Some 15 years from now, the construction industry will have been transformed into a highly efficient unified process, which integrates design ideas that are fully informed by exhaustive, iterative engineering analysis and simulation, with a seamless, factory-based, optimized manufacturing and assembly process. Construction will have become a holistic organism in which all stages in the process inform each other through a technologically enabled network of collaboration and information exchange shared by man and intelligent machines alike.

Projects will become ever more cost effective, because designs will be informed by optimized real-time planning and logistics strategies. Energy efficiency, life cycle costs and safety will be greatly improved. There will be one all-encompassing, 3D global asset database of built form, which will be integrated with other rapidly evolving information databases that will be instantly accessible to all.

Through the integration of new paradigms of material and structural analysis and simulation, business management, parametrically driven design analysis and auto-mated generative computer design, construction projects will be more efficient and will take advantage of forms of innovation that are only possible through the use of these new media and tools. The syntaxes of architecture and construction will manifest a higher level of artificially conceived and functionally driven elements and systems that result from this enhanced process. Rather than being limited to semi-arbitrary form making, the beauty of sophisticated logic-driven ideas and explorations will become the driving force in design.

Individuals will be seamlessly connected to the vast network of global collaborators and will become ever more specialized as the never-ending renewal and evolution of the entire process accelerates exponentially. Construction process dashboard controls will become increasingly simple and yet powerful at the top of the decision-making chain, and increasingly granular, numerous and automated at the bottom. There will be substantial 'feedback' from the site and post-construction processes back into the design and manufacturing cycle.

Ultimately, a perfect union will be achieved between the imagination of the individual designer and a massive interconnected symphony of planning, production and life cycle management. The process itself will gain an increased life of its own and will become, in itself, a recognizable revolutionary presence in future society, which will help to empower the political structures that currently define the limits of the industry.

5.8. Conclusions

As the improvements in the construction process on pioneering implementations like the One Island East project spread to the regional and global construction industry as a whole, significant gains will be realized. In future, the implementation of new collaborative enabling technologies and working practices will greatly enhance the breadth and sophistication of the iterative construction process. Design and construction will become a singular, fully unified process that will not only be more efficient, but will in itself open up a new world of design possibilities.

Acknowledgements

An abridged version of parts of this chapter can be found in the *Southeast Asia Building Magazine*, March/April 2006 issue. Special thanks are due to SEAB for permission to reprint parts of the article. Portions of this paper also appeared in the *Proceedings of the Australasian Universities Building Educators Association (AUBEA) Conference 2006*. Special thanks are due to the conference organizer for permission to reprint parts of the article in this chapter.

References

Seletsky, P. (2005) Digital design and the age of building simulation. *AECbytes* **31** October.
Tse, T. C., Wong, K. D., Wong, K. W., Leung Y. M. and Riese, M. (2006) Building information modelling – a case study of building design and management in One Island East, Hong Kong. *Proceedings of the Australasian Universities Building Educators Association (AUBEA) Conference 2006*, 11–14 July, University of Technology, Sydney.

6 The virtual building: A designer's perspective

Martin Simpson

Have we made any significant advances in the last 25 years? Is it right to assume that the construction industry of 2030 will be any different to today's?

Since the publication of *Rethinking Construction* (Egan, 1988) there has been significant investment in White Papers at regular frequency by the construction industry. These White Papers rightly tend to deal with the big picture and not so much with the specific tools or philosophies required to deliver the reports' conclusions. However, these Papers do base much of their conclusions and recommendations on the ideals that have proved successful in other manufacturing industries. It makes sense to begin the search for our solutions in these industries and then adapt, modify and amend them so that they fit the business models of the construction industry.

Virtual prototyping and lean manufacturing are not new concepts. They have been applied with great success in many manufacturing industries: automotive and aerospace in particular.

This chapter will argue that the construction industry must adopt lean manufacturing as one of its key philosophies in order to reach the stated goals of the recent White Papers. Many technological solutions can provide improvements to the construction industry; however, this chapter will focus on the advantages of the virtual prototype both as a philosophy and as a tool.

6.1. How do we define a virtual prototype?

As this chapter is written to encourage discussion about the impact and development of virtual prototypes, we should start by looking at the *Compact Oxford English Dictionary of Current English* (Soanes and Hawker, 2005) definitions of 'virtual' and 'prototype':

- virtual – (*adj.*) not physically existing as such but made by software to appear to do so
- prototype – (*noun*) the first or preliminary form from which other forms are developed or copied

However, in its truest sense a virtual prototype is not only a prototype but also a philosophy and a set of procedures (D'Adderio, 2001):

- A prototype, in that it contains links to all data, information and codified knowledge required to design, engineer and manufacture a product

- A philosophy, in that it outlines the technological and organizational path that organizations have to follow to integrate their development processes
- A set of standard procedures, in that it is created in such a way to facilitate the physical manifestation of the digital model to be produced from the digital model and will be based on industry best practice

In today's society where the term 'virtual' is overused, it is easy to confuse a virtual prototype with a computer visualization. A virtual prototype must have intrinsic links to engineering design and manufacturing knowledge. Therefore multi-physics simulation is an essential part of the prototype methodology. However, in order to realize the full potential of the virtual prototypes, they must be used in conjunction with design and construction philosophies such as lean manufacturing.

6.2. Lean manufacturing philosophy

Lean manufacturing is a holistic and comprehensive management philosophy (Wikipedia, 2006). It is designed to be implemented into the core philosophy of an enterprise, with the primary focus on reducing or eliminating waste. The three principles of lean manufacturing are:

- Improvement in quality
- Reduction in production time
- Reduction in cost

Lean manufacturing is not a new concept and many wise industrialists from Ben Franklin to Henry Ford have strived to employ its principles. When lean manufacturing is successfully implemented it can dramatically improve the process of design and manufacturing to create superior products. A good example of this is the way Toyota revolutionized the automotive industry in the 1980s.

The principles of lean manufacturing can also be found at the heart of key philosophical reports for the construction industry, such as *Rethinking Construction* (Egan, 1998), *Modernising the Construction Industry* (Bourn, 2000) and *Accelerating Change* (Department of Trade & Industry, 2002).

Many people equate the improvement of efficiency with cost-cutting. The problem with most designs within the construction industry is that the true cost is determined very late in the process, and sometimes only after the project has been completed. Engineers and architects create design solutions based on familiar specifications, materials and methodologies. Feedback of the real construction costs into design is essential. We have design tools to optimize for strength and weight, but we rarely get sufficient cost data to enable the design team to optimize for cost.

Having worked on several major projects around the world it is most unusual if the design team does not go through late 'value-engineering' exercises. These exercises rarely bring value and are in reality 'cost-cutting' because the projected cost of the project exceeds the client's budget. The later these changes occur, the more the redesign costs (Figure 6.1). It seems fairly obvious, but unless we can optimize and improve cost certainty at the early stages of the project we can never fully reduce waste and efficiencies in the design and construction process.

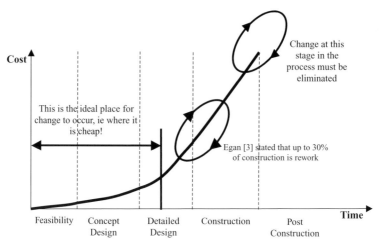

Figure 6.1   The cost of change throughout the construction life cycle.

Examples of the duplication of tasks between design team members are very common, but only through using share digital models and employing intelligent collaboration tools can we streamline the processes involved and reduce the occurrence of errors and misunderstandings. It can be argued that one of the key drivers for inefficiencies and cost increases is the occurrence of late design changes. Figure 6.1 illustrates that the cost of change can increase exponentially as the design moves through the construction life cycle. It is accepted that post-construction changes cost an order of magnitude more to implement when compared to implementing the same change during construction.

The virtual prototype has a significant role to play in the implementation of cost certainty. By embracing greater simulation earlier in the design phase and developing technology that allows cost data, along with engineering and manufacturing knowledge, to be implemented much earlier in the life cycle, we dramatically reduce the risk of changes required during the later stages.

6.3. Whole-life costs

The true cost of a product is not embodied in the manufacturing process alone but across the complete life cycle: from cradle to grave. For the western world to remain competitive in a global economy we may go one step further and adopt 'cradle-to-cradle' as our preferred working ethos (McDonough and Braungart, 2002). Cradle-to-cradle is a new philosophy where deconstruction and recycling become key aspects of the design and manufacturing process. One man's waste is another man's food. But that is a debate outside the scope of this chapter.

However, if we concentrate on the immediate goals of eliminating waste and optimizing the manufacturing processes then we can find suitable arguments in favour of investment in simulation at the design stage by examining a typical 'design–build–finance and operate' (DBFO) contract. These are usually subject to a 25-year concession and are now seen as the standard model for large public schemes such as hospitals,

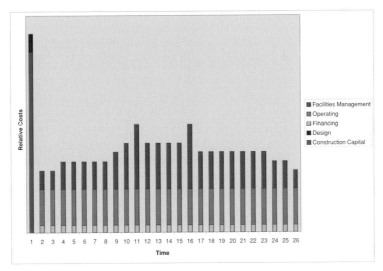

Figure 6.2   Possible costs for a typical 25-year DBFO contract.

schools and roads. It is generally accepted that the continuing costs of facilities maintenance (FM) outweigh the initial capital expenditure. Some reports estimate that FM can cost up to about five times the capital construction value (Figure 6.2).

Early stage design up to the detailed design stage can have a significant effect on the upfront capital cost of the project, and yet the continuing FM costs only account for about 10% of the overall construction cost. In other words, design costs about one-sixtieth of the sum it can affect.

Again the virtual prototype has a significant role to play in improving the efficiency of design when the whole-life costs are taken into account. The ability to reduce clashes between services and structure is important during construction, but it is the ability to design spaces and link component prototypes to multi-physics simulation that will ultimately yield the optimal designs. Once a prototype has been produced for the purpose of clash detection, it requires only slight modifications to be able to use the same information to create an asset management tool that can be used for the whole life of the project.

6.4. Why change?

Obviously the elimination of waste should not be limited to the design and manufacturing process alone, but rather should be taken into account on the whole-life costs of a project. There are eight areas where waste can occur. All can be reduced at the design stage, but only with adequate access to knowledge of the costs, methods and processes involved in the construction and continuing maintenance of the project.

- *Defects*: -The effort involved in inspecting for and fixing defects
- *Over-production*: Making more than what is needed, or making it earlier than needed
- *Waiting time*: Products waiting on the next production step, or people waiting for work to do

- *Transportation*: Moving products further than is minimally required
- *Inventory*: Having more inventory than is minimally required
- *Motion*: People moving or walking more than minimally required
- *Scrap*: Having goods or materials left over after the process
- *The process itself*

> It is unwise to pay too much, but it's worse to pay too little. When you pay too much you lose a little money – that is all. When you pay too little, you sometimes lose everything because the thing you bought was incapable of doing the thing it was bought to do. The common law of business balance prohibits paying a little and getting a lot – it can't be done. If you deal with the lowest bidder, it is well to add something for the risk you run. (Ruskin, 1872)

This quote, by John Ruskin, is still as true today as when it was first written. Invariably, clients are always tempted by the lowest bidder. It is sometimes possible to argue that a superior product should cost more and that the quality of a product is worth paying more. This should apply both to the physical products and also design services.

By adopting lean manufacturing and eliminating waste whilst improving efficiencies in the process it will be possible to:

- Provide a superior product at the same price as an inferior product, or
- Provide the same product at a reduced price, or
- Make more profit by supplying the same product at the same prices as competitors but employing a more efficient process.

In any of these cases the law of business would favour those who adopt the new technology or philosophies (with the assumption that companies that make more profit can continue to invest in improvements, while those that do not eventually go out of business).

6.5. Barriers to change

So far this chapter has concentrated on the drivers for change. However, we cannot fully appreciate the scale of the challenge within the construction industry until we also examine some of the barriers to change and suggest how these may be overcome.

The construction industry is often perceived as slow to adopt new technologies. In many cases this is an unfair observation. It is easy to look at other manufacturing sectors and forget that the amount of research and development (R&D) expenditure in the construction industry is several orders of magnitude smaller than that of the aerospace or automotive industries. This lack of resource and other factors result in both a reluctance to invest in the tools and the training necessary to change. Consequently, the construction industry is not quick to adopt wholesale changes to technology, philosophies or methodology as soon as they become available.

Barriers to the implementation of new technology and philosophy can be grouped into the categories shown in Table 6.1 and will be discussed in the following sections.

Table 6.1 Barriers and possible solutions to implementation of changes of philosophy and technology in the construction industry.

Barriers to change	Possible solutions
The client's business model – The client's business model has a significant effect on the product created by the design and construction teams. For example, a client who will own and operate the building for its lifetime will be more receptive to investing more in the initial design and production costs than a client who wants to minimize capital cost to sell it on for the most profit.	Though legislation to disclose the ongoing F&M costs is already on its way, ultimately it will be the consumers and end-clients who will change the market dynamics.
	Once it becomes obvious (and quantifiable) that investment in capital expenditure can have significant improvements, either in terms of rental or resale value, then market forces will allow change.
	Until this time, as an industry we need to ensure that data on projects, where whole-life costing is a key driver, is published to allow metrics and business cases to be developed.
The client's financial model – Just as a business model can affect the desire to change the design and production methodologies, the client's internal financial model can also create barriers. Ironically this is the case for public clients who are the biggest advocates of change.	The industry needs to lobby major clients, especially at national policy level, to amend the financial operating structure to enable a more holistic approach to the finance of projects. It is true that the PFI methodologies do allow maintenance and operation to be taken into consideration, but there are many projects tendered under other methods that do not.
At a recent discussion with a client we were told that the budgets for capital construction costs and running costs were totally separate. It was not possible to borrow from the maintenance pot to pay for improvements to the design and construction.	The industry needs to publish metrics that can prove a financial case for altering national and local policy, as well as allowing private clients the confidence that changing the financial structure for the projects will yield dividends.

(continued)

Table 6.1 *(continued)*

Barriers to change	Possible solutions
Time to handover (and therefore income!) – The construction industry is unique in the manufacturing world in the fact that we do not build prototypes. Every project is unique and, by definition, is a prototype. Many design decisions are made very late in the day and it is common for construction projects to be delayed due to changes in specifications and on-site re-designs. It is very common for foundations and underground drainage to start on site before the designs for the superstructure and building services are complete. In some cases the completion date is fixed due to a particular event, but in most cases the design and construction process is set out without a complete assessment of the financial model over the project's lifetime.	Great reductions in waste can be achieved by delaying the start on site until the design has been completed. Build it once and build it right. It is not possible for the construction industry to create full prototypes, but this actually strengthens the case for virtual prototypes. Combined with component-based design the speed of the design process can actually be increased. Egan (1998) stated that many projects overrun in time and budget, but clients invariably think that it will not happen to them.
Resistance to investment in technology – Investment in the hardware and software necessary for virtual prototyping is viewed as expensive when compared to standard CAD software.	The upfront cost (and sometimes annual costs) of the software is expensive compared to CAD. But the advantages that the software brings can be offset against staff costs and the costs inherent in late design change, on-site clashes and errors due to misunderstanding and insufficient information. The additional benefits of working in a 3D environment also highlight a lack of information and design incompatibilities that are not always revealed by over-laying 2D drawings. Further savings and optimizations can be offset against the reuse of knowledge and components from project to project.

The design management and tendering process – The current system of tendering means that a significant amount of redesign takes place late in the project. A concept design may be developed by engineers based on assumptions that are not realized until the contractor appoints the actual supplier systems much later in the process.

Professional indemnity – The climate of individual professional indemnity is divisive and opposed to coordinated team-working. In many cases problems that arise in design need to be solved in a multi- disciplinary and coordinated design sessions. Yet in many cases ideas cannot be developed because it may require one specialist taking on more responsibility beyond their traditional scope.

Aversion to innovation – Many construction industry clients are first-time and one-time projects. As a rule they are not construction professionals. They employ a project manager to manage the project on their behalf. Professional project managers sometimes perceive innovation as risk.

Framework agreements between clients, designers, contractors and supply chains can reduce much of the uncertainty of tendering. The move to component-based design can also assist in allowing ranges of solutions and parts that can be used to create unique designs that do not compromise buildability. Cost certainty and supply chain management can also be incorporated into the information flow.

Collaborative and project indemnities do provide the framework for greater teamwork and coordination, though they are more expensive than relying on individual indemnity.

The industry needs case studies and metrics to prove the benefits of changing the way we work.

Another solution is to create a profession that is able to deal with the technological considerations of modern design – the technological manager.

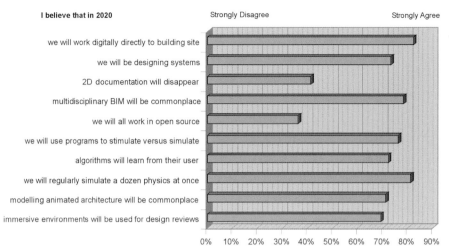

Figure 6.3   Changing practice in 2020.

6.6. Where may the future take us?

So far this chapter has argued that not only does the construction industry need to implement new philosophies of design and construction, but that it also needs to embrace new technology that is modified and adopted from other manufacturing industries. If this is the case then we should look at the current technologies that are in use and try to extrapolate where these may lead in the future.

In 2006 Arup undertook an international survey of over 20 leading academics and industrialists within the construction industry (Luebkeman and Simondetti, 2006). The purpose of the study was not to predict 'what the future practice may bring' in terms of specific tools but rather 'how general practice may change over the next 15 years'. Figure 6.3 shows the response to ten statements.

The research highlighted inconsistencies of vision and scepticism within the construction industry. However, the conclusion was that the industry should expect to see the emergence of four digital specialists:

- *The toolmaker*: Professionals who are able to create bespoke digital tools such as programs or scripts. Toolmakers would have sufficient knowledge of engineering principles and existing digital tools to enable them to create seamless and efficient processes that involve several distinct packages.
- *The modeller*: There are two types of modellers:
  - professionals who can distil engineering principles into the governing criteria so they can be modelled digitally (i.e. the mathematics governing geometry or optimization);
  - professionals who can model geometry and embed the necessary engineering information for use by analysis or design software.
- *The technology manager*: Similar in scope to the modern-day project manager, the technology manager is an individual with solid experience in 3D or 4D modelling, preferably in several different software packages. This person would have

an understanding of how to set up modelling protocols, have a broad grounding in construction techniques and a good understanding of multiple design disciplines (Froese, 2004).

- *The researcher*: In many ways the research is both an attempt to close the gap between academia and industry and also a method for increasing the rate of technology update. One way of satisfying this requirement is to embed the work of the research student, such as an EngD or a PhD, with the industry.

Even though this survey is looking 15 years ahead, it is possible to use recent examples and experiences from past projects, with which Arup has been involved, to show that these professions are already starting to evolve.

6.6.1. Beijing Olympic Stadium

The complex nature of the design concept by ArupSport with Herzog & De Meuron and CAG required a new approach to geometric modelling (Figure 6.4). The solution was to adopt virtual prototyping software from the manufacturing industry to support the process. The whole design team adopted CATIA as the main geometry platform. The use of the software and the specific approach to this complex project were essential to understand the construction and fabrication processes. This gave the contractors and the architect the confidence that the project could be built and that the aesthetics would

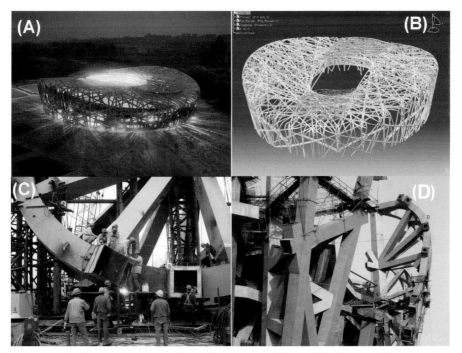

Figure 6.4 Beijing Olympic Stadium: (a) concept; (b) model; (c) and (d) construction. (Reproduced with permission from Gehry Technologies.)

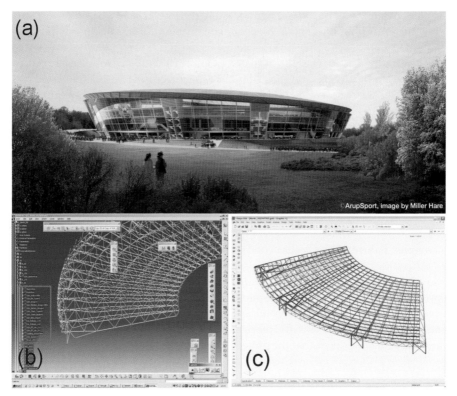

Figure 6.5   FC Shakhtar: (a) concept; (b) geometric model; (c) analytical model. (Figure provided by ArupSport.)

not be diluted by the construction process, respectively. The success of this project was entirely due to the inclusion of specific geometry modellers and an attempt to embed manufacturing constraints into the connection and element design. This project also laid the foundation for further work in linking geometry to analysis and design tools.

6.6.2. FC Shakhtar Stadium

Building on the skills developed on the Beijing Stadium, the ArupSport design team wanted to create a close link between design and geometry software to facilitate both geometric and structural optimization. This was done using a bespoke applet created to link the geometry created in CATIA, the analysis in a structural analysis package (GSA by OASYS), custom design routines and optimization algorithms. This project reinforced, within ArupSport, the need to develop the 'toolmaker' as an integral part of the design team. There were significant advantages gained by the ability to seamlessly integrate and improve the interoperability of various software packages that were not originally developed to fit together (Figure 6.5). It should also be noted that the method employed for transferring digital information was the software interface COM (Component Object Model) rather than the IFC (Industry Foundation Classes) system as this was not adopted by either of the software vendors.

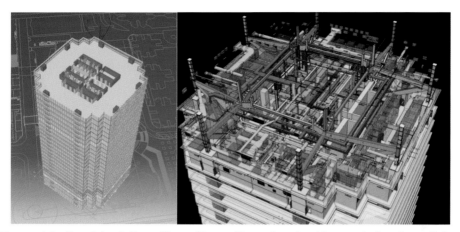

Figure 6.6 One Island East, Hong Kong. (Reproduced with permission from Gehry Technologies.)

6.6.3. One Island East

The project is a 194-m, 69-storey tower in Hong Kong (Figure 6.6). The design and construction programme only allowed six months to produce structural tender drawings. The client, Swire Properties, were aware of the advantages of ensuring interoperability between all project team members from the outset and insisted on using a common platform. A total of 25 team members – including the client's project manager, cost consultant, architect, structural engineer (Arup) and building services engineers – were trained by a Building Lifecycle Management team from Gehry Technologies. The project team worked together to produce a building information model (BIM) that was used for coordination and clash detection. Preliminary feedback has been very positive from the client and project members, both in terms of savings and improvements in efficiency.

One of the key reasons for the success was the use of a model coordinator resident in the project office. In addition to the use of a model coordinator, Gehry Technologies also provided a toolmaker to develop new functionality and enable interoperability in the form of scripts and macros.

6.7. Future extrapolation of current trends

It is entirely possible that, by the year 2030, significant advances in computer software and hardware will render any predictions we currently propose to seem obsolete. However, the construction industry, if it remains true to type, will significantly slow the pace of implementation so that at least some of these ideas may still be in use.

Without doubt, virtual prototyping technology will be used as standard throughout the industry, just as CAD (computer-aided design) is used today. With the advantages of prefabrication we should assume that the construction industry will be almost entirely a factory-driven process.

Building components will be created in factory environments with assembly on-site. This creates some interesting issues as the ability to drive a factory process will be an important feature of the software and philosophies adopted. There are very few packages on the market today that can deal with this aspect of manufacturing.

Even with a prefabrication/manufacturing bias it will still be necessary to complete some tasks on-site, such as the actual assemblies of components, and some trades involved in fit-out will require human intervention. It is entirely possible, but also unlikely, that in 25 years' time we may see some plastering and painting as automated robotic tasks.

The key questions are not whether this will happen but why and how soon? We can answer 'why' by looking at the other manufacturing industries. There is simply a balance point at which employing humans to do a task that can be automated and streamlined is financially uneconomic. How soon? This is already happening. Some of the major building services' contractors pre-assemble several different systems onto a single rig for easy transportation and erection on site. The precast concrete industry and steel industry are factory-based processes. Even reinforcement meshes can be robotically assembled for entire floor plates, delivered to site and rolled out.

Although it is almost impossible to predict exactly how far this technology will get in the next 25 years, the next stage of evolution is clear. Designers must make the transition from thinking about the building as an individual discipline to starting to think in terms of components that are both multidisciplinary and capable of being pre-assembled.

Over the next few years we will see the convergence of analysis and geometry. Many component producers for mechanical systems already produce catalogues, some are in the form of 2D model files and some are even 3D model files. However, engineers do not have the full range of tools and analysis techniques to enable metadata integration between geometry and engineering design. This is a prerequisite to rapid, full systems' comparisons and optimizations. Further integration of space and multi-physics analysis, such as computational fluid dynamics, would be the next stage.

It is quite likely that the legal requirements of planning consent will force designers to set out the predicted energy consumption of a building over a series of standard scenarios; this is already beginning to be seen through the recent revisions to Part L of the UK Building Regulations. At the moment it is not really possible to optimize and test a range of building systems at the concept stage of the design. Yet crucial design decisions are taken on building form and orientation at the very beginning of the conceptual phase with very little engineering input.

Once we are able to link systems together in terms of geometry and design then this will enable the development of intelligent building management systems (BMS). If we look at the level of sophistication between the component management system of a modern aircraft and a modern building there is no real comparison. However, the increase in the complexity of aircraft systems has not solely been driven by safety but also by the efficiency and optimization of maintenance schedules. Though current BMS systems collect sensor data that are not self-aware – i.e. they can tell you that an element has failed but not why – the next stage in evolution will be a BMS dashboard that brings together the sensor data, maintenance schedules and engineering information. This has the potential to show where individual components are in real time, and also whether they are operating in the most efficient manner and how their performance can be improved. For example, it may be able to tell that it is more economical to replace a generator five years earlier than planned because the reduction in energy costs will pay for itself.

We should assume that the consumer of our products will change in the construction industry in much the same way as in other industries. Reliability, energy usage and maintenance determine the whole-life cycle costs, which can be dramatically reduced by correct investment in simulation and technology at the design and construction stages. The production of asset monitoring and management tool will be a standard part of construction contracts in the future. Most of the information is already produced at the design stage anyway; it just isn't brought together in an interactive environment.

6.8. Conclusions

As the construction industry strives to improve quality, reduce overall construction time and costs it will search for alternative methods and philosophies. In many cases it is more efficient to borrow and adapt than to create from scratch. Therefore, many of the principles of lean manufacturing will be adopted and made to work in the construction industry.

A technology platform will be required to support and develop the transition towards a manufactured component-based design and product delivery system. The virtual prototype has the potential to become the backbone of the construction industry, just as CAD is today.

Virtual prototype technology can deliver significant improvements in the construction industry providing that certain barriers to innovation and change can be overcome. In addition to the tools and philosophy, the industry needs professionals to drive these changes forward. Perhaps the most crucial of these emerging professions is the technological manager.

The only certainty we can have about the future is that it will be different to how we imagine it to be. However, it does have the potential to be a very interesting journey.

References

Bourn, J. (2000) *Modernising Construction*. London, National Audit Office. Available at http://www.nao.org.uk/publications/nao_reports/00-01/000187.pdf

D'Adderio, L. (2001) Crafting the virtual prototype: How firms integrate knowledge and capabilities across organisational boundaries. *Research Policy* **30** (9): 1409–1424.

Department of Trade & Industry (2002) *Accelerating Change*. A Report by the Strategic Forum for Construction (Chaired by Sir John Egan). London, Rethinking Construction.

Egan, J. (1998) *Rethinking Construction: The Report of the Construction Task Force to the Deputy Prime Minister*. Norwich, Department of the Environment, Transport and the Regions.

Froese, T. (2004) Help wanted: Project information officer. *European Conference on Product and Process Modeling (ECPPM-2004)*, 8–10 September, Istanbul, Turkey.

Luebkeman, C. and Simondetti, A. (2006) Practice 2006: Toolkit 2020. *Intelligent Computing in Engineering and Architecture* **4200**: 437–454.

McDonough, W. and Braungart, M. (2002) *Cradle to Cradle: Remaking the Way We Make Things*. New York, North Point Press.

Ruskin, J. (1872) Munera Pulveris: *Six Essays on the Elements of Political Economy*. London, Smith, Elder [S.l.].Soanes, C. and Hawker, S. (2005) *Compact Oxford English Dictionary of Current English*, 3rd edn. Oxford, Oxford University Press.

Wikipedia (2006) *Lean Manufacturing*. Available at: http://en.wikipedia.org/wiki/Lean manufacturing

Part 3
Challenges for visualization and simulation

7 Planning and scheduling in a virtual prototyping environment

Andrew Baldwin, C. W. Kong, T. Huang, H. L. Guo,
K. D. Wong and Heng Li

Virtual prototyping extends building information management and product modelling to allow construction organizations to model the construction process, to explore different construction methods and to enable both the client and supplier organizations to visualize how and when construction will take place. This chapter describes how this new way of working may impact on the planning and scheduling of construction work, on how the skills and abilities of those engaged in these processes may change and on possible changes in the management of the planning and scheduling process. An outline is also given of the development of current planning methods and planning systems, together with the principles of planning and scheduling.

7.1. Introduction

The *Collins English Dictionary* (1986) defines planning as: 'a detailed scheme, method, etc for obtaining an objective'. Mawdesley *et al.* (1997) extend this definition to define planning as:

> a general term which is used to encompass the ideas which are commonly referred to as programming, scheduling and organising. Its aims may be defined as: to make sure that all work required to complete the project is achieved in the right order, in the right place, at the right time, by the right people and equipment, to the right quality, and in the most economical, safe, and environmentally acceptable manner.

Planning should not be confused with scheduling: understanding and producing a set of sequenced construction activities. All projects require both planning and scheduling if their design and production is to be successfully monitored and controlled. Prior to Henry L. Gantt, planning and scheduling was undertaken without formal processes or methods. Gantt introduced the 'Gantt Chart', now commonly referred to as 'the bar chart', as a basis of scheduling production activities; and this form of communicating the schedule of construction continues to be used for most, if not all, construction projects. Even the advanced planning and modelling techniques that are incorporated within today's project management systems adopt the bar chart as a means for presenting their results. These systems are invariably based on the network analysis techniques introduced in the mid-1950s and developed in the subsequent years.

The introduction of network analysis for planning and scheduling construction activities was at first limited to major construction projects and large construction organizations. These organizations had the resources, in terms of planning staff and access to computing power, that enabled them to take advantage of the techniques. The introduction of the microcomputer in the early 1980s and the plethora of software that was produced in the following decade provided all organizations with cheap access to computer power and the software to model the construction process. This, however, did not result in the widespread adoption of these modelling techniques on all construction projects. Even with the advent of cheap, powerful computer-based modelling software, it is only on major construction projects that a detailed construction model is maintained throughout the construction phase of the project. For most construction organizations the use of the technique has become primarily limited to three distinct phases within the construction process: in the pre-construction phase (to prepare the tender), at the contract phase (to finalize the contract programme) and the post-contract phase, where the technique has become important in the preparation of claims for the extension of time and additional payments.

To the range of planning and scheduling techniques currently available may now be added the technique of virtual prototyping. To understand how virtual prototyping may impact on construction planning and scheduling it is necessary to review some of the elements of the planning and scheduling process.

7.2. Construction planning and scheduling

Construction project planning is a critical process in the early project phases that determine the successful implementation and delivery of project. During this stage, project planners need to develop the main construction strategies, to establish construction path and assembly sequences, and to arrange construction methods and resources required for the execution of work packages (Koo and Fischer, 2000; Jaafari *et al.*, 2001; Waly and Thabet, 2002). Traditionally, construction documentation has been prepared in a standard two-dimensional (2D) format: consisting of plans, elevations, details, schedules and specifications (Goldberg, 2004). The traditional 2D drawings and paper-based delivery processes, however, limited the capability of visualizing and understanding the design and subsequent construction work involved (Waly and Thabet, 2002). Members of project teams may develop inconsistent interpretations or imagination of the project images when viewing the 2D drawings, and this could result in ineffective communication (Koo and Fischer, 2000). On the other hand, the critical path method (CPM) and bar charts have still been widely employed by project teams as a main tool to express the project schedules and coordinate the activities of members of the project team (Koo and Fischer, 2000). Many project planners have continually relied on these traditional ways in selecting construction equipment, reviewing constructability and arranging construction methods and site layout. These approaches impose a heavy burden on project teams due to the large amount of information and the interdependence of different elements (Waly and Thabet, 2002). In addition, tougher building codes, higher performance requirements, tighter construction schedules and the need to deploy innovative construction methods and technologies have forced project teams to seek new tools to facilitate the better planning and management of contemporary building design and construction.

Such shortcomings of traditional communication tools, together with the advances in digital technologies, have stimulated various research and development efforts to generate new innovative construction process planning techniques in order to enhance the visualization of the construction sequence and finished product. The development of the three-dimensional (3D) computer-aided design (CAD) systems have reduced the burden on verbal and written communication, allowing product designs to transcend differences in location and time (Jim, 2004). A number of software products have been designed to accomplish the digital design (i.e. Bentley Architecture, Graphisoft ArchiCAD, VectorWorks ARCHITECT, Digital Project and Autodesk's Revit and Architectural Desktop). The latest research development relates to the development of a graphical presentation of the construction plan via the four-dimensional (4D) geometric models (i.e. 4D-Planner) (Waly and Thabet, 2002). A 4D CAD model is generated from the combination of 3D graphic images and the time. The 4D visualization technique provides an effective means for communicating temporal and spatial information to project participants (Koo and Fischer, 2000). Finished projects are visualized and spatial configurations directly shown. Visualization of construction plans allows the project team to be more creative in providing and testing solutions by means of viewing the simulated time-lapse representation of corresponding construction sequences (McKinney and Fischer, 1997), and prompting users to think about all the missing details (e.g. site access) (Jaafari *et al.*, 2001). Despite such advancements, 4D CAD models rely heavily on the availability of full plan or schedule information to provide a graphical simulation of the project schedule. The planner mostly uses these tools as a means of visualizing and comparing, rather than for implementing different decision alternatives (Waly and Thabet, 2002). In addition, 4D CAD systems cannot effectively simulate construction processes in which various resources are used to transform construction from one stage to the next of the time-lapse.

Virtual prototyping (VP) is a CAD process concerned with the construction of digital product models ('virtual prototypes') and realistic graphical simulations that address the broad issues of physical layout, operational concept, functional specifications and dynamics' analysis under various operating environments (Pratt, 1995; Xiang *et al.*, 2004; Shen *et al.*, 2005). Dedicated VP technology has been extensively and successfully applied to the automobile and aerospace fields (Choi and Chan, 2004). For instance, an automobile can be fabricated virtually via VP technology, which allows various team members to view the 3D image of the finished product, evaluate the design and identify the production problems prior to the actual start of mass production. However, to date, the development and application of VP technology in the construction industry (i.e. construction process simulation) has been limited. This is probably because each construction project is unique in terms of its conditions, requirements and constraints. A production line in manufacturing industry is almost as constant and stable as the machine's operation is predictable; whilst a construction project is human-dominated, involving various construction parties and uncertainties.

Given its successful implementation in manufacturing industries, various research efforts have attempted to apply the VR concept in forming effective dynamic construction project planning and scheduling tools. Researchers at the University of Teesside (UK) developed VIRCON (VIRtual CONstruction) as a prototype application for the evaluation, visualization and optimization of construction schedules within a virtual reality interface (Jaafari *et al.*, 2001). The Virtual Design and Construction (VDC) method was also designed as a model for integrating the product (typically a building or plant) so that the contractor can design, construct and operate based on the model (Kunz and Fischer, 2005). Virtual Facility Prototyping (VFP) was another interesting method developed at Penn State for visualizing the building facilities during the construction planning phase; and Immersive Virtual Environment (IVE) was designed to improve the project planning process by generating and reviewing construction plans in a virtual environment (Yerrapathruni *et al.*, 2005). Waly and Thabet (2002) developed an integrated virtual planning tool called the 'Virtual Construction Environment' (VCE), which allows the project team to undertake realistic rehearsals of major construction processes and examine various execution strategies in a near-reality sense before the real construction work.

7.3. Virtual prototyping for the construction of high-rise residential buildings

High-rise construction is the predominant form of residential building in Hong Kong and many other cities. Construction techniques for this type of building continue to evolve, and the use of prefabricated components in residential building in Hong Kong is increasing. This increase is driven by better build quality, faster construction and government support. But there are difficulties. The shift from on-site fabrication to off-site factory production increases the risk of miscommunicating requirements and design changes, as the inclusion of suppliers of prefabricated components increases communication layers. The use of prefabricated components also introduces additional construction joints, which require higher design accuracy. The traditional 2D drawings as a means for communicating building design and construction operations are frequently found to be insufficient. Construction planners often have to rethink the construction project many times because there has been no way to capture best practices and carry them over from one project to another. Thus, there is a need to present building components in 3D and to present construction operations in a virtual environment so that the ideas of planners can be captured, communicated and reused.

By implementing virtual prototyping for construction operations' evaluation, constructability data can be evaluated and captured. The data can be used by the construction manager to check design feasibility and provide feedback to the design team. It allows the discovery of constructability problems early in the design stage to minimize cost of change. During the construction stage, constructability data can be used to produce a detailed process plan and generate 3D construction operation instructions for the workers and foreman. The data can also be used for future maintenance planning: 3D maintenance and repairs' instructions can be built based on the data.

7.4. Construction planning in a virtual prototyping environment

To overcome the deficiencies of existing systems, new virtual prototyping techniques have been developed at The Hong Kong Polytechnic University in the form of a construction virtual prototyping (CVP) system. The CVP is a construction process simulator based on the 'Dassault Systemes®' software.

The CVP system was used on a residential building project that implemented virtual prototyping for planning and scheduling the construction work, comprising the

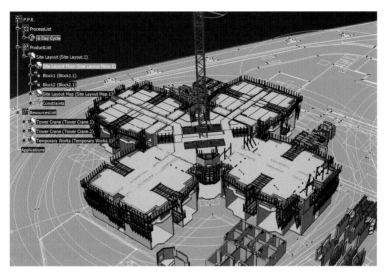

Figure 7.1 Building block layout in a virtual prototyping environment.

construction of two, 41-storeyfloor building blocks with 16 apartments per floor. The structure of these buildings included more than 60% of precast components, ranging from load bearing walls and slabs to toilet and kitchen units. Figure 7.1 shows the layout of a building block in the virtual prototyping environment. This project is pioneering by adopting a large percentage of precast construction in a residential building in Hong Kong. The contractor had two main concerns: the capability of the tower cranes in lifting all the precast components, materials and temporary work facilities; and the sequence of installing precast components. The objectives of using virtual prototyping in this building project were to verify and optimize the 6-day floor construction cycle in the areas of resources' utilization, space allocation, sequence of works and the design of temporary work facilities and precast components. The expected outcomes of this virtual prototyping were reduced construction schedule variances, avoidance of unbuildable or ergonomically unsafe conditions, minimized change orders after design completion and shortened response time to changes and unplanned conditions in the construction site.

7.5. The implementation of virtual prototyping

There are three main phases in implementing virtual prototyping: the project requirement collection phase; the 3D models' building phase; and the process simulation phase. Table 7.1 depicts the tasks, information and people involved in these three phases. They are discussed in detail in the following sections.

7.5.1. Project requirement collection phase

In the project requirement collection phase, major project challenges are identified and these become the basis for defining the scope of works. The challenges can be

Table 7.1   The three phases of virtual prototyping implementation.

	Project requirement collection phase	3D models' building phase	Process simulation phase
Tasks	Identify design and construction challenges Define scope of works	Create 3D models of building components, temporary work facilities and plants Carry out digital mock-up to check dimensional conflict	Simulate planned construction process Validate construction sequence Find time–space conflict Check and optimize resources' utilization Try alternative construction plan
Information required	Preliminary construction method statement Architectural drawings of major building components Master construction programme Preliminary floor cycle programme	Workshop drawings and layouts of building components, temporary work activities facilities and plants Detailed floor cycle programme showing divided working bays	Detailed process programme Productivity rates of different activities Safety plan
People involved	Architect Engineer Project manager of main contractor Representatives from major subcontractors	Project planning team of the main contractor Suppliers of building components and temporary work facilities	Project planning team of the main contractor Representatives from major subcontractors

divided into those of design-related and those of construction-related. Design-related challenges come from the coordination of building components and design details of temporary work facilities to ensure harmonized building design and construction operations. The major design challenges in the Hong Kong residential building project mentioned above were the coordination of welding joint and reinforcement layout of precast components, and the coordination of working platform design. Construction-related challenges come from uncertainty in the method of construction, the duration of an activity and the level of resources' utilization. The major construction challenges in this project were ensuring that the tower cranes were not overloaded, and preparing the best sequence of installing precast components to streamline other construction works, like concreting, rebar fixing, formwork fixing, etc. The information required during this phase includes a preliminary construction method statement, architectural drawings of major building components, a master construction programme and a preliminary floor cycle programme. The people usually involved in this phase include the architect, the engineer, the main contractor's project manager and representatives from the major subcontractors.

7.5.2. 3D models' building phase

In the 3D models' building phase, 3D CAD models of building components, plants and temporary work facilities are built according to the need for tackling design and construction challenges. Building components, including both precast and in-situ parts, have to be broken down into smaller units that suit the simulation of construction operations. For instance, the slab reinforcement and in-situ concrete have to be broken down into units that fit the divided working bays. The level of detail of the 3D models has to be discussed in this phase so that the models can reflect the situation that needs to be examined. It is obvious that it is not feasible to build 3D models down to the detail of every nut and bolt. The models just need to be built to the detail that can reflect both dimensional and space conflicts, and sometimes enough for examining safety issues. In virtual prototyping, construction plant can have its physical properties, like a degree of freedom in movement, speed and acceleration of movement and association between different mechanical joints. Physical properties are set for evaluating reachable area, travelling time and viewing angle of the plant operator. The 3D models prepared for the Hong Kong residential building project include precast components, reinforcement and in-situ concrete components, conduits and boxes, steel wall and beam formworks, shoring, struts, working platforms and tower cranes. After creating 3D models, a digital mock-up will be arranged for checking dimensional conflict between building components and between temporary work facilities. Figures 7.2 and 7.3 show dimensional conflict-checking between working platforms and between wall formworks and precast slabs, respectively. The various tasks in this phase have to be done after producing the first version of workshop drawings and before the manufacturing of building com-

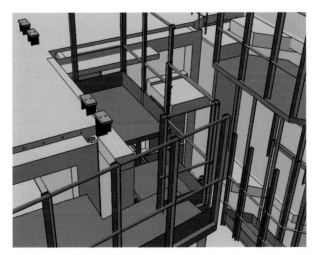

Figure 7.2   Digital mock-up to check dimensional conflict between working platforms.

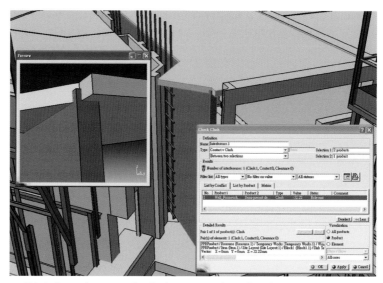

Figure 7.3   Digital mock-up to check dimensional conflict between wall formworks and precise slabs.

ponents and temporary work facilities, so that dimensional conflicts can be rectified before manufacturing to reduce reworks in the actual construction. The information required in this phase includes workshop drawings and layouts of building components, temporary work facilities and plants, and a detailed floor cycle programme that shows divided working bays. People involved in this phase include the main contractor's project planning team and the suppliers of building components and temporary work facilities.

7.5.3. Process simulation phase

After building 3D models and validating the design of building components and temporary work facilities, the next process – the simulation phase – is to simulate the planned construction process, validate the construction sequence, find the time/space conflict, check and optimize resources utilization and try alternative construction plans. Construction processes are simulated to tackle identified construction challenges. The process is detailed, as every movement of a human or plants can be simulated in the virtual prototyping environment. Figure 7.4 shows a simulation of a worker moving a table formwork to its final position for slab concreting, and Figure 7.5 shows the mobilization of workers when installing precast building components. Sometimes it is unnecessary to simulate the behaviour of workers or every detailed work process, but just to highlight building components that are under construction. Figure 7.6 shows a simulation of fixing a wall reinforcement by using a blue coloured highlight. The level of details of the simulation depends on the nature of the construction challenges that need to be tackled. For tackling workspace-related problems, we probably need to simulate every step of a construction process and the movements of both humans and

Figure 7.4   Human movement simulation in detailed process analysis.

plant involved. But for reviewing the overall work sequence or the resources' utilization of a floor cycle, the use of colour highlighting is sufficient to reflect the physical conditions of the construction site. Higher levels of detail require more effort from the virtual prototyping team to produce the simulation, and also from the contractor's project team to provide more detailed productivity and planning information, which are sometimes difficult to obtain due to limitations in time and information. The information required in this phase includes a detailed process programme, the productivity

Figure 7.5   Simulation showing mobilisation of workers.

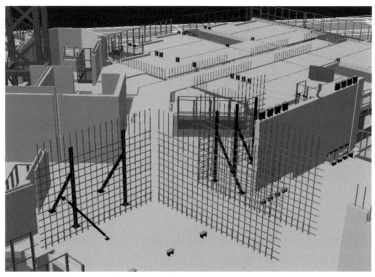

Figure 7.6   Simulation of reinforcement fixing process.

rates of different activities and a safety plan. Those people involved in this phase are the contractor's project team and the representatives of major subcontractors. (A floor construction cycle involves the work of different subcontractors.) It is necessary to collect information on the deployment of resources, productivity rate and work sequence from them in order to produce the simulation.

The process simulation phase usually starts at the construction project planning stage and stops at the actual construction stage. Process simulation helps contractors to tackle construction challenges in the planning stage. The simulation can then be used as 3D work instructions for communication among the project team members and for giving guidelines to workers. During the construction stage the actual productivity data are recorded and compared with that of the simulation. Adjustment to the simulation will be made if there is any discrepancy. The adjusted simulation will serve as a knowledge base for reference by future projects.

7.6. Benefits of virtual prototyping

Through the process of virtual prototyping, the 6-day cycle used for the residential building in Hong Kong (see Section 7.4), which was initially planned by the main contractor, was incrementally improved and optimized through an iterative trial and error process. The finalized plan was a 4-day cycle with every aspect of constructability and safety examined and confirmed by the virtual prototyping process. During the process of constructing 3D models of all prefabricated cast-in-situ components, over ten design errors, which could contribute to over 40% of rework, were identified and corrected. In addition, the virtual prototyping process also identified a number of 'unsafe' spots where possible human–machine interaction could occur and result in severe accidents. In addition to the tangible benefits, the main contractor accepted that the virtual prototyping system was an ideal platform to capture and manage the knowledge and expertise of the main contractor. As the virtual prototyping system provides a rich environment to capture all properties and attributes needed to represent construction processes and sequences, the simulated construction processes eventually became the most valuable knowledge asset to the contractor, and the virtual prototyping system became an effective knowledge management tool for the firm.

With the incremental accumulation of simulated construction projects, it is possible for a contractor to try different 'what-if' scenarios within the virtual prototyping environment, and to achieve incremental improvements in efficiency and productivity over time. Arguably, this is how the car industry has managed to achieve over 10% productivity improvement annually over the last decade or so.

7.7. The difficulties of implementing virtual prototyping

The difficulties of implementing virtual prototyping in a construction project come from three major areas:

- computer hardware requirement
- information collection and dissemination, and
- communication of virtual prototyping ideas

Virtual prototyping applications compute the sequence of activities and illustrate those activities in a real-time, 3D virtual environment. This requires considerable computational power to drive the simulation. Even a workstation computer with a very fast CPU (central processing unit) and graphics acceleration card can just simulate the construction of a typical building at an acceptable speed. The contractor has to purchase dedicated workstations for virtual prototyping. This can become a hurdle for contractors as they need to invest in the start up hardware. Currently the virtual prototyping applications are run on a 32-bit version of Microsoft® Windows® XP, which allows it to use 2 GB of physical memory at maximum. This amount of physical memory is sometimes not enough for simulating a complicated scenario. Adjustment to the 3D models is needed to reduce memory usage. An alternative is to use a 64-bit version of Windows® XP, which does not have this memory limitation.

Difficulties in information collection can come from a conflict of interest. The main contractors in Hong Kong usually sublet major works like concreting, reinforcement fixing, building services' installation, system formwork, etc. They usually do not have a comprehensive database of productivity rates of different trades. To accurately simulate the construction process, the productivity rates and the number of workers for different trades have to be collected from the subcontractors. However, the subcontractors usually refuse to give this information due to confidentiality and protection of their business interest. The main contractor has to make their own estimate based on experience in the planning stage and then measure it during actual construction. Difficulties are also encountered in disseminating simulation information to the workers, as the communication chain starting from the main contractor ends at the leading subcontractors, but not necessarily the workers. It is up to the work leader whether to disseminate the information to the workers. As in the manufacturing industry, the effectiveness of virtual prototyping will be largely hampered if not all the people involved in the construction works understand the results of simulation.

In order to produce a practical and thorough simulation of construction works, input from the planning personnel of the main contractor and the subcontractors is necessary. Their ideas have to be collected in the initial simulation planning stage and also in the simulation review stage. However, it has been found that calling up all the parties to attend meetings for virtual prototyping works is difficult due to the subcontractors' lack of interest. It is necessary to persuade subcontractors to accept the benefits of virtual prototyping at the very beginning.

7.8. The future of virtual prototyping for construction planning and scheduling

The virtual prototyping techniques described above, together with other similar systems currently available, have already been successfully applied to a number of different construction projects by several construction organizations keen to utilize the available tools for construction planning. Current use of the software by such organizations is heavily reliant on construction planners working with consultants or software experts to produce the simulation models. (This method of adoption reflects the early methods of working when network analysis techniques were introduced. As with the introduction of network analysis software this way of working may be expected to change as the techniques available mature and become more widely adopted.)

Feedback from current users of virtual prototyping software is generally positive; the software functions that are considered to be the most beneficial are the visualization functions and the ability to detect clashes in construction activities. The latter facility is particularly important when reviewing construction activities in confined areas and construction that is heavily dependent upon prefabrication. Where the construction work is repetitive (e.g. on high-rise residential or commercial buildings), the cost savings identified by virtual prototyping in the construction process can be substantial. Savings in the cycle time for each floor of the building considerably reduce the overall construction time, leading to reduced site administration costs and an earlier handover of the project to the client.

The disadvantages in the use of current systems are the time taken to input data and the difficulties in modifying the simulation process. Construction planners find this task too time-consuming, particularly when faced with the normal time-scales permitted within the tendering process. Unless the information on the building product is supplied in the form of an electronic data model that can be easily accessed, it may be necessary to develop detailed models of the product before the simulation of the construction process can proceed. Modelling the construction process is more complicated than modelling a production process based on a production line environment. This inevitably demands that new software routines be produced.

All this may be expected to change with the continued development of these products and the customization of systems to suit the construction environment. How will this affect the work of the construction planner and construction organizations? A fundamental change in the role of the construction planner as a result of building information modelling and virtual prototyping is considered unlikely, the basic function of the planner may be expected to remain the same. Construction planners may be expected to engage with virtual prototyping when and where it provides additional information and additional facilities to enable the planner to meet the demands of the client within the time-scales available. The main use of virtual prototyping for the construction planner will initially continue to be at the pre-contract stage of construction. Virtual prototyping will provide additional advantages to the construction planner in the analysis of complex construction activities, particularly repetitive activities, and the analysis of space requirements, including the allocation of working space for different trades. There would also appear to be opportunities to use virtual prototyping to assist in making improvements in site safety. To achieve this it will be necessary to develop new simulation and modelling techniques to meet the demands of construction. The existing techniques developed for the production industries are incapable of meeting the needs of the architectural engineering and construction industry.

For the construction planner to become more engaged with virtual prototyping there is a clear need for simulation tools to be easy to use and require a minimum of input. The functionality of these tools must supplement the functionality provided by existing project management software and other discrete event simulation software tools that are currently available. With such tools it will be easier for the construction planner to quickly model the construction process and to fully understand the implications of decisions made within the uncertainties of pre-tender planning. There is also a clear need to develop data libraries to support the simulation and decision-making involved. These libraries are unlikely to be based on traditional libraries of resource performance data and costs. Within 3D design, parametric modelling has become the accepted basis for model development. Parametric construction cost estimating with embedded method statements will need to be further developed as a replacement for traditional resource-based estimating.

As with the emergence of a specialist role in the construction team for modelling the construction process using network analysis, a specialist role may be expected to emerge for individuals with both building information modelling and virtual prototyping skills and construction experience. This resource will initially be used by the construction organization in a specialist consultant or specialist support role to the main construction team. For an increase in the use of building information modelling and virtual prototyping, construction organizations will need to directly employ specialists with these skills within the construction team. These people will need direct construction experience to enable them to work as planners and not simply as operators or modelling specialists.

The roles of the professions within the design, procurement and construction of buildings and infrastructure are well established, having been developed over centuries. However, it may be confidently predicted that, with anticipated advances in building information modelling and virtual prototyping, these roles will change; the only debate is about how much and when. Experience from projects where virtual prototyping has been introduced have identified changes in the procurement process as construction clients become aware of the potential of the technologies, not only during the design and construction of the building but as a platform for maintaining the building product throughout its lifetime.

The benefits of building information modelling and virtual prototyping will only be fully realized if there are fundamental changes in the practices and procurement of buildings and infrastructure to permit the advantages of the technology to be fully utilized, e.g. through the widespread adoption by clients of design and build projects. Unless this happens then adoption will at best be piecemeal and at worst non-existent. Gann (2000) identifies both the driving forces from changing supply and the driving forces from changing demand which are determining the reinvention of the construction industry. He argues that:

> supply side forces include the availability of equipment and systems, the availability of specialized software and new materials and components and that changing demand is being driven by changes in work patterns, changes in attitudes and changes in regulation. Overall this is leading to interactive design, flexibility in the organisation of work and new forms of computer aided management. The demands from the clients of construction who within their business world will continually seek new ways of working to meet their business demands will drive such change in construction.

If construction organizations are to achieve similar savings and similar benefits then they must have a clear strategy to fully develop their business relationships with preferred suppliers, and include the provision to share working expertise in building information modelling and virtual prototyping. Much of the decision-making made by the construction planner at the pre-tender stage of construction relates to the work of specialist suppliers. With increased prefabrication and smart materials, the specialist suppliers may be expected to have an increasing part to play in both the design of the building product and the construction process. If construction organizations do not adapt and adopt building information modelling and virtual prototyping to meet the demands of the client's design team then they will find their role changes. So too may the construction planner, who may find that the designer, assisted by the tools and techniques of building information modelling and virtual prototyping, is able to usurp their role in the production process.

Acknowledgements

Funding for this research was supplied from the Research Grants Council of Hong Kong through allocation from the Competitive Earmarked Research Grant for 2005-2006 under Grant Number: PolyU 5103-05E, PolyU 5157-04E and PolyU 5209-05E. The authors also wish to acknowledge the support obtained from the Housing Authority of Hong Kong and Yau Lee Construction Company Ltd for their assistance in providing information and comments.

References

Choi, S. H. and Chan, A. M. M. (2004) A virtual prototyping system for rapid product development. *Computer-Aided Design* 36: 401–412.
Collins Dictionary of the English Language (1986). London, Collins.
Gann, D. (2000) *Building Innovation*. London, Thomas Telford.
Goldberg, H. E. (2004) *Is BIM the future for AEC design? AEC From the Ground Up: The Building Information Model.* Available at http://aec.cadalyst.com/aec/article/articleDetail.jsp?id=133495
Jaafari, A., Manivong, K. K. and Chaaya, M. (2001) VIRCON: Interactive system for teaching construction management. *Journal of Construction Engineering and Management* 127: 66–75.

Jim, B. (2004) *Digital Manufacturing – The PLM Approach to Better Manufacturing Processes*. Available at http://www.aberdeen.com/c/report/tech_clarity/Digital_Manufacturing.pdf

Koo, B. and Fischer, M. (2000) Feasibility study of 4D CAD in commercial construction. *Journal of Construction Engineering and Management* **126**(4): 251–260.

Kunz, J. and Fischer, M. (2005) *Virtual Design and Construction: Themes, Case Studies and Implementation Suggestions*, CIFE Working Paper #097. Available at http://www.stanford.edu/group/CIFE/Publications/index.html

Mawdesley, M., Askew, W. and O'Reilly, M. (1997) *Planning and Controlling Construction Projects*. Harlow, Essex, Addison Wesley Longman.

McKinney, K. and Fischer, M. (1997) 4D analysis of temporary support, *Proceedings of the 1997 4th Congress on Computing in Civil Engineering*, ASCE, 16–18 June, Philadelphia, PA, pp. 470–476.

Pratt, M. J. (1995) Virtual prototypes and product models in mechanical engineering. In: *Virtual Prototyping – Virtual Environments and the Product Design Process* (ed. J. Rix, S. Haas and J. Teixeira). London, Chapman and Hall, p. 113–128.

Shen, Q., Gausemeier, J., Bauch, J. and Radkowski, R. (2005) A cooperative virtual prototyping system for mechatronic solution elements based assembly. *Advanced Engineering Informatics* **19**: 169–177.

Waly, A. F. and Thabet, W. Y. (2002) A virtual construction environment for preconstruction planning. *Automation in Construction* **12**: 139–154.

Xiang, W., Fok, S. C. and Thimm, G. (2004) Agent-based composable simulation for virtual prototyping of fluid power system. *Computers in Industry* **54**: 237–251.

Yerrapathruni, S., Messner, J. I., Baratta, A. J. and Horman, M. J. (2005) *Using 4D CAD and Immersive Virtual Environments to Improve Construction Planning*. CIC Technical Reports in Penn State #45. Available at http://www.engr.psu.edu/ae/cic/publications/TechReports/TR_045_Yerrapathruni_2003_Using_4D_CAD_and_Immersive_Virtual_Environments.pdf

8 Reshaping the life cycle process with virtual design and construction methods

Martin Fischer

For the foreseeable future, construction executives face significant challenges in delivering projects profitably. Virtual design and construction (VDC) technologies are already addressing some of these challenges by providing a better method for multi-disciplinary project teams to coordinate the design and their work on a facility. In the future, more integrated information will enable project teams and facility professionals to learn more about the facility's performance over its life cycle and help to automate many routine design and fabrication tasks. The result should be facilities that respond better to their users' business needs and their environmental and social context.

8.1. Construction business challenges today

When asked, many construction executives cite the following challenges for their businesses today:

- Competitive market conditions
- Lack of qualified human resources, which impedes sustained company growth and consistent delivery of successful projects
- Lack of industry data standards, which create inefficiencies both within companies as well as between multi-company project team members
- More complex projects that require greater cost reliability and also demand increased schedule and scope flexibility, while addressing demanding legal, regulatory and environmental requirements
- Challenging conditions to sell and execute projects, including new market conditions, new delivery methods, new market entrants and new business models

For the foreseeable future, these influences will continue to challenge the traditional roles and processes of firms. Many firms have started to leverage VDC technologies to differentiate themselves in the market-place, to leverage the human resources, data and knowledge they have and to manage the risks of complex projects proactively.

Figure 8.1   Conceptual design review with a virtual reality model. (Majumdar *et al.*, 2006.)

8.2. State-of-the-art use of virtual design and construction models today

Many project teams routinely review and coordinate their designs and the corresponding schedules with 3D and 4D models in many phases of a facility's life cycle.

For example, Figure 8.1 shows a judge, his staff, the developer and the architect reviewing the conceptual design of a courtroom with/through a virtual reality model. This collaborative session cut the conceptual design review time from 2 days to just over 2 hours, and produced more consistent and complete feedback on the proposed design to the architect than the traditional physical mock-up model would have done.

Mechanical, piping, electrical and fire protection subcontractors co-located for the detailing phase of a new medical office building in Mountain View, California. They used 3D models to detail and coordinate their respective areas of responsibility (Figure 8.2). This approach reduced the time taken from one discipline asking a question to receiving an answer from another discipline to minutes instead of the typical weekly review cycle. Wasted detailing efforts were greatly reduced, the detailing phase was completed more quickly and the output of the detailing phase was improved. As a result, the subcontractors were able to prefabricate a much higher percentage of their work. They also saw productivity in the field increase dramatically. For example, from the dimensionally accurate and coordinated 3D model, the piping subcontractor prefabricated and delivered materials daily, and sometimes twice daily, in kits for specific areas for a crew (Figure 8.3). Furthermore, the number of change orders and requests for information during construction was reduced enormously.

On a multi-family housing project in Charleston, South Carolina, the concrete contractor's superintendent and field engineers planned the tasks for their workers for the next day interactively with a 4D model displayed on a SMART Board in the site trailer (Figure 8.4). This planning session allowed the site management team to: verify that the workers, materials, space and access needed for the planned work was

Figure 8.2   Co-located sub-contractors, using 3D models, for the detailing of mechanical, electrical and piping systems. (Khanzode, 2006.)

indeed available; create clear work instructions for the workers for the next day; and understand, mitigate and communicate safety concerns.

These examples show that building professionals now use visualization methods combined with some data integration and tools like clash detection with confidence and with considerable impact in all phases of developing, designing and building a facility. Building virtually first is one of the best risk management methods available to project teams today. Companies are now working on the widespread deployment of

Figure 8.3   Kit of prefabricated pipes for one day's work in one area. (Khanzode, 2006.)

Figure 8.4   Planning the next day's construction work using 3D and 4D models on a project by Accu-Crete, Alexandria, VA.

work processes that leverage VDC methods. Taking advantage of VDC visualization methods often requires a different organization of the design and construction (or virtual and real making) phases of a project to ensure that the appropriate expertise is available when it matters most. For example, to leverage virtual building methods in the early project stages requires the input of experienced builders.

In my experience, widespread adoption is currently hindered most by a shortage of project engineers familiar with VDC tools and of executives who fail to recognize the operational and strategic value of VDC-based processes in a timely manner.

Nevertheless, many project teams coordinate several building systems (most commonly the structural, mechanical, piping and fire-proofing systems) with visual 3D models, using some automated clash detection, and plan some of the construction with 4D models. Given coordinated building systems, some subcontractors then fabricate close to 100% of the components just-in-time directly from their 3D model. Given this state-of-the-art use of VDC technologies, many firms are also working on advancing these technologies to enable more computer-based predictions of the performance of facilities and the corresponding organizations and processes, to integrate data use across disciplines and project phases, and to automate design, fabrication and construction work.

8.3. Emerging virtual design and construction methods

Methods to predict facility performance from 3D building information models are being developed and used in practice by some companies (Kam *et al.*, 2003). For example, Granlund (http://www.granlund.fi) routinely calculates a building's energy consumption, life cycle cost and other performance metrics with input from a building information model (Figure 8.5). BIM-based (building information model) cost estimating tools, e.g. D-Profiler by Beck Technology, are also emerging (Carroll, 2006). The

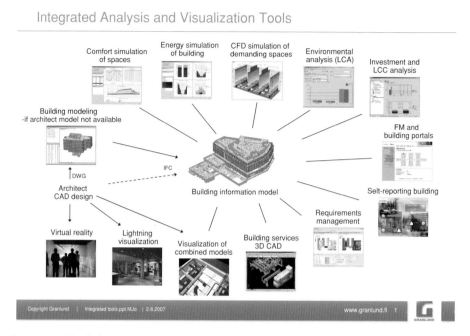

Figure 8.5 Tools for analysis and visualization integrated through shared product models are emerging as cornerstones of integrated, performance-based, life cycle-focused facility design. The figure illustrates the current capabilities and offerings of the mechanical design firm, Granlund (Helsinki, Finland). (Figure from Hänninen, 2006.)

General Services Administration is implementing a BIM-based spatial program valida-tion method (GSA, 2006). Such BIM-based, automated facility performance, prediction tools require significant research and development (R&D) as well as investment to de-velop and test them. They make it possible, however, to predict the performance of a proposed design quickly and consistently, speeding up and improving the comparison of alternatives and decision-making.

Another area that requires significant development is the user interface for groups of multi-disciplinary stakeholders and for field workers. Early in a project, many pro-fessional and non-professional stakeholders should provide well-informed input to the design of a facility. The user interfaces that exist today do not facilitate this input, which leads to a long design process and the discovery of design criteria and stake-holder requirements sequentially. Better user interfaces that focus on engaging groups of stakeholders from many different backgrounds could help to achieve a more effec-tive and efficient early project design phase. BIMs then hold great promise to carry the project requirements determined by the stakeholders forward throughout the design process and create a more transparent design and design process (Kiviniemi, 2005). Later in the project, the design decisions need to be communicated to the field, in particular to the crews installing the various components. Construction companies are experimenting with handheld devices to make design information available in the field when and where needed (Means, 2007). Nevertheless, the user interface to retrieve and interact with project information with such devices in the field still requires significant development. Furthermore, the updating and sorting of project information to make it available when needed also needs to be simplified and improved.

8.4. The facility life cycle in 15 years' time

How will all these developments shape the facility life cycle in the future? I foresee four main phases needed to develop a facility and two main phases to run a facility.

(1) *Four, facility development phases*:
   - making the business case for the facility (feasibility study, conceptual design, setting performance targets, links to the client's business case)
   - designing the facility and its life cycle, including its systems, parts and materials as well as the related processes and organization
   - detailing the facility, getting it ready for fabrication and construction
   - making the facility, including fabrication and the installation of parts, components, and subsystem, and documenting what is actually part of the building and how it was put together; also practising the facility start-up and use while it is being built (starting it up virtually)
(2) *Two, main facility use phases*:
   - starting up and commissioning the facility and its systems, really understanding how the facility works and comparing how it functions with the design assumptions and targets
   - using, reusing, tracking performance, keeping an inventory of the building contents and make-up

VDC technologies will support the development of the business case for a facility by supporting the representation of a 4D model that shows the major components and phases of a project in its expected context, including the relevant physical and digital infrastructure and other facilities related to the project and the environmental, social and economic context. Such a 4D model will serve as the 'defending champion' of the scope and schedule for the proposed facility, and provide the initial project-specific data for analysis of its income (economic and other) and costs (first cost, operating costs, including energy, carbon, etc.). Rapid digital prototyping methods (e.g. Carroll, 2006) will enable project teams to develop project scenarios and corresponding high-fidelity analyses rapidly, and to support the project pro-forma with clear facility design and phasing assumptions. The conceptual 3D model will also provide the visualizations necessary to enable all the relevant stakeholders to visualize the facility over its major life-cycle phases (see, for example, Figure 8.1). Furthermore, methods that relate the client's business strategy and requirements to the facility design (e.g. Pennanen *et al.*, 2005; Haymaker and Chachere, 2006) will be linked to the conceptual 4D model to provide transparency between the users' business goals and the design of the facility.

In summary, VDC technologies will enable project teams to leverage performance data much earlier and more comprehensively than is possible today, and link the business, stakeholder and facility requirements with its design and desired performance. They will allow more stakeholders to understand a design from more perspectives and criteria far more quickly than is possible today, which should, in turn, lead to a set of design criteria that is specified much better.

In the design phase, VDC technologies will enable project teams to carry forward the 'defending champion' design from the business case development phase, and to design the products, organizations and processes of the project's life cycle concurrently and in a coordinated way (Kunz and Fischer, 2005). VDC technologies will also automate many design development and analysis tasks and continue to provide high-fidelity visualizations, as necessary, to inform and engage stakeholders. During the design phase, VDC technologies support the testing of many design versions over the entire (simulated) facility's life cycle with a cradle-to-cradle perspective (McDonough and Braungart, 2002). VDC technologies will dramatically reduce the response time for questions that affect two or more disciplines. This faster response time will reduce the amount of information that needs to be coordinated and so speed up design cycles. Finally, VDC technologies will provide a continuous information platform, similar to the product life-cycle management systems in manufacturing today (Grieves, 2005).

Once the key decision-makers are happy with the design, which includes the design of the processes and organizations to make and operate the facility, the project team will move into the detailing and fabrication phases. Many detailing and fabrication tasks will be automated, since there should be no interference in the facility's design. Already today, several firms can fabricate and deliver assemblies and subassemblies to the site within three days: i.e. three days after the final design decision has been made the design is realized and installed in its final place. The Center for Integrated Facility Engineering (CIFE), Stanford University, CA, and its member companies have set the goal that almost any facility should be built in six months or less using automated detailing and fabrication and well-coordinated parallel fabrication and construction efforts. During the construction phase, the life cycle simulation of the completed facility will continue, and the commissioning team will start up the facility virtually many times. Similarly, operators will undergo training in the facility, in much the same way as pilots use flight simulators today. This phase will not only deliver the physical artefact, but also the digital information necessary to operate and retrofit the facility efficiently.

Finally, in the operations phase, VDC technologies will help to validate the design assumptions and models against the observed facility's performance, including its occupancy, environmental conditions and comfort, as well as the energy consumption of the various building systems. However, the development of methods to compare observed and predicted performance requires significant research.

8.5. Conclusions

Through more timely and comprehensive modelling, simulation, analysis and visualization, VDC technologies will enable a dramatic improvement in construction performance. Not only will it be possible to design and build complex facilities very quickly (12–18 months), but the facilities will also perform better because, over time, the underlying engineering models will become more precise and appropriate. The formalization of the knowledge necessary to allow such improvements will, however, require significant effort over the next decade or two. Furthermore, simpler and robust methods to integrate and manage multi-disciplinary project data are needed, as well as more intuitive user interfaces for professional and non-professional stakeholders for the vast amount of data. Organizationally and culturally, firms in the construction industry are already evaluating their ability to assume risks differently than is possible today. In addition, many executives realize the importance of developing innovation processes to shape the life cycle and facilities of their projects so that their firm maintains or expands its competitive position.

Acknowledgements

I thank Julius Chepey, Micky Doner, David Hudgens, Atul Khanzode, John Kunz, Frank Neuberg, Dirk Schaper and Mike Williams for the discussions and insights that led to the work, concepts and ideas presented in this chapter.

References

Carroll, S. (2006) VDC to sustain and to disrupt. *Presentation at the CIFE Summer Program*, 21 June, Stanford, CA.

General Services Administration (GSA) (2006) GSA BIM guide for spatial program validation. *Version 0.90, GSA Building Information Modeling Guide Series*. Washington, DC, General Services Administration, Public Buildings Service, Office of the Chief Architect.

Grieves, M. (2005) *Product Lifecycle Management: Driving the Next Generation of Lean Thinking*. New York, McGraw Hill.

Hänninen, R. (2006) Building lifecycle performance management and integrated design processes: How to benefit from building information models and interoperability in performance management. Invited presentation: *Donald R. Watson Seminar Series on Construction Engineering and Management*, 18 January, Stanford University, CA.

Haymaker, J. and Chachere, J. (2006) Coordinating goals, preferences, options, and analyses for the Stanford Living Laboratory Feasibility Study. In: *Intelligent Computing in Engineering and Architecture, 13th EG-ICE Workshop, LNAI 4200* (ed. I. F. C. Smith). Heidelberg, Germany, Springer, pp. 320–327.

Kam, C., Fischer, M., Hänninen, R., Karjalainen, A. and Laitinen, J. (2003) The product model and Fourth Dimension project. *ITcon (Electronic Journal of Information Technology in Construction)* [Special Issue: IFC – Product Models for the AEC Arena] **8**: 137–166.

Khanzode, A. (2006) 3D model based integrated concurrent detailing at DPR. *Presentation at the CEE320 Seminar on Integrated Facility Engineering, CIFE*, 28 November, Stanford University, CA.

Kiviniemi, A. (2005) *Product Model Based Requirements Management*. PhD thesis, Dept of Civil & Env. Eng., Stanford University, CA.

Kunz, J. and Fischer, M. (2005) *Virtual Design and Construction: Themes, Case Studies and Implementation Suggestions. Working Paper Nr. 97, CIFE*. Stanford University, CA.

Majumdar, T., Fischer, M. and Schwegler, B. R. (2006) Virtual reality mock-up model. *Building on IT, Joint International Conference on Computing and Decision Making in Civil and Building Engineering* (ed. H. Rivard, E. Miresco and H. Mehlhem), 14–16 June, Montreal, Canada, Canada Research for Sustainable Building Design, pp. 2902–2911.

McDonough, W. and Braungart, M. (2002) *Cradle to Cradle: Remaking the Way We Make Things*. New York, North Point Press.

Means, J. (2007) Experience of using tablet PCs in the field. *Presentation at the CIFE-Tekes Virtual and Mobile Real Estate and Construction Symposium*, 26 January, Stanford, CA.

Pennanen, A., Whelton, M. and Ballard, G. (2005) Managing stakeholder expectations in facility management using workplace planning and commitment making techniques. *Facilities* **23**(13/14): 542–557.

9 Virtual prototyping from need to pre-construction

Robin Drogemuller

The most important decisions about a building or structure are made during the early stages of design when the least amount of detailed information is available. Analysis tools that can help designers assess the implications of their decisions across a range of important parameters are needed. These should all be capable of operating from a single shared representation of the project. Once these initial decisions are made, designers need to be able to track the results as the building design becomes more comprehensive. This allows deviations to be identified, and then either corrected or the earlier decisions reconsidered. This process needs to continue until the building design is completed and then fed into the building usage and management phase of the facility's life.

This chapter provides a vision of how such systems will operate in the future and examines the current state of integrated, whole-life cycle analysis systems. This allows areas where gaps exist to be identified.

9.1. What is virtual prototyping?

For the purposes of this discussion, virtual prototyping is defined as: the use of computer-based tools to model a proposed construction project and to run a series of analyses against the model to iterate towards an acceptable configuration of the model, before starting to physically construct the proposed facility. Better results can be obtained where a number of factors are analysed to move towards a solution that satisfies a range of parameters.

This definition deliberately allows for a range of methods of interaction and coordination between the users, the information models and the mathematical models. There is no specific requirement for a visual representation of the shape of a proposed facility. At the early stages of planning a new shopping complex, for example, the information model may consist of breakdowns of proposed rental mixes combined with a catchment model. The analytical models may consist of formulae that describe returns to the complex owner, which are based on projections of income factored on the mix of tenancies derived from the population mix in the catchment and competition from other shopping complexes. There is no explicit requirement to see the possible shape and number of floors of the proposed complex, although the representation of various catchments within a GIS (geographical information system) would be useful.

The definition excludes systems where the user can not significantly influence either the information model(s) or the analytical model(s), such as VRML (Virtual Reality

Figure 9.1   Three tiers of built environment.

Modeling Language) environments and movies of results. These may be outputs from a virtual prototyping system, but are not considered virtual prototyping *per se*.

9.2. Vision for virtual prototyping

Predictions of the future are fraught with difficulty as there are no guarantees that all of the myriad requirements for a particular vision to come to fruition will actually occur in the necessary timeframe and sequence. Consequently, the predictions of this chapter are divided into a number of factors: technical, social and political.

In 20–30 years' time we will have the technical capability to assess the impacts of proposed buildings and infrastructure designs from a comprehensive, triple bottom-line perspective. We will be able to analyse these impacts at multiple levels of detail: urban, neighbourhood and facility level (Figure 9.1). We will be able to visualize, generate and manipulate the relevant information in various formats, where changes in one 'view' of the information will automatically flow through to the other 'views'. We will be able to share this information with others virtually to enable geographically dispersed collaboration. It will also be possible to share this information widely to allow public comment on proposals. Issues of using defaults for missing information, continuity through time, versioning, etc. will be handled semi-automatically by the IT systems supporting designers, constructors and maintainers.

The types of user interfaces that we use to interact with these virtual models will range from a selection of existing interfaces: plans, sections, etc. of the model itself, Gantt charts of the proposed processes, schedules and textual documents through to sophisticated immersive environments such as those portrayed in the movie *Minority*

Phase Zero	Phase One	Phase Two	Phase Three	Outline Planning Approval	Phase Fourr	Phase Five	Detailed Planning Approval	Phase Six	Phase Seven	Phase Eight
Demonstrating the need	Conception of Need	Outline Feasibility	Substantive feasibility study & outline financial authority		Outline conceptual design	Full conceptual design		Production design, procurement & full financial authority	Production information	Construction
Establish	Finalise									
Outline					Revise full			Finalise		
Initial stakeholder list	Assess impact & requirements									
Establish									Finalise	
	Prepare feasibility		Prepare concept design brief						Finalise	
	Consider								Finalise	
		Undertake outline for each option	Assess most "feasible" options							
Establish										
Compile inial register										
Prepare										
			initial						Finalise	
			Prepare initial assessment						Finalise	Manage
					Prepare initial Outline	Full			Finalise	
						Prepare				Revise & implement
								Produce	Finalise	Develop as-built information
									Prepare	Update & implement
								Start		
										Manage & undertake

Figure 9.2 Process protocol from a document perspective.

Report (Twentieth Century Fox, 2002) and the 'Future Workspaces' (Fernando *et al.*, 2003) project. All these interfaces will work off a shared model, with view-dependent and protected intellectual property maintained in separate, but linked, data stores.

The best way to explain this vision is to put together a storyboard to explain how it all may work. The vision is set out following the 'Process Protocol', developed by SCRI (Salford Centre for Construction Innovation) (Cooper *et al.*, 1998). The idea for this story comes from an actual proposal for a new multi-storey residential development proposed for Mitcham, a suburb of Melbourne, Australia, but the details used within the story bear no relationship to the actual proposal. The issues will be discussed in a bottom-up manner, starting with the Process Protocol, which deals with the facility level.

Figure 9.2 shows the Process Protocol from a document perspective. The shading indicates where a document is initiated and where development of the document ceases. Text is added to shaded boxes where there is a significant change. Non-annotated shaded boxes indicate that the document is further refined without significant change.

Using 'documents' as the basis for a discussion of virtual prototyping may seem retrograde. However, it can be justified if each 'document' is considered as a 'view' of the information describing a project that is useful for specific groups within the process at particular times. Implicit in this understanding of the project procurement process is the concept of the unified project description, which is one of the major steps along the road to a fully integrated virtual prototype of building and infrastructure projects.

9.2.1. Statement of need

The statement of need is developed during the first two phases of the Process Protocol and covers the process of identifying and scoping an opportunity. For many types of developments the information needed for such analyses can be presented in GIS-type environments. A driver for the example project could be the construction of the new Eastlink ring road through Melbourne's eastern suburbs (Thiess John Holland, 2007). A GIS-level query to identify new opportunities might be:

> Find locations where there are two major arterial roads running near orthogonally, with good public transportation and access to local services, where the population density is low to medium and the local council is friendly to higher density development.

Looking up 'Maroondah Hwy, VIC, Australia' in Google Earth™ (latitude – 37.816561, longitude >145.19231) shows the area under consideration. The new Eastlink ring road is visible running north/south, while the Maroondah Highway, a major artery into the Melbourne central business district (CBD), runs east/west. A railway line is also visible running into the CBD. The large development to the right of Eastlink is the Eastland shopping centre.

9.2.2. Stakeholder impact and requirements

Identification of stakeholders can also benefit from a GIS-type approach to presentation. Overlays can be used to indicate the likely level of stakeholders' interest and their return on investment from the business case.

9.2.3. Business case

The business case is built around identifying the stakeholders, the benefits that each group can expect, from a triple bottom-line perspective (TBL), and how much they will be willing to pay. This, then, provides an estimate of the likely benefit to the developer, which needs to be balanced against the TBL cost to the developer. As noted above, the use of a GIS-based approach that could combine the various factors across stakeholders and benefit to the developer would be useful.

While GIS systems present a 2D view of the map and results, other software can present deformed surfaces to show the value of the function being considered.

Major issues here lie in how to access the information that is available for use in such analyses, how to generate information that is not currently available and how to display more subjective factors (i.e. 'friendly to development' local government authorities in the query above).

9.2.4. Communication strategy

The communication strategy is developed, modified and extended continuously through the project. Virtual environments could be used in conjunction with the stakeholder identification information to target the various stakeholder groups in appropriate ways.

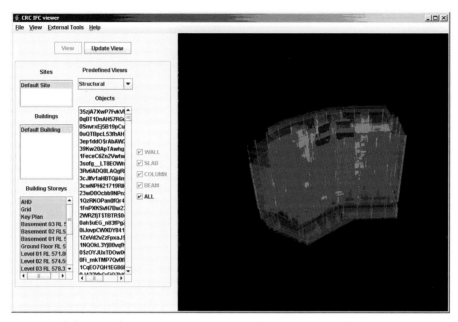

Figure 9.3   Selection of particular components within a virtual model.

The use of virtual tools could be a significant part of the communication strategy and could inform public participation in developments if it were desired.

9.2.5. Project brief

The term 'project brief' has been retained here to retain the terminology used within the Process Protocol, but 'client and user requirements' is a preferred term. This information is generated in 'Phase 1, Conception of need', and is then maintained through to 'Phase 7, Production information', in the Process Protocol. It could be argued that this should continue throughout the life of the building or structure.

Kiviniemi's work (Kiviniemi, 2005) on 'client requirements' management' explains how they can be maintained in a product model or virtual environment. If there were software tools that could identify mismatches between the client requirements and the current design then they could be highlighted within the virtual prototype of the project (Figure 9.3).

Solibri Model Checker (Solibri, 2007) and DesignCheck (CRC CI, 2005a) are two tools that could be used for this task, if there was a generally usable method of linking requirements to rules that could be used to check that the requirements were met.

9.2.6. Site and environmental issues

Access to reliable geographic and environmental information is necessary before virtual prototypes can be used to present this information in new ways. Overlays of data in a GIS could be used to highlight areas that need attention. Such representations

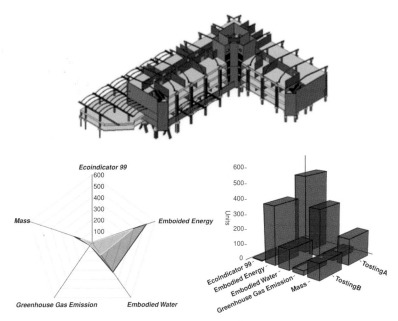

Figure 9.4   LCADesign – input model and LCI graphs.

could also be used as part of the communication strategy to stakeholders to overcome misconceptions due to a lack of understanding, or gaps in understanding, of the issues.

The relationship between environmental issues and design is important given the significant contribution of the built environment to environmental degradation. While the design stages are yet to be discussed, this is an appropriate place to introduce some of the environmental analysis tools and methods.

While there are a number of internationally used environmental rating systems, the lack of reliable, consistent information on environmental impacts and agreed methods of using the available information make rigorous application difficult. A number of tools is available that link environmental performance to design, but most of them are only applicable to the conceptual design stages and do not link in seamlessly to other representations of the project information.

However, one system that does link directly to virtual (CAD) representations is LCADesign (Tucker *et al.*, 2005). LCADesign extracts the required materials and volumes directly from an IFC (Industry Foundation Classes) model of the building, combines it with LCI (life cycle impact) data and then presents the results in graphical form (Figure 9.4). It also provides the ability to substitute alternative construction types for components in the structure and to generate graphs comparing performance against selected parameters.

9.2.7. Feasibility studies

Feasibility studies are undertaken in Phases 2 and 3 of the Process Protocol. Much of the information covered in the previous descriptions is used in undertaking the studies and in presenting the results.

9.2.8. Risk analyses

Risk analyses and the risk register are used throughout the procurement process. Outputs from other processes are used in generating this information. The use of risk considerations is demonstrated in the 'statement of need' section where 'friendly to development' is used as part of the GIS query.

A risk knowledge base could be built up by matching cases of previous projects, and the potential and actual risks that arose, to the characteristics of new projects. This could be built on abstractions of the relevant project characteristics for each risk type.

9.2.9. Execution plan

This is another area where the effective use of virtual prototyping is the consideration rather than virtual prototyping being an input itself. However, there is some work on software tools that can be used to assess the 'health' of projects and to act pre-emptively on problems. The Project Diagnostics system (CRC CI, 2005b) uses a 'critical success factors → contributing factors approach'.

9.2.10. Procurement plan

Procurement planning can be linked with virtual prototypes in the same manner as suggested under Section 9.2.8 (Risk analyses) above, through matching previous procurement decisions against perceptions of their success or otherwise to build a knowledge-based case tool. One tool that gathered procurement histories was the Project profiler (CRC CI, 2004), which matched the key characteristics of proposed projects against recommended procurement methods (Figure 9.5).

9.2.11. Occupational health, safety and the environment

On site OHS & E issues can be addressed through the application of 4D CAD techniques. Legal pressures are forcing the consideration of on-site safety issues during design. The increased interest in environmental issues is also increasing the interest in design for deconstruction. This indicates that automated systems could assist designers in considering constructability, OHS & E and design for deconstruction. Such software will have a market within the timeframe under consideration.

The Automated Scheduler project examined some of the issues of storing project planning information and could be extended to cover this area (Figure 9.6).

9.2.12. Cost plan

The cost plan is first generated in conjunction with outline conceptual designs and then refined until completion of the production documentation. There are a number of systems available that support the extraction of quantities from 3D models, but most of these require some form of user input for them to work successfully. Within

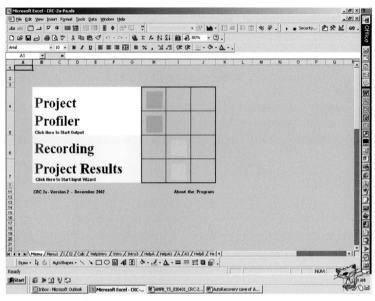

Figure 9.5   Project profiler user interface.

our 20-year timeframe, the issues of automating the extraction of quantities are likely to be largely resolved. A greater challenge lies in bringing automated cost planning forward into the design process where the information is much less detailed. Another related issue is in the explicit representation of information that is not currently added to building models because it is too time-consuming to represent it explicitly. These will be addressed in the next section.

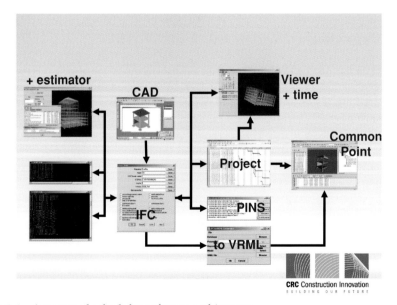

Figure 9.6   Automated scheduler software architecture.

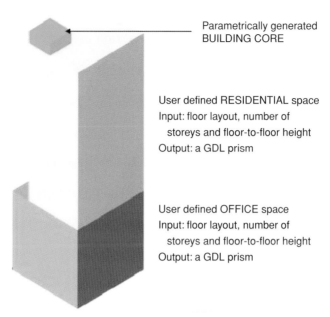

Parametrically generated
BUILDING CORE

User defined RESIDENTIAL space
Input: floor layout, number of
    storeys and floor-to-floor height
Output: a GDL prism

User defined OFFICE space
Input: floor layout, number of
    storeys and floor-to-floor height
Output: a GDL prism

Figure 9.7   Massing model of multi-use building.

9.2.13. Concept design and product model (production design)

The representations of the building or structure are normally considered as the core of virtual prototyping. The discussions above should indicate that there are many other opportunities in integrating information about projects and facilities. It is also clear that many of the 'pre-design' documents define and constrain the proposed building or structure design. For a fully integrated system this pre-design information needs to be captured, linked into the representations of the facility and updated as decisions are made throughout the project processes.

The major opportunities to apply this wealth of information lie during conceptual design, where decisions have the most impact and 'good' decisions will have the most benefit. The Parametric Design project within the CRC for Construction Innovation (CI) examined the issue of scoping the engineering systems in buildings during the massing-model design stage (Figure 9.7). This massing model was then used to generate estimates of the requirements for architectural spatial layout, structural, fire protection, water supply, electrical (power, lighting), environmental (LCA: life-cycle assessment) and mechanical (vertical transport, HVAC (heating, ventilation and air conditioning)) systems (Figure 9.8).

For each system and subsystem the user was able to enter appropriate key parameters to allow the impacts of the changes to parameters and the interrelationships between different systems to be explored. There has been a range of projects within the CRC-CI looking at the issues involved in supporting more detailed design representations, and these are indicated in Figure 9.9.

Some of these have been discussed previously, but the important general issues arising from the development of a suite of software that will impact on how design is undertaken in the next 20 years are:

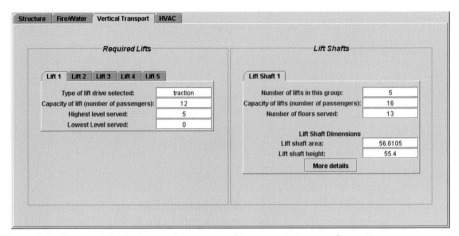

Figure 9.8   Parametric design system: vertical transportation configuration.

- Shared services attached to the virtual data model that provide functions required by a number of user services. For example, the ability to take off quantities is shared across LCADesign, Automated Estimator and Automated Scheduler. The way the quantities are structured varies for each application, but the underlying service is the same.
- The ability to analyse a design at various stages of development. 'DesignCheck', for example, can check conformance to AS1428.1 – Design for access and mobility

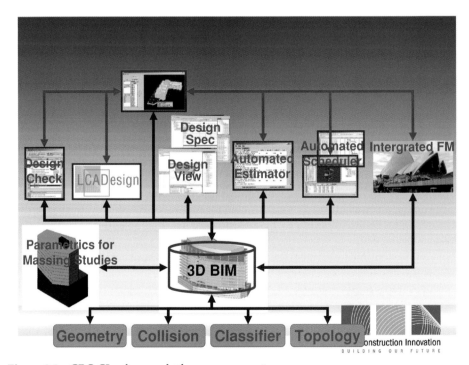

Figure 9.9   CRC-CI software platform components.

(disabled access) – from sketch design stage through to complete documentation. This allows the identification of potential problems from the spatial layout, before hard decisions are made about the provision of stairs, lifts and ramps.

- Linking client requirements into the representation of the facility so that variances can be identified and issues addressed as soon as they arise.
- Sharing of common data so that the different participants can ensure that clashes are identified within the virtual model rather than on site.
- Issues of versioning control, not only on complete models but also within models.

9.2.14. Operational policy and maintenance plan

The models developed during design could be used as the basis for analyses of the maintenance processes on equipment and components within and outside the buildings and structures. The addition of permanent small cranes to assist in cleaning complex buildings is one example where such analyses are useful. Examples exist of opposite extremes of planning for operation and maintenance. Some buildings need walls to be partially demolished in order to replace HVAC plant for example, while other buildings provide hatches and lifting points to make access to HVAC equipment easy and non-disruptive.

9.2.15. Handover plan

Developing handover plans within a virtual prototype could exploit many of the facilities already described. If scheduling information is stored appropriately, staged handover of large projects or refurbishments could be automatically generated and visualized, so that the client can more readily understand what is becoming available and when. The availability of a full representation of the facility would make the handover of 'as built' documentation trivial and could mean that it was actually used. An opportunity lies here in the designers providing a service to their clients to continue to maintain the information throughout the facilities' life, either in a facility management role, or in partnership with the facility managers. Whether the design professions are interested in this will be seen in the near future.

9.2.16. Planning approval

An interesting scenario that could emerge with the integration of building information, urban planning and infrastructure (utilities, transport, etc.) is the use of more definitive planning tools and a more rational approach to 'head works' allocation. Planners could, for example, apply a maximum population to an area. If a developer wanted to increase the density of development, then an approval cost could be calculated based on the increase from the current total population to the new population once the development was completed. This would reflect the benefit received by the developer and the 'loss' to the local government authority from the reduction in planned capacity. This type of system could be used to return some of the windfall gains achieved when areas are re-zoned from one use to another.

Another useful facility at the planning level would be the ability to feed information on planning proposals through to the utility providers: water, power, gas, transportation, etc. This could then be fed into virtual models of the utility networks for the prediction of demand, forward work planning and even headworks' cost estimates.

9.3. Links to urban issues

Buildings and structures do not exist in isolation. The surrounding environment has a significant impact on their performance. While some consideration is given to this in various codes, this area has not been heavily studied. Another CRC-CI project examined the environmental issues at the local level. It analyses the local wind speed, precipitation and daylighting around a building and the expected temperatures in the intra-building spaces within a virtual model of the local area (Figure 9.10). The analyses are performed at two levels – analyses are done on the area as a whole, which take around 2 days for each run, and then analyses of individual buildings can be performed to assess the impacts of changes. These small-scale analyses use the whole of model analysis to set boundary conditions.

This type of project needs to support both the IFC model and GIS capabilities. Neither of these systems is appropriate for all of the data, so some merging of the capabilities of the two types of representation can be expected over the next few years.

Another area where some integration needs to occur within the timeframe under consideration is the links between urban planning, subdivisional layout and building design. Several projects within the CRC-CI are examining the effects of subdivisional layouts on the levels of performance that buildings can achieve. To give an example, in hot climates, a site with its long axis aligned north–south is going to impose a similar orientation on any building placed on it. This will mean that the long axes

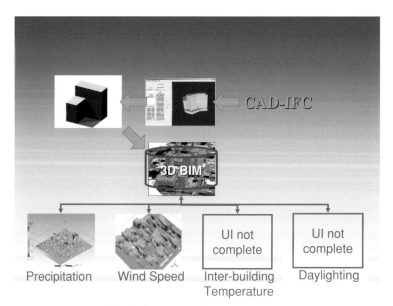

Figure 9.10   Assessment of local climates within a CBD.

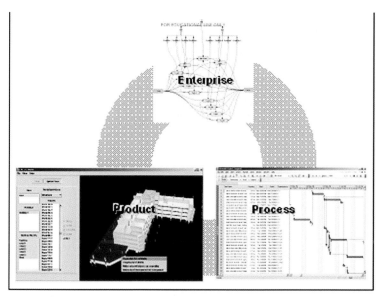

Figure 9.11   Product, process and enterprise as interrelated concepts.

are oriented to maximize heat gain at the times of day when it is most difficult to protect the building. A CSIRO (Commonwealth Scientific and Industrial Organization) project (Schevers and Drogemuller, 2006) is examining the relationships between urban planning and subdivisional planning and various KPIs (key performance indicators) used by planners. This work is using the semantic web technology to provide advanced inferencing features.

9.4. Unifying the representations

The various representations of information described above have been presented under a chronological structure. If the content and use of the information is examined it can be broken down in other ways. A useful breakdown of the information is into:

- *Product*: Describing the resulting facility and its intermediate states
- *Process*: Describing the means of transforming the product between states
- *Enterprise*: Describing the organizational structure(s) used within the project

These all form part of a continuum that needs to be supported across the facility life cycle if virtual prototyping is to be successful (Figure 9.11).

Breaking the documents down into these categories leads to the following structure:

- *Product*:
    - statement of need
    - stakeholder impact and requirements
    - project brief
    - site and environmental issues

- feasibility studies
- risk process: technical
- *Process*:
  - communication strategy
  - risk process: project management, health, safety and environment
- *Enterprise*:
  - business case
  - risk process: commercial

9.5. Technical issues

A number of components are necessary before a virtual prototype can be built. Firstly, some type of data model is required which describes the building or infrastructure that is being analysed, the relationships of the object within its environment and the relationships of the parts of the object. Secondly, the processes that lead from one state of the facility data model to the next must be described. Finally, the organizational structure that performs the processes to construct the virtual and then the real facility must be stored.

No single software vendor can afford to support the wide range of applications that are necessary to meet the vision for virtual prototyping described above. The sharing of information will need to be seamless between vendors; this is interoperability, across product, process and enterprise descriptions.

Bringing the various pieces of the 'virtual prototype puzzle' together will require complex information models and modelling capabilities, which will need to be intuitive and seamless for industry users. These will support analysis at various stages of design (levels of detail) and will also support multiple versions of a project being developed in parallel. This is likely to lead to a diminution of the influence of CAD and geometrical representations to a situation where CAD is just another 'view' of the project.

The ability to link constraints – i.e. legal requirements, such as building regulations, and also self-imposed constraints, such as client requirements – to the information models that support the procurement process will also be critical.

The integration of GIS and facility information will also be necessary to ensure that all possible dimensions of the facility are examined at the relevant levels of detail.

One issue that has technical as well as social implications is that some technical analyses are well supported by accepted algorithms and data, while other technical areas are still prone to very subjective assessments. A comparison between thermal analysis and environmental impact assessment illustrates this point. The ASHRAE (American Society of Heating, Refrigerating and Air-Conditioning Engineers) handbooks consist of four volumes that are accepted internationally as the definitive description of the science behind the thermal assessment of buildings. In contrast, there are no widely accepted, rigorous methods of assessing environmental impact, and the available data and methods of obtaining the data are still very much up to the individual. While national and regional environmental impact assessment methods are defined, any two people analysing the same structure are likely to derive significantly different results. Consequently, to provide high quality results from virtual prototyping environments in 20 years' time, work needs to start in the very near future on defining the underlying science in these more subjective areas.

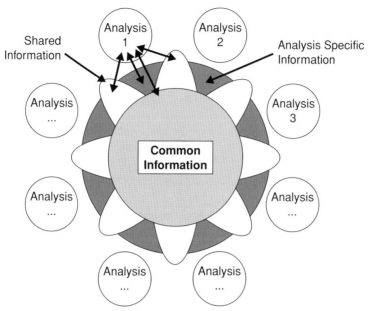

Figure 9.12   Tiers of information.

9.6. Information as the basis for virtual prototyping

The discussion above indicates that integrated information systems are necessary for virtual prototyping of buildings and structures to progress. This leads to three tiers of information within projects (Figure 9.12).

Common information is required by everyone and is available to everyone. Shared information is needed for two related processes to cooperate but is not needed by others, while private information (not shown in Figure 9.12) is only used internally. Private information may be private because it is of no use to any other analysis, or it may be core intellectual property for some group.

One aspect of information use and sharing that came out strongly in the previously explained Parametric Design project was the need to identify the information output/input dependencies. This information was captured by the use of 'Perspectors' (Haymaker *et al.*, 2003) as shown in Figure 9.13.

9.7. Social and cultural issues

A number of social and cultural issues will affect the rate of uptake or even the viability of virtual prototyping.

The major area will be information ownership and access. This is a constraint within projects, where participants need to be willing to share more information than is necessary with document-centric processes. However, this issue is even more critical when new analytical methods, such as environmental impact, are considered. On the one hand, qualified, reliable and unbiased people need to generate new data, and need to be supported in some way while doing so. On the other hand, new analytical techniques

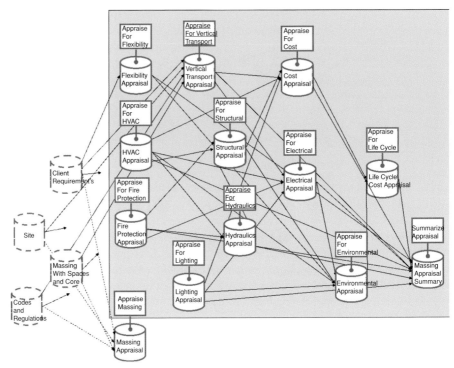

Figure 9.13   Perspectors for building services design.

will struggle for widespread acceptance until such information is widely available at a comparatively small cost. New techniques like environmental impact assessment will most likely require some level of government intervention before they are widely used. Fortunately in this area, government intervention is occurring.

Some early users of technology will use the technology for market advantage, either because it makes their own processes more efficient or for a perceived marketing advantage.

One impediment to the move to virtual prototyping is that some interests will perceive a threat to their business from new technologies. An example is the introduction of national energy rating schemes for housing in Australia and the raising of the required rating to 5 stars in the state of Victoria. This has meant that for the temperate parts of Australia it is more expensive to gain a required rating with a timber floor rather than a concrete slab on the ground. This has caused some angst for the timber industry and significant political lobbying.

One major advantage of virtual prototyping is that it will lead to a better informed market and hopefully to a better standard of design and construction.

9.8. Conclusions

A number of possible scenarios for the use of virtual prototyping through the AEC-FM (Architecture Engineering & Construction – Facilities Management) development process have been described. There is a plethora of opportunities to improve the way that information is generated, displayed and manipulated within our industry.

A number of technical, social and cultural issues will impact on the rate and extent of uptake of virtual prototyping. However, given the current rate of 3D technology uptake within the AEC-FM sector, virtual prototyping is likely to be the norm in 20 years' time.

Acknowledgements

The figures used throughout this chapter are derived from a number of actual projects funded either through the Cooperative Research Centre for Construction Innovation or through CSIRO. The research groups, besides CSIRO, that have contributed to these projects include QUT, RMIT, Stanford University, the University of Newcastle and the University of Sydney. The industry organizations who have contributed in-kind and models are Arups, John Holland, Queensland Department of Public Works, Rider Hunt and Woods Bagot. The author was employed by CSIRO as a principal research scientist while undertaking most of the work described above.

References

Commonwealth Research Centre for Construction Innovation (2004) *Achieving Value Alignment in Project Delivery.* Available at http://www.construction-innovation.info/images/pdfs/Achieving_Value_Alignment_final.pdf

Commonwealth Research Centre for Construction Innovation (2005a) *Design Check: Automated Code Checking, CRC for Construction Innovation.* Available at http://www.construction-innovation.info/images/pdfs/Brochures/Design_Check_brochure.pdf

Commonwealth Research Centre for Construction Innovation (2005b) *Project Diagnostics: Providing Successful Project Outcomes.* Available at http://www.construction-innovation.info/images/pdfs/Research_library/ResearchLibraryC/Brochure/Project_diagnostics_April05[1].pdf [Accessed 9 March, 2007]

Cooper, R., Hinks, J., Aouad, G., Kagioglou, M., Sheath, D. and Sexton, M. (1998) *A Generic Guide to the Design and Construction Process Protocol.* University of Salford, UK.

Fernando, T., Wilson, J., Dalton, G., Dangelmaier, M., Cros, P-H., Baudier, Y., Skjaerbaek, J. and Varalda, G. (2003) *Future Workspaces: A Strategic Roadmap for Defining Distributed Engineering Workspaces of the Future, Reports.* Available at http://www.avprc.ac.uk/fws/ [Accessed 9 March, 2007]

Haymaker, J., Suter, B., Kunz, J. and Fischer M. (2003) *PERSPECTORS: Automating the Construction and Coordination of Multidisciplinary 3D Design Representations, CIFE Technical Report #145.* Stanford University, CA.

Kiviniemi, A. (2005) *Requirements Management Interface to Building Product Models.* Finland, VTT **Technical Research Centre of Finland**. Available at http://www.vtt.fi/inf/pdf/

Schevers, H. and Drogemuller, R. (2006) Ontology driven concept modeller for urban development. *Proceedings of the AUSWEB 2006Conference*: Australasian World Wide Web Conference, 1–5 July, Noosa Beach, Queensland, Australia.

Solibri (2007) http://www.solibri.fi/ [Accessed 9 March, 2007]

Thiess John Holland (2007) http://www.thiessjohnholland.com.au/home/designing and building_eastlink [Accessed 9 March, 2007]

Tucker, S. N., Seo, S., Ambrose, M. D., Johnston, D. R. and Newton P. W. (2005) Eco-assessment of commercial buildings. *Proceedings of the Fourth Australian Conference on Life Cycle Assessment, Sustainability Measures for Decision Support*, February, Sydney, Australia.

Twentieth Century Fox (2002) http://www.minorityreport.com [Accessed 9 March, 2007]

10   The need for creativity enhancing design tools

Souheil Soubra

The construction sector, which is labour-intensive and has a low entry level, was (and still is to a large extent) very profitable. Economic activity has been strong with profits at their highest levels for the last five years, despite increased costs associated with materials and more stringent building regulations. Hence, we consider that the 'buzzwords' used over the last few decades in the Construction-IT (information technology) research agenda, such as 'improved productivity' and 'additional market segments', are not and have never been the real drivers of the in-depth change within the sector.

Nevertheless, the context is changing rapidly. European society is facing an overwhelming number of challenges: demography changes and an ageing population (in industrial countries), climate change, unemployment and, in particular, the end of low cost fossil resources.

At the same time, the construction sector is expected, more than ever, to obtain better living and working conditions: conditions that are accessible and comfortable for all, safe and secure, durably enjoyable, efficient and flexible to the changing demands, available and affordable.

Obviously, current business models and working methods have reached their limits, and it appears there is an urgent need for creativity enhancing tools supporting an 'out of the box' approach to design, aiming for:

- Environmentally sustainable construction (in a context of limited resources – such as energy, water, materials and space)
- Meeting clients' and citizens' needs in terms of health (following indoor and outdoor exposures), security (against natural and industrial hazards), accessibility and usability for all (including disabled and elderly) and enhanced life quality in the urban environments

It would appear that future research challenges in this sector address:

- Seamlessly integrated 'visualization/simulation' packages that allow simulation of physical products and processes in order to support complex decision-making processes throughout a project's life cycle. This needs to be done through a visual representation of the project coupled with underlying knowledge related to design, construction and operation.
- Standards and interoperability at various scales (geographical information, buildings, components).

- Communication media to support collaborative design between stakeholders (client, architect, engineer, citizen, etc.) and across disciplines (structural, thermal, acoustics, etc.).

10.1. Introduction

Among the industrial sectors, the construction sector is one of the biggest consumers of energy, either directly (e.g. heating and air conditioning) or indirectly (e.g. production of building materials). It contributes greatly to the massive use of some critical resources (such as energy, water, materials, space, etc.) and is responsible for a large portion of greenhouse gas emissions.

Buildings can also be dangerous to their occupants and users: domestic accidents kill, each year, up to 180 000 and injure up to 500 000 in France alone (the victims are mostly children and the elderly). Poor air quality and some emissions can also cause a large number of diseases and be detrimental to the public health, even if the effects can only be proven several decades later (e.g. the use of asbestos which was only prohibited in 1997 in France, the known effects of legionella, radon and various other pollutants).

The mutual interactions between cities and buildings are becoming more and more complex. Obviously cities, buildings and their various elements need to be integrated as complex systems of material, energy and information flows. However, due to the lack of adapted models and tools, such a holistic model can not yet be achieved.

Ken Yeang (1996) considers that bioclimatology, in architectural terms, is the relation between the form of a structure and its environmental performance with respect to its external climate. Although such an approach has higher start-up costs, it produces lower life cycle energy costs, as well as providing a healthier and more human environment within the structure. Other issues, he considers vital to bioclimatic consideration, are those of place-making, preserving vistas, creating public realms, civic zones, physical and conceptual linkages, and the proper massing of built forms. One of his salient quotes is: 'The global economy today is increasingly aware of energy as a scarce resource; the need to conserve energy and design for a sustainable future is becoming imperative for all designers' (Yeang, 1996, p. 9).

Despite that, assessments of both the environmental performance of buildings and the impact of a building in its own context are still largely based on assumptions supported by simple, disconnected mono-disciplinary models that cannot support the holistic approach needed.

We consider that addressing these complex issues of:

- Environmentally sustainable construction in a context of limited resources
- Meeting the ever increasing and changing users' needs such as health, safety, accessibility, flexibility, affordability

requires sophisticated 'visualization/simulation' packages that allow a holistic approach to design by integrating:

- Various disciplines (technical performances of the envelope, comfort conditions, health and safety, cost analysis, etc.)
- Different scales (materials, components, buildings and cities)
- The overall life cycle (inception, design, construction, operation and demolition)

10.2. Sustainable design on building and urban levels

Environmentally sustainable construction is rooted in an understanding of the natural ecosystem and the building, along with the interaction between the two. A definition might include the use of innovative materials, the recycling of existing materials, the use of natural energy sources, energy efficient power plants and pre-design consultation with clients and users.

But environmentally sustainable construction cannot be limited to buildings alone, even though the challenges represented by buildings are extensive and cover a wide range of issues. Even the best designed buildings would malfunction if integrated in an uncontrolled city plan. Hence, taking into account the urban dimension of sustainability is of paramount importance.

Several studies concerning the rate of urban growth show a commonly agreed phenomenon. The last century witnessed the most intensive period of urbanization our planet has ever experienced. The urban population increased from 150 million (in 1900) to about 3 billion today. Also, estimates indicate that up to 61% of the world's population will live in cities by 2030 (Bocquier, 2005). The urban growth rate is expected to be high, particularly in developing countries, since the urbanization level in developed countries already tends to stabilize at a rate around 75%. These figures highlight the importance of city planning as an increasingly complex issue involving numerous actors (e.g. planners, developers, citizens, neighbourhood communities, local and national government agencies and civil security organizations). Each of these parties represents a set of varying interests that often conflict.

In the area of technology development, the past two decades have witnessed a real revolution related to the use of geographical information. We have now reached an impressive situation in which computer hardware and software have made it possible to record, display, store, analyse, deconstruct or simulate very different kinds and scales of geographical information. The available technologies and tools include: high-resolution satellite imagery accessible through the web; standards for handling geographically referenced data; GIS (geographic information systems) for making cadastral plans and topographic maps; image processors for analysing remotely sensed images; geo-statistical packages for spatial interpolation; and many more. Moreover, several attempts have been made to link models of planning, of hydrological, meteorological or geological processes to spatial data in an attempt to capture the interactions between pattern and process in geographical space. There is still, however, a lot to deal with, especially concerning the modelling of the interaction between urban planning and sustainability related issues, e.g. energy consumption, travel requirements, parking needs or social diversity and the use of space. Nevertheless, it is now commonly agreed that the research agenda must not solely be concerned with technology, as social, organizational and human issues also need to be considered in an interdisciplinary manner.

Curwell *et al.* (2003) have argued that the ability of stakeholders to share their understanding of the present and visions for the future is an important prerequisite to informed planning and, through this, to building a consensus on difficult issues such as sustainable development. It is believed that by developing advanced 'virtual planning' technologies and understanding how they can be coupled to the planning process, we might be able to strengthen local governance and thus support more sustainable urban development.

As an application, the research carried out on the virtual planning issue in the Intel-cities[1] integrated project (Intelligent Cities – IST no. 507860) identified advanced environmental visualization and simulation as a powerful means to support a more inclusive planning process that ensures more rapid and consensual urban (re)development decision-making (Soubra, 2006). Indeed, new technologies for virtual planning provide intuitive methods to interact with relevant data so that experts and non-experts will be able to assess the environmental consequences of urban (re)development projects. It is only by visualizing the outputs of environmental simulations in various 'what-if' scenarios that actors can be expected to comprehend the consequences of a planning proposal. The research addressed the seamless integration of multiple sources of digital data. As these data have been amassed by many different organizations over several decades, a preferred solution was to adopt open data standards for handling data towards which legacy systems can then migrate. Fortunately, several industry-led consortia have been promoting standards relevant to urban planning. The mission of the Open Geospatial Consortium (OGC), for example, is to deliver interface specifications for geographically referenced data. To this end, they have developed several specifications to describe, encode and transport information pertaining to spatial entities. In contrast to Industry Foundation Classes (a common building information model developed for the construction industry by the International Alliance for Interoperability), the OGC specifications apply to generic spatial entities such as points, lines and areas and need to be contextualized before they can be used for a particular application. Figure 10.1 shows an application schema developed for virtual planning according to the OGC Geography Mark-up Language (GML[2]) specifications.

GML-encoded information can be easily transported over the internet. It can be used to communicate descriptions of feature types or transport actual data for a particular feature. A parser was developed allowing GML files to upload through a MapServer implementation and the data delivered as a Web Feature Service to a Mozilla Firefox browser or displayed in the visualization engine of the environmental simulators (Figure 10.2). CSTB (Centre Scientifique et Technique du Bâtiment) noise and air quality simulators were adapted to read a GML2 document. The simulator then generated 3D fields for noise and air quality and a visualization engine displayed these data with 2.5D building data (Figure 10.3).

The environmental simulators operate as a spatially enabled component within the Intelcities service-oriented architecture. The final stage of the technical developments has been to connect the simulators to the Intelcities core and declare the service. Declaring the service refers to the possibility that other services could also operate with the environmental simulators. In this case, a geobroker service, developed elsewhere in the Intelcities project, act as a geodata server to the simulators. In the visualization engine,

---

[1] Intelcities is a research and technological development project to pool advanced knowledge and experience of electronic government, planning systems and citizen participation from across Europe.
  The project was led by the city of Manchester and brought together 18 European cities, 20 ICT companies and 36 research groups.
  The project was part of the *European Union's Sixth Framework Programme*, with €6.8m of the €11.4m budget from the EU's Information Society Technologies programme.

[2] GML is an XML-based mark-up language that is used to describe geographical phenomena and encode information about them. These features have geometry and non-geometry properties but feature types, such as roads, houses or municipal boundary, are not defined in GML itself. They are defined in application schemas.

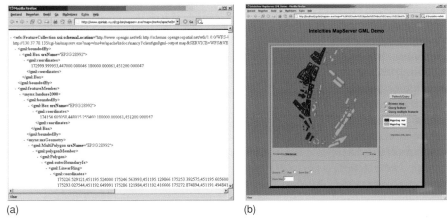

(a)   (b)

Figure 10.1   (a) Extract from a PlanningGML data file; (b) building data displayed in a Web Feature Service using MapServer.

a user is able to define an area of interest on a georegistered aerial photograph and then upload the specific GML data required by a simulator using WFS (Web Feature Service) (Figure 10.4). The user, authenticated by the system, is then enabled to create 'what-if' scenarios by changing the weather or traffic conditions (Figure 10.5).

10.3. Design requirements: comfort, accessibility, flexibility

The ever-changing ways of life bring about new comfort requirements as well as new tools. For example, from the nineteenth century came the collective distribution of water

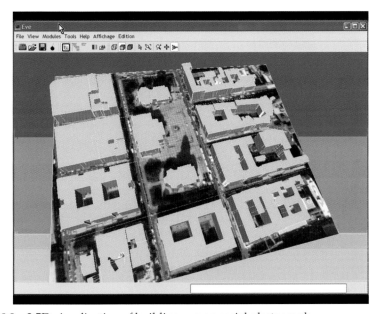

Figure 10.2   2.5D visualization of buildings on an aerial photograph.

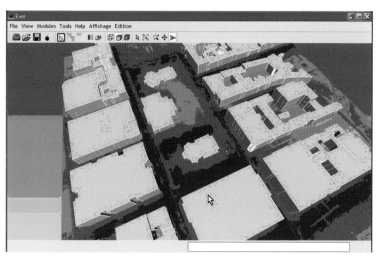

Figure 10.3   Visualization of noise levels using colour coding.

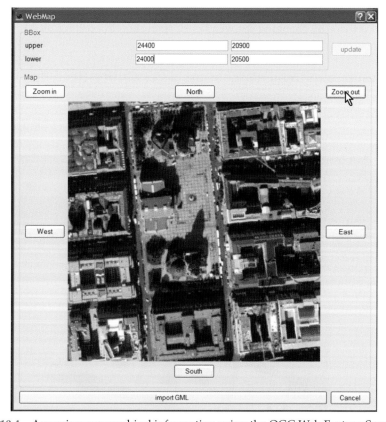

Figure 10.4   Accessing geographical information using the OGC Web Feature Service.

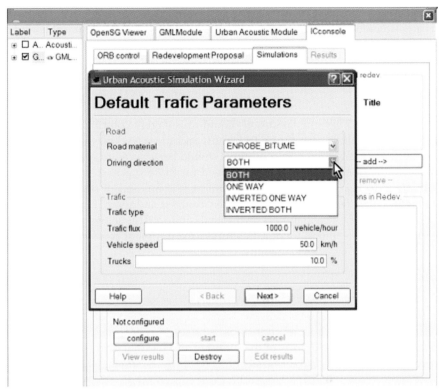

Figure 10.5   User-defined traffic conditions for the noise simulator.

and sewerage systems; in the twentieth century there came ducts for air extraction, mechanical air ventilation and the wide distribution of electricity and gas. Today, there is broadband access to triple play services for telephone, television and the internet. The response to the new comfort requirements has always resulted in the development of new products and new services. Recent times have seen the emergence of users' requirements related to thermal comfort (in winter and now in summer), protection against external and interior noise, visual comfort, air quality, etc. Comfort is a decisive criterion in design since occupants will obviously react to any discomfort by taking actions to try and restore their comfort. These actions can, in fact, induce a bigger discomfort or even a dysfunction for the building. In any case, it alters the operation conditions that were envisaged by the designers.

10.3.1. Comfort

10.3.1.1. Thermal

Thermal comfort is critical to energy efficiency. It is composed of several parameters, among which are the energy efficiency of the building envelope and the heating systems, as well as psychological (how the brain handles information stimulus coming from skin sensors about warmth and cold and how these are integrated as an overall sensation of pleasure or displeasure) and behavioural aspects (Salat, 2006). The interaction between occupants and buildings is an area that still lacks systemic and adaptive models that could be used in the design process.

10.3.1.2. Visual

Visual comfort and lighting conditions could also largely contribute to energy use. Poor lighting conditions can cause discomfort and stress, and play an important role in domestic accidents. Lighting conditions should be adapted to space requirements to facilitate the undertaking of tasks reserved for those spaces, bearing in mind the 'visual' needs of the occupants. Visual comfort is composed of several parameters, among which are the luminance of surfaces, contrast and glare. Lighting (including solar radiation) has a strong impact on thermal comfort. This shows the importance of integrated interdisciplinary models to support the design of comfortable spaces that match basic principles of sustainability.

10.3.1.3. Acoustic

Acoustic comfort takes into account sound sources (e.g. transport, human activities) and the technical characteristics of the building and its components. Exposure to excessive sound levels may have a variety of adverse effects, among which are the disturbance of speech communication, disturbance of concentration, sleeping problems and even hearing impairment. Annoyance caused by noise may also have social effects and may affect other comfort conditions (e.g. opening windows, using balconies).

10.3.1.4. Air quality

Bad air quality inside and outside the buildings can cause a large number of diseases, since people may be exposed to potentially toxic, allergenic or infectious pollutants for up to 16–24 hours per day in their homes, offices, schools and the like. It is now clear that efforts towards energy savings have, as a side effect, reduced air quality by improving the insulation of buildings and reducing the air renewal rates.

10.3.2. Accessibility

The quality of the accessibility of the buildings is also an increasingly strong concern of building owners and of the other stakeholders, especially designers. It addresses the question of the space used by the occupants. From the point of view of the users, accessibility is appreciated in their various movements in and around the building, including:

- *From outside towards inside*: to locate oneself and to reach the target (e.g. entrance)
- *Inside the building*: to have the ability to locate oneself at any point inside the building, and to be able to circulate

- *From inside toward outside*: to locate oneself inside the building to be able to reach the exit

Although each of the above applies to any user, they become more challenging and take on an extra dimension for people with a disability, such as those with reduced mobility (including the elderly), with partial or no sight, or with impaired hearing.

10.3.3. Flexibility

Flexibility lies in the capacity to authorize different practices and being able to ensure their long term capacity for change. This topic is still greatly underestimated during the design phase, which is probably due to its complexity and the lack of adapted design tools. Flexibility should be considered both at space and usability levels (such as how the space can be adapted to users' requirements as these evolve through different life periods). Flexibility could also manifest itself at the structural level. Does it make sense to invest heavily in building structures assuming they need to support loads for a long period, or would it not be more relevant to adapt the structure progressively depending on the current need and then to have a progressive investment policy?

10.4. The need for innovative design methods and tools

Environmentally sustainable design is a holistic process where architects and engineers must share a solid knowledge of local conditions, existing resources and the possibilities of using renewable forms of energy and materials. Several issues need to be addressed:

- As stated earlier, the city and the building need to be integrated as an overall complex system of material, energy and information flows
- All the stakeholders need to share an in-depth knowledge of all functional, technical and design relationships
- At all stages, design should be 'testable' through simulation, allowing assessment of performance and comfort conditions in a clear and comprehensive form

Collaboration is much more than just exchanging information since it also involves the participants' thinking processes and their networking, dealing with a wide range of issues and systems, which needs to be at the heart of the design process. Environmental performances of the building, environmental protection of the construction site, careful consideration of occupants' and workers' health through all the phases, the emphasis put on reducing construction and demolition waste, are all complex issues with a high interdisciplinary aspect that require innovative design methods and tools. 'Testable design' is therefore important to deal with this complexity, which would be particularly useful at the very early stages of the design process. Indeed, wrong or insufficiently mature decisions made at the early stages of the design often cannot be subsequently corrected, and an interdisciplinary approach requires the ability to confront options as early as possible in order to identify and handle conflicts.

Early concept evaluation (obviously through virtual prototyping) needs to be on the research agenda along with following issues:

- The ability to handle uncertainty in the design (i.e. being able to assess a project even when a clear definition of some of its parts is not yet available).
- The potential of keeping design options open as long as possible to avoid falling into the same traditional patterns and to support an 'out of the box' approach to design.
- The ability to document design decisions (when, by whom and especially why a decision has been made) and eventually backtrack some of these decisions to handle new circumstances or emerging ideas.
- The ability to make links at a meta-level between the evaluation models used. Indeed, it is necessary to be able to use these models to create interdisciplinary simulations, which allows searching not only for local optimums (each found in a separate discipline) but also, and especially, for a global optimum for the overall project taking into account the various and usually contradicting constraints of each discipline.

In this emerging paradigm, the client's brief and design can run parallel, which will necessitate a strong collaboration between the design teams and the client/users. The emerging bi-directional influence that emerges out of such collaboration is difficult to monitor and support, if not impossible.

It is also evident that enhancement of the design process is needed in order to deal with the increased complexity of the user environment in a better manner. This complexity comes from several sources, including regulation, clients' expectations and their changing needs through life, emerging technologies and the growing importance of services.

10.5. From collaborative work to cooperative design

Collaborative environments have been on the Construction-IT and EU IST (Information Society Technologies programme) research agendas for several years now. It is still a basic requirement, but it needs to incorporate more than just an exchange of information through videoconferencing systems. More needs to be done to move towards cooperative design, while avoiding a simple continuation of past work.

Hanser (2003) argues that design in the construction sector and, more particularly, design project management are social and professional activities characterized by a specific cooperation context. However, the available collaborative tools are not totally adapted to this context. Indeed, the cooperative design (or co-design) – contrary to the collaborative or distributed design – requires a cognitive synchronization based on 'unplanned' activities that he qualifies as 'implicit' activities. Autonomous actors, belonging to one or more entities, realize these activities with varied and complementary competences. In this group, the actors cooperate in order to achieve a common objective, which can be the production of a document, a product or a building.

In this cooperation, the role of a tool is to allow the actors to not only have a vision of their work development but also to increase their action's potential and widen the scope of each action. The existing tools, commonly referred to as 'workflow', are based on cooperation and coordination models where the activity must be explicit. They can be adapted to some industrial sectors or administration processes but fail to meet the specificities of the construction sector.

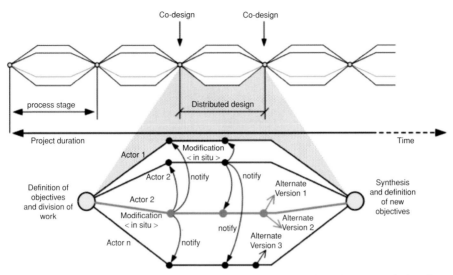

Figure 10.6 The collaborative design process composed of co-design and distributed design steps. (Hanser, 2003.)

Hanser presents cooperative design in the construction sector as a cyclic process composed of two main phases:

- *Co-design*: In this step, designers synchronously communicate in order to define a common goal and dispatch work. They bring their own competences to reach this goal together. Specific interdisciplinary issues need to be solved here. Global optimums need to be decided at this stage.
- *Distributed design*: This step characterizes a collective design where the tasks are clearly defined and distributed to each actor at the beginning of the design process. Each actor tries to reach a goal (or 'sub-goal') by participating simultaneously (not jointly) in the realization of a final common goal known and agreed by everyone. In other words, each actor optimizes and details his version of the design (searching for local optimums) and can notify the others about his modifications. Communication becomes asynchronous (Figure 10.6).

Depending on whether the collaboration takes place in the same place or in different places, at the same time (synchronous) or at different times (asynchronous), four types of interactions can be identified (Table 10.1; David, 2001). The differentiation between synchronous and asynchronous collaboration is very important since these two modes are applied, sequentially, in the cooperative design process. During an asynchronous cooperation, the exchange is done by whole entities (numerical model, complete documents, etc.). The more synchronous the communication, the smaller the granularity of the exchanged information (e.g. discussions about specific building components, etc.), thus the importance of the access management becomes strong. This imposes an important set of constraints on collaborative tools for the construction sector since these tools necessitate going very quickly from the bigger picture to more specific details, while allowing modifications at various different scales. Arthaud (Arthaud and Soubra,

Table 10.1   Different types of interactions to support collaboration.

	Same time	Different times
Same space	Traditional meetings Decision-making space Electronic blackboard Brainstorming and 'what-if' tools	Shared resource centre
Different spaces	Audio- and videoconferencing Virtual meetings Chat Shared applications	Email Mailing list Shared workspaces: (BSCW, CVS, etc.)

Abbreviations: BSCW, Basic Support for Cooperative Work; CVS, Concurrent Versions System.

2006) considers that this is what is missing in the model-based approach, which was never widely used in the construction sector. Therefore, he developed a tool allowing the differences between STEP[3]-based product models (among which are the IFC) to be tracked. Starting with two versions of the product model, the changed, added and removed elements are quickly and visually detected, facilitating the transition between the distributed design phase (where designers work on their own version of the model) and the co-design phase (where these versions are merged after the decision-making phase) (Figure 10.7).

In the Construction-IT R&D agenda, special emphasis should be placed on the move to collaborate beyond the videoconference mindset of today to make it the means by

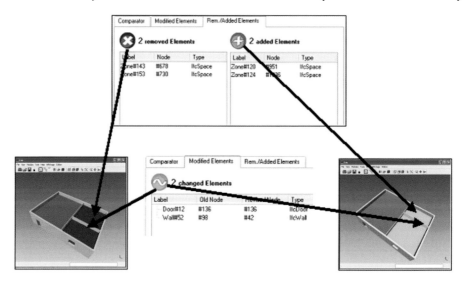

Figure 10.7   Tracking differences between STEP-based product models. (Arthaud and Soubra, 2006.)

---

[3] STEP, the Standard for the Exchange of Product Model Data, is a comprehensive ISO standard (ISO 10303) that describes how to represent and exchange digital product information.

which it becomes possible to work together on product and process representations, across organizational and enterprise boundaries. Collaboration needs to be based on virtual co-location, and multi-user, multi-discipline and immersive environments. Virtual prototyping tools should be developed to help collaborative teams reach negotiated solutions, or a global optimum. Cooperative design should also provide facilities for modelling, decision-making and client-oriented simulations. These design environments should also be location- and context-aware. Filters that direct selected information and knowledge to different groups, depending upon their needs and activities, should be provided.

10.6. Conclusions

The Construction-IT research agenda should be revisited in order to move away from the irrelevant buzzwords used in recent decades, for example 'improve productivity', to include new societal challenges such as:

- Sustainable construction in a context of limited resources
- The end of the low cost fossil energy
- Demography changes and an ageing population
- Climate change

These are obviously complex issues with a high interdisciplinary aspect, which ask for creativity enhancing design methods and tools that allow the investigation of 'out of the box' solutions that are now abandoned due to their complexity. More specifically, research objectives include the following:

- Buildings and their various elements need to be integrated as complex systems of material, energy and information flows. Models linking city planning issues and sustainability related issues on both the urban and the building levels need to be developed.
- Various disciplines (technical performances of the envelope, comfort conditions, health and safety, cost analysis, etc.) need to be integrated in interdisciplinary and holistic approaches supporting a 'global optimum' search that allow designers to confront options in order to identify and handle conflicts.
- Consultation with clients and citizens is a definite must in order to put the users in the centre of the process and to envisage a sound operation of the building. New design paradigms need to emerge where the client brief and design are conducted in parallel and can inform each other as they progress.

Some ideas of R&D topics that should be on the Construction-IT research agenda include:

- Seamlessly integrated 'visualization/simulation' packages that allow the simulation of physical products and processes in order to support complex decision-making throughout the project's life cycle. This needs to be done through a visual representation of the project coupled to underlying systemic and adaptive models related to technical aspects and user-centred issues.

- Early concept evaluation through virtual prototyping that support:
  - the ability to handle uncertainty in the design (i.e. being able to assess a project even when a clear definition of some of its parts is not yet available), thus allowing a comparison of design options from the very earliest stages
  - the potential of keeping design options open as long as possible to avoid failing in the same traditional patterns, and to support innovative design options
  - the ability to document design decisions (when, by whom and, in particular, why a decision has been made) and eventually backtrack some of these decisions to handle new circumstances or emerging ideas
- New collaborative environments that go way beyond the videoconference mindset of today towards tools that allow stakeholders to share an in-depth knowledge of all functional, technical and design relationships. Special emphasis should be put on taking into account the specialisms of the construction sector, and especially on developing tools adapted to the design process where both co-design and distributed design tasks are conducted.

References

Arthaud, G. and Soubra, S. (2006) Tracking differences between IFC product models: a generic approach for fully-concurrent systems. *European Conference on Product and Process Modelling (ECPPM)*, 13–15 September, Valencia, Spain.

Bocquier, P. (2005) L'urbanisation mondiale? Pas si galopante que ça! *La Recherche* **389**: 5.

Curwell, S., Hamilton, A. and Soubra, S. (2003) *The Intelcity roadmap. Report of the EU Intelcity project IST-2001-37373.* European Commission.

David, B. (2001) IHM pour les collecticiels. *Réseaux et Systèmes Répartis* **13** (Nov.): 169–206.

Hanser, D. (2003) *Proposition d'un modèle d'auto-coordination en situation de conception, application au domaine du bâtiment.* PhD thesis, Institut National Polytechnique de Lorraine.

Salat, S. (2006) *The Sustainable Design Handbook.* Paris, Hermann.

Soubra, S., Marache, M. and Trodd, N. (2006) Virtual planning through the use of interoperable environmental simulations and OpenGIS® geography mark-up language. *eChallenges Conference 2006*, 25–27 October, Barcelona.

Yeang, K. (1996) *The Skyscraper Bioclimatically Considered – A Design Primer.* West Sussex, Wiley–Academy.

Part 4
Challenges for information and knowledge modelling

11 Context-aware virtual prototyping

Chimay J. Anumba and Zeeshan Aziz

Virtual prototyping has evolved in response to the requirement for designers and others involved in product development to have a better feel for the end product, at the earliest possible opportunity and without the expense of building a physical prototype or making the product itself. This is increasingly seen as necessary within the construction industry and various approaches are being adopted for its implementation. One of the key limitations of existing efforts relates to the lack of context-specificity in the development and use of the virtual prototypes. This results in both information overload (where some people are provided with more information than they need) and information starvation (where the prototype does not provide other people with the information that they do need).

This chapter explores how awareness of user context (such as user profile/role, preferences, task, location, existing project conditions, etc.) and other project contexts can be incorporated into virtual prototyping in order to ensure the delivery of pertinent information and enhance construction collaboration. It argues that parallel developments in web services, which provide the ability to dynamically discover and invoke services regardless of the operating system or programming language, can be leveraged to enable construction professionals to access, in real time, different corporate back-end systems and multiple inter-enterprise project resources. This chapter discusses how context awareness and web services can be integrated with virtual prototyping to facilitate the creation of an intelligent modelling environment, within which intelligent decisions can be taken based on the interpreted context. A framework is presented, which facilitates context capture, context brokerage and integration with legacy applications using a web services-based architecture. Aspects of the framework implementation, which was based on a pocket PC platform, are used to illustrate the use of context-aware computing technologies in realistic construction scenarios. Conclusions are drawn about the possible future impact of context awareness and web services in virtual prototyping.

11.1. Introduction

The importance of models and prototypes in the product development process has long been recognized, and has ranged from very simple drawings and plans to sophisticated 4D and nD models. In fact, the earliest architectural drawings are thought to date back to 2000 BC. Since then, considerable progress has been made and, until very recently, considerable effort was devoted to creating physical models and prototypes. These models and prototypes offer useful perspectives on a product or project prior to its actual development or implementation. However, these were limited in a number of respects, particularly with regard to being expensive and providing primarily static information with inadequate capacity for assessing several behavioural and other important characteristics. These have largely been replaced by virtual models and prototypes (3D, 4D, nD and VR (virtual reality)), which have enhanced facilities for the exploration of further attributes and behaviours, and the investigation of 'what-if' scenarios. Models that incorporate a temporal dimension are also able to model dynamic information by illustrating changes to the basic model or prototype over time. A major limitation of these models is that they are designed primarily to deliver a pre-programmed functionality with no consideration of the user context and/or other project contexts. This often leads to a mismatch between what an application can deliver and an end-user's data and information requirements.

In contrast to the existing static and pre-programmed information delivery approaches, project information by its very nature is dynamic and incorporates the multiple perspectives of the project team members. Thus, the value of the virtual model or prototype is highly dependent on the perspective of the end-user, which changes from time to time in line with changes in the user context. Thus, virtual prototyping systems need to be able to understand changes in the context of the user (such as having a knowledge of who they are, where they are located, what tasks they are involved in, what stages/aspects of the model they are interested in or responsible for, etc.) and to deliver the right information at the right time on an as-needed basis.

This chapter introduces the concept of context-aware virtual prototyping and discusses how a combination of context awareness (by enabling better understanding of the user's context) and web services (by supporting resource discovery and integration in a context-aware environment) can be used to facilitate virtual prototyping. It starts by presenting an overview of context-aware computing and web services. It then presents a framework for a context-based information delivery system for construction workers, which facilitates context capture, context brokerage and integration with back-end systems using a web-services model. Aspects of the implementation of the framework in a construction context are illustrated and its possible extension to virtual prototyping discussed. Conclusions are drawn about the possible future impact of context awareness and web-services technologies in virtual prototyping and the wider construction industry.

11.2. Context-aware computing

Context-aware computing is defined by Burrell and Gay (2001) as the use of environmental characteristics such as the user's location, time, identity, profile and activity to inform the computing device so that it may provide information to the user that is relevant to the current context. Context-aware computing enables an application to leverage knowledge about various context parameters, such as who the user is, what the user is doing, where the user is and what mobile device the user is using (if any). Pashtan (2005) described four key partitions of context parameters, including

- User static context: includes user profile, user interests, user preferences
- User dynamic context: includes user location, user current task, vicinity to other people or objects

- Network connectivity: includes network characteristics, mobile terminal capabilities, available bandwidth and quality of service
- Environmental context: includes time of day, noise, weather, etc.

The awareness of user context (such as user profile/role, preferences, task, location, existing project conditions, etc.) can enhance communications and collaboration in the construction industry by providing a mechanism for determining information relevant to a particular context. In recent years, the emergence of powerful wireless web technologies, coupled with the availability of improved bandwidth, has enabled mobile workers to access in real time different corporate back-end systems and multiple inter-enterprise data resources to enhance construction collaboration. Context-aware information delivery adds an additional layer on top of such real-time wireless connectivity (Aziz *et al.*, 2005) offering the following benefits:

- Delivery of relevant data based on the worker's context, thereby eliminating distractions related to the volume and level of information.
- Reduction in the user interaction with the system by using context as a filtering mechanism. This has the potential to increase usability by making mobile devices more responsive to user needs.
- Awareness of mobile worker's context, through improved sensing and monitoring, can also be used to improve security and health and safety practices on the construction site. At the same time, it is possible to use the knowledge of on-site activities to improve site-logistics, site-security, accountability and health and safety conditions on the site.

While the user context is often the most well-known context to be considered, it is important to highlight that there are other project contexts that may need to be provided for. These are discussed further later in the chapter.

11.3. Using web services in a context-aware environment

Web services are self-contained, self-describing, modular applications that can be published, located and invoked across the web using standard internet protocols. Once a web service is deployed, other applications (and other web services) can discover and invoke the deployed services regardless of the operating system or programming language. As identified by Fensel (2001), the key to web services is on-the-fly software creation through the use of loosely coupled, reusable software components. In contrast, previously used systems based on established infrastructures such as CORBA (Common Object Request Broker Architecture), RMI (Remote Method Invocation) and EDI (Electronic Data Interchange) were tightly coupled, each with their own transport protocol and inability to communicate with TCP/IP (Transmission Control Protocol/Internet Protocol). Using web services, data is free to move about the web without the constraints imposed by tightly coupled transport-dependent architectures (Coyle, 2002). Thus, the web-services technology ensures standards-based, low cost integration.

Existing virtual prototyping applications often have limited facilities for integration with other applications and services. In many cases, the integration issue is being addressed through the use of proprietary or open standard API (Application Programming Interfaces), with each application requiring a specialized API to communicate with other applications. Some of the most obvious problems of using such an approach to system integration include lack of consistency, scalability and robustness. Sophisticated interfaces are usually required to overcome the interoperability problems and reduced versatility.

In contrast, the web services approach to application integration addresses some of these limitations. A web services-based architecture is based on modular components, with each component representing a specific function. These modular components can be composed into solutions to offer the exact set of features required by a particular context. This raises the prospects for enhancing the versatility of virtual prototyping applications in the construction industry, as these applications can be combined in real time with complementary third party services depending on the project requirements or user context under defined constraints. Also, it is possible to discover services and use them on an as-needed basis, based on the context of the end-user. Aziz *et al.* (2003) discussed various application scenarios of using context awareness and web-services technology to support construction workers. However, this focused on the construction stage of the project delivery process and did not explore their potential in virtual prototyping. The framework developed by Aziz *et al.* (2003) is presented below, prior to a discussion of its applicability to virtual prototyping.

11.4. Context-aware service delivery framework

Figure 11.1 presents a framework that combines context awareness and web services to create a pervasive, user-centred environment, which has the ability to deliver relevant information to mobile construction workers by intelligent interpretation of their context so they can make better informed decisions.

Key layers of the framework are discussed below:

- *Context capture tier*: Helps in context capture and provides access to the system. It also supports mobile workers by providing context-relevant information through a human and a software interface layer. The human interface layer allows intuitive user interaction, by ensuring that data is delivered according to the worker's device type. The software interface layer ensures integration of software operating on the mobile worker's device with the back-end systems. User agents allow the personalization of contents and services in line with workers' preferences. The task-specific agents help mobile workers in accomplishing a specific task by understanding the task context, and by identifying, filtering and accessing the services taking into account the physical conditions on the construction site.
- *Access tier*: Provides the vital communication link between the wired back-end and the wireless front-end. The use of IP-based technologies enables handover and seamless communication between different wireless communication networks, such as wireless wide area networks, local area networks and personal area networks. The access tier supports both push and pull modes of interaction, i.e. information can be actively pushed to mobile workers (through user-configured triggers), or a worker can pull information through *ad-hoc* requests, on an as-needed basis.

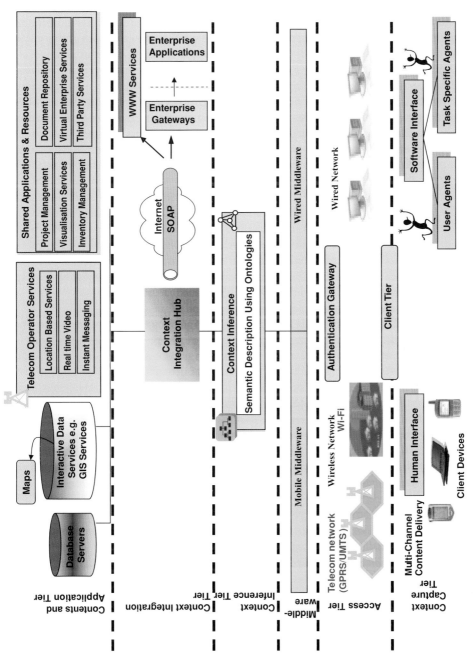

Figure 11.1   Context-aware services delivery framework.

- *Middleware tier*: Ensures the separation of data from presentation and applications. This separation allows for the reuse of the same middle tier for services delivery to wired and wireless clients. Mobile middleware connects desktop-based, back-end systems with different mobile networks, addressing the limitations imposed by mobility, e.g. device limitations, bandwidth variation, etc. Use of XML transformation technologies allow support for a wide range of mobile devices with varying form factors.
- *Context-inference*: Provides the ability to reason about the captured context using the semantic web-based model to describe a knowledge model for a corresponding context domain, thereby helping context description and knowledge access (by supporting information retrieval, extraction and processing) based on the inferred context. The understanding of semantics (i.e. meanings of data) enables the creation of a relationship between the context parameters and the available data and services. The output from the context-inference layer is fed into applications to make them aware of events on the site.
- *Context integration tier*: Is intended to help construction workers (or their agents) in service discovery. The service hub uses semantic mark-up for resource and service type description. Semantic mark-up will also allow users (or agents) to make intelligent decisions about when and how these resources and services should be used, based on the interpreted context. Adherence to open standards technologies will allow applications to share data and dynamically invoke capabilities from other applications in a multi-domain, multi-technology, and heterogeneous remote collaboration environment.
- *Contents and applications tier*: Contains construction project data and applications. The latter may be provided by project partners or application service providers (ASP).

11.5. The framework implementation

The framework implementation involved a proof-of-concept demonstrator and provided an initial working model of a large, more complex entity. Figure 11.2 presents the deployment architecture, while its key features are discussed below:

11.5.1. Context capture

In the implementation, context was drawn from the following sources (see Figure 11.3 for a variety of possible context dimensions):

- *Current location*: via a wireless, local area, network-based positioning system. A client application running on a user's mobile device or a tag sends constant position updates to the position engine over a WiFi link. This allows real-time position determination of users and equipment. It is also possible to determine user location via a telecom network-based triangulation.
- *User device type* (e.g. Personal Digital Assistant, TabletPC, SmartPhone, etc.): via W3C CC/PP standards (CC/PP, 2003). These standards allow the description of capabilities and preferences associated with mobile devices.

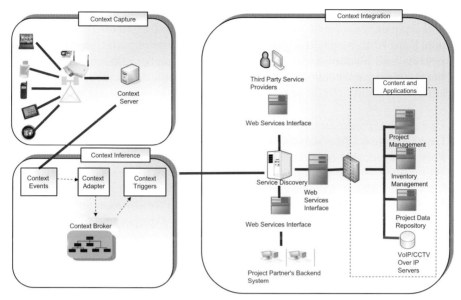

Figure 11.2   The deployment architecture.

- *User identity* (e.g. foreman, electrician, site supervisor, etc.): via the unique IP address of the mobile device. User profile was associated with user identity.
- *User's current activity* (e.g. inspecting the work, picking up skips, roof wiring on the ground floor, etc.): via integration with project management application.
- *Visual context*: via a CCTV (closed circuit television camera) over an IP camera.
- *Time*: via the computer clock.

11.5.2. Context inference and integration

Various context events are fed into the context engine. The context adapter converts the captured context events (e.g. user identification and location, time, etc.) into

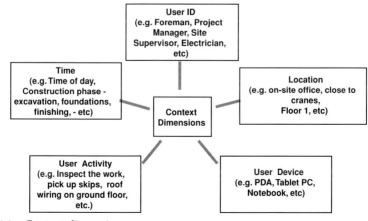

Figure 11.3   Context dimensions.

semantic associations. The Resource Description Framework (RDF) schema (RDF, 2005) was used to provide vocabulary and structure to express the gathered contextual information. Being XML-based, RDF also ensures the provision of context information in an application- and platform-independent way. Also, using the RDF schema, the context broker maps captured contextual information to available data and services. Mapping includes:

- *User profile to project data*: Mapping of information, based on the role of user on site
- *Location to project data*: Mapping user location to project data (e.g. if electrician is on floor 3, he probably requires floor 3 drawings and services)
- *User task to project data*: Mapping information delivery to the task at hand

RDF was also used as a meta-language for annotating construction project resources and drawings. Such a semantic description provides a deeper understanding of the semantics of construction documents and an ability to flexibly discover required resources. A semantic view of construction project resources logically interconnects project resources, resulting in the better application of context information. At the same time, semantic description enables users to have different views of data, based on different criteria such as location and profile. Changes in the context prompt the context broker to trigger the pre-programmed events, which may include pushing certain information to users or an exchange of information with other applications using web services, to make them aware of the events on the site. As the user context changes (e.g. change of location, tasks), the context broker recalculates the available services to the users in real time. Some of the implemented services illustrate the use of context-aware service delivery in realistic construction scenarios and are discussed below:

Profile-based task allocation

In the implementation, tasks were allocated to site workers based on their identity, profile and location. The task list (Figure 11.4) specifies the activities a site worker must perform, and it also includes the associated method statement, describing how tasks need to be performed. Using an administration application, the site manager assigns tasks and method statements to the site worker. As the site worker arrives for work, the site server detects the unique IP address of their mobile device and prompts the worker to log-in. On a successful log-in, the server pushes the worker's task list and associated method statement. The client application running on the site

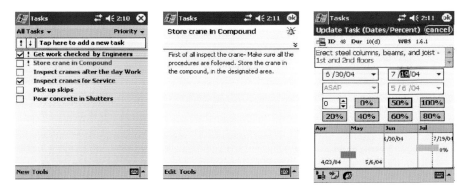

Figure 11.4   Profile-based task allocation and updates.

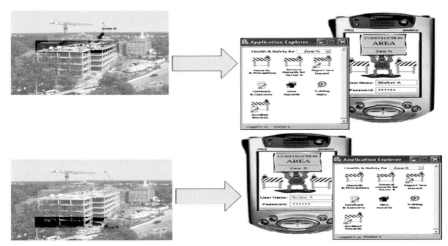

Figure 11.5   Context-aware access to project data.

worker's mobile device detects that data on the server has been changed. Modified files are then synchronized using WLAN-based synchronization. Synchronization is a two-way process (i.e. synchronizing files between the mobile device and the server application). This way, the completion of tasks can be monitored in real time and an audit trail maintained.

Context-aware access to project data

Mobile workers were provided with access to project information based on their context (i.e. location, profile and assigned task) (Figure 11.5). The context broker played the key role of capturing the user context and mapping the user context to project data, at regular intervals.

Context-aware inventory logistics support

WLAN tags were used to store important information about a bulk delivery item. XML schema was used to describe the tag information structure. As the delivery arrives at the construction site, an on-site wireless network scans the tag attached to the bulk delivery and sends an instant message to the site manager's mobile device, prompting him/her to confirm the delivery receipt. The site supervisor browses through the delivery contents and records any discrepancies. Once the delivery receipt is confirmed (Figure 11.6), data is stored locally on the site manager's mobile device. Local

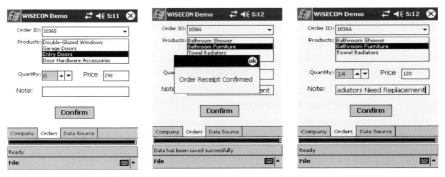

Figure 11.6   Inventory logistics support.

information stored on the mobile device is subsequently synchronized with the site server, resulting in an update of the inventory database.

11.6. Towards context-aware virtual prototyping

While existing virtual prototyping applications provide considerable benefits in terms of being able to model a facility such that it can be evaluated from a variety of perspectives, a serious limitation of these systems is that the same information is made available to all users irrespective of their context. In order to be more effective, virtual prototyping systems need to be able to understand changes in the context of the user (such as having a knowledge of who they are, where they are located, what tasks they are involved in, what stages/aspects of the model they are interested in or responsible for, etc.) and to deliver the right information at the right time on an as-needed basis. This is the goal of context-aware virtual prototyping (CAVP).

The integration of context awareness and web services into virtual prototyping applications offers considerable potential for enhancing their versatility and ensuring that end-users are provided with access to context-specific data, information and services. The context parameters that need to be incorporated can be defined by the project team based on the requirements of each project. These will include some of the user-context parameters previously outlined (e.g. role, discipline, interests, preferences, etc.) as well as project parameters (e.g. project stage, client requirements, project location, procurement/contract type, technical/design requirements – with respect to issues such as sustainability, cost, quality, function, etc.). Whatever the chosen context parameters, an appropriate context broker is needed that is able to track changes in context and trigger the appropriate response from the virtual model/prototype and/or invoke any required web services. In undertaking this, it is important to have ontologies that provide a common understanding of key concepts, such that reasoning can be undertaken based on the interpreted context and intelligent action taken.

Integrating context awareness and web services with virtual prototyping can be undertaken at a variety of levels. At a simple level, the context could relate to the user's profile (e.g. architect, structural engineer or project manager) and enable navigation of the virtual prototype to be based on highlighting and visualizing features of the prototype that are relevant to that particular context. For example, a 'walk-through' of a given floor could show the structural elements to the structural engineer and the finishes to the architect. A more sophisticated implementation of context awareness in virtual prototyping would enable the embedding within the virtual model of appropriate context triggers, ones that enable the user's context to be captured in real time and the visual representation to be modified to reflect any changes. For example, as the user moves from exploring the model from a design point of view to a construction perspective, the information provided could be modified to reflect the relationship between the 'as-built' facility and the virtual model. There is also considerable scope to make linkages between context triggers in the physical world and those in the virtual environment. Thus, contextual changes in the physical environment can trigger modifications to the virtual model and vice versa. This requires effective mechanisms for bi-directional consistency evaluations, although the degree to which automatic updates to the model would be allowed can be explicitly defined by the project team.

The role of embedded web services in enhancing virtual prototyping systems should not be underestimated. The growth in the number and sophistication of web services means that, increasingly, useful applications will be available on the internet that can be invoked directly from virtual prototyping systems. For example, it should be possible to invoke a web-based, whole-life costing application, a finite element analysis package or a sustainability assessment program to assess the evolving virtual prototype. These invocations can be triggered by changes in the context of the users or the project, and can significantly enhance the utility of virtual prototypes.

11.7. Conclusions

In this chapter, a vision for the development of a new generation of context-aware virtual prototyping applications has been presented. The key concepts were introduced and supported by the detailed description of the architecture and implementation of a context-aware information delivery system for mobile construction workers. Context-aware virtual prototyping (CAVP) has the potential to cause a paradigm shift in the design and construction process, by allowing project team members and other end-users of virtual prototypes access to context-specific information and services on an as-needed basis. The framework developed for context-aware information delivery to mobile construction workers can be adapted to different implementations, including virtual prototyping. However, considerable work still needs to be done to make this a reality. With further developments in ICT (information and communication technologies), it will be possible to capture and make provision for a wide range of context variables that can be built into virtual prototypes. The potential benefits of this new range of context-aware virtual prototyping applications are considerable and will usher in a new era of collaboration in construction.

References

Aziz, Z., Anumba, C. J., Ruikar, D., Carrillo, P. M. and Bouchlaghem, N. M. (2003) Semantic web based services for intelligent mobile construction collaboration. *ITcon* **9**, [Special Issue: *Mobile Computing in Construction*]: 367–379.

Aziz, Z., Anumba, C. J., Ruikar, D., Carrillo, P. M., Bouchlaghem, N. M. (2005) Context-aware information delivery for on-site construction operations. *22nd CIB-W78 Conference on Information Technology in Construction*, Institute for Construction Informatics, Technische Universitat Dresden, Germany. CBI Publication No: 304, pp. 321–327.

Burrell, J. and Gay, K. (2001) Collectively defining context in a mobile, networked computing environment. *Short Talk Summary in CHI (Computer–Human Interaction) 2001 Extended Abstracts*, May 2001.

CC/PP(2003) Available at http://www.w3.org/TR/2003/PR-CCPP-struct-vocab-20031015/

Coyle, F. (2002) Mobile computing, web services and the semantic web: opportunities for m-commerce. *Proceedings of the Workshop at ISMIS'02 (International Symposium on Methodologies for Intelligent Systems)*, 26 June, Lyon, France.

Fensel, D. (2001) *Ontologies: Silver Bullet for Knowledge Management and Electronic Commerce*. Berlin, Springer-Verlag.

Pashtan, A. (2005) *Mobile Web Services*. Cambridge University Press.

RDF (Resource Description Framework) (2005) Available at http://www.w3.org/RDF/

12   nD modelling, present and future

Ghassan Aouad, Song Wu and Angela Lee

This chapter describes the concept of nD modelling and combines the findings of five workshops to form an nD roadmap of the future. The roadmap has contributed to the ongoing debate on the value of nD modelling and, as such, presents a research and a technology framework. The authors anticipate that the roadmap will be used to formulate agendas towards global nD-enabled construction. nD modelling is now a concerted idiom associated with ICT (information and communication technologies) in the AEC/FM (architectural, engineering, construction and facilities management) industry. This chapter discusses current and future work.

12.1. The concept of nD modelling

A building information model (BIM) is a computer model database of building design information, which may also contain information about the building's construction, management, operations and maintenance (Graphisoft, 2003). Thus, an nD model is an extension of the BIM, incorporating all the design information required at each stage of the life cycle of a building facility (Lee *et al.*, 2003). From this database, different views of the information can be generated automatically; views that correspond to traditional design documents such as plans, sections, elevations and schedules. As the documents are derived from the same database, they are all coordinated and accurate – any design changes made in the model will automatically be reflected in the resulting drawings, ensuring a complete and consistent set of documentation (Graphisoft, 2003). nD builds upon the concept of 2D, 3D and 4D.

2D and 3D modelling in the construction industry takes its precedence from the laws governing the positioning and dimensions of a point or object in physics, whereby a three-number vector represents a point in space: the x and y axes describing the planar state and the z axis depicting the height (Lee *et al.*, 2003). Further, 3D modelling in construction goes beyond the object's geometric dimensions and replicates visual attributes such as colour and texture. This visualization is a common attribute of many AEC design packages, such as 3D Studio Max and ArchiCAD, which enable the simulation of reality in all its aspects or allow a rehearsal medium for strategic planning. Combining time sequencing in visual environments with the 3D geometric model (x, y, z) is commonly referred to as '4D CAD/modelling' (Rischmoller *et al.*, 2000). Using 4D CAD, the processes of building construction can be demonstrated before any real construction activities occur (Kunz *et al.*, 2002). This will help users to find the possible mistakes and conflicts at the early stage of a construction project, and enable stakeholders to predict the construction schedule. Research projects around the world have taken the concept and developed it further, so that software prototypes and commercial packages have now begun to emerge. In the USA, the Center of Integrated Facility Engineering (CIFE) at Stanford University has implemented the concept of the 4D model on the Walt Disney Concert Hall project. In the UK, the University of Teesside's VIRCON project integrates a comprehensive core database designed with standard classification methods (Uniclass) with a CAD package, a project management package (MS Project) and graphical user interfaces as a 4D/VR (virtual reality) model to simulate construction processes of an £8 million, three-storey development for the University's Health School (Dawood *et al.*, 2002). Commercial packages are also now available, such as 4D Simulation from VirtualStep, Schedule Simulator from Bentley and 4D CAD System from JGC Corporation.

nD modelling builds on the concept of 4D modelling by integrating an *n*th number of design dimensions into a holistic model, which would enable users to portray and visually project the building design over its complete life cycle. nD modelling is based upon the building information model (BIM), a concept first introduced in the 1970s and the basis of considerable research in construction IT (information technology) ever since. The idea evolved with the introduction of object-oriented CAD (computer-aided design); the 'objects' in these CAD systems (e.g. doors, walls, windows, roofs) can also store non-graphical data about the building in a logical structure. The BIM is a repository that stores all the data 'objects', with each object being described only once. Both graphical and non-graphical documents, such as drawings and specifications, schedules and other data, respectively, are included. Changes to each item are made in only one place and so each project participant sees the same information in the repository. By handling project documentation in this way, communication problems that slow down projects and increase costs can be greatly reduced (Cyon Research, 2003). Leading CAD vendors, such as AutoDesk, Bentley and Graphisoft, have promoted BIM heavily with their own BIM solutions and demonstrated the benefits of the concept. However, as these solutions are based on different, non-compatible standards, an open and neutral data format is required to ensure data compatibility across the different applications. Industry Foundation Classes (IFCs), developed by the International Alliance for Interoperability (IAI), provide such capabilities. IFCs provide a set of rules and protocols that determine how the data representing the building in the model are defined, and the agreed specification of classes of components enables the development of a common language for construction. IFC-based objects allow project models to be shared, whilst allowing each profession to define its own view of the objects contained in that model. This leads to improved efficiency in cost estimating, building services design, construction and facility management: IFCs enable interoperability between the various AEC/FM software applications, allowing software developers to use IFCs to create applications that use universal objects based on the IFC specification. Furthermore, this shared data can continue to evolve after the design phase and throughout the construction and occupation of the building.

12.2. 3D to nD modelling project

The 3D to nD research project, at the University of Salford, has developed a multi-dimensional computer model that will portray and visually project the entire design

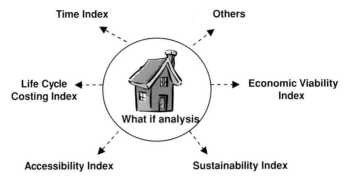

Figure 12.1   'What-if' analysis indexes of the 3D to nD modelling project.

and construction process, enabling users to 'see' and simulate the whole life of the project. This, it is anticipated, will help to improve the decision-making process and construction performance by enabling true 'what-if' analysis to be performed to demonstrate the real cost in terms of the variables of the design issues (Figure 12.1). Therefore, the trade-offs between the parameters can be clearly envisaged:

- Predict and plan the construction process
- Determine cost options
- Maximize sustainability
- Investigate energy requirements
- Examine people's accessibility
- Determine maintenance needs
- Incorporate crime deterrent features
- Examine the building's acoustics

The project aimed to develop the infrastructure, methodologies and technologies that will facilitate the integration of time, cost, accessibility, sustainability, maintainability, acoustics, crime and thermal requirements. It assembled and combined the leading advances that had been made in discrete ICTs and process improvement to produce an integrated prototyping platform for the construction and engineering industries. This output will allow seamless communication, simulation and visualization, and intelligent and dynamic interaction of emerging building design prototypes, so that their fitness of purpose for economic, environmental, building performance and human usability will be considered in an integrated manner. Conceptually, this will involve taking 3D modelling in the built environment to an *n*th number of dimensions. The project was funded by the EPSRC (Engineering and Physical Sciences Research Council), UK, under a Platform grant for the amount of £0.5 million for four years. The unique nature of this grant encouraged blue-sky innovative research, international collaboration and supports future funding opportunities.

The developed nD tool's system architecture consists of:

- *nD knowledge base*: A platform that provides information analysis services for the design knowledge related to the various design perspective constraints of the nD modelling (i.e. accessibility requirements, crime deterrent measures, sustainability requirements, etc.). Information from various design handbooks and guidelines on the legislative specifications of building component will be used, together with physical building data from the BIM to perform individual analysis.

- *Decision support*: Support for the decision-making process has proved to be problematic. Traditionally, a whole host of construction specialists are involved in instigating the design of modern buildings. With so much information coming from so many experts, it becomes very difficult for the client to visualize the design, the changes applied and the subsequent impacts on the time and cost of the construction project. Changing and adapting design and planning schedules and cost estimates to aid client decision-making can be laborious, time-consuming and costly. Each of the design parameters that the stakeholders seek to consider will have a host of social, economic and legislative constraints that may be in conflict with one another. Furthermore, as each of these factors vary – in the amount and type of impacts they can have – they will have a direct impact on the time and cost of the construction project. The criteria for successful design, therefore, will include a measure of the extent to which all these factors can be coordinated and mutually satisfied to meet the expectations of all the parties involved. Multi-criteria decision analysis (MCDA) techniques have been adopted to tackle the problem. The solution for the combined assessment of qualitative criteria (i.e. criteria from the Building Regulations and British Standard documents that cannot be directly measured against in their present form) and quantitative criteria (e.g. expressed in geometric dimensions, monetary units, etc.) has also been investigated. Analytic Hierarchy Process (AHP) is used to assess both qualitative criteria (i.e. criteria that cannot be directly measured) and quantitative criteria (e.g. expressed in dimensions, monetary units, etc.). However, the common understanding and definition of the similar concepts in the construction project from different specialists is yet to be developed; it is required to make the decision-making process more interactive and intelligent.

So far, the nD prototype tool incorporates whole-life cycle costing (using data generated by Salford's Life-Cycle costing research project), acoustics (using the R_w weighted sound reduction index), environmental impact data (using BRE (Building Research Establishment)'s 'Green Guide to Specification' data), crime (using the Secured by Design Scheme standards) and accessibility (using BSI: 83001). Technology for space analysis is also developed to support the accessibility analysis.

12.3. nD modelling roadmap

The nD modelling roadmap was developed as part of the 3D to nD modelling project, and draws its results from a number of research team workshops (academic), a national workshop (industrial), and two international workshops (both industrial and academic) that took place over a period of 3 years:

- The first academic workshop sought to gain a consensus, amongst the 20+ strong group of the multi-disciplinary research team, of the boundaries of the vision: applied or blue-sky, short- or long-term industry implementation? An electronic voting tool was used to ascertain the consensus of the participants in their own grappling of what they perceived would be the nD model.

- The second academic workshop set about defining the need and scope of nD modelling using a case study exemplar. Research topics, such as data quality/availability and decision-making mechanisms, were identified.
- At the national workshop, the findings of the academic workshops were presented to enable a consensus with both industry and academia across the UK. Those attended included contractors, clients, suppliers and architects, ranging from one-person to large multi-national organizations. The spectrum of participants was to reduce any inherent exclusion of industry players in the vision development, and to gain interest and acceptance of the work. The positioning of IT in the industry was established.
- The 1st International Workshop, held at Mottram Hall, Cheshire, UK, in February 2003, brought together world-leading experts (both from the industrial and academic communities) within the field of nD modelling. There were 52 participants from 32 collaborating organizations, from 10 countries. The workshop developed a business process and IT vision for how integrated environments would allow future nD-enabled construction to be undertaken. A summary of all the aforementioned four workshops is published in *Developing a Vision of nD-Enabled Construction* (Lee *et al.*, 2003).
- The aim of the 2nd International nD Modelling Workshop was to build upon the findings of the preceding events, concentrating on tackling the strategic and operational issues confronted by the widespread application of nD-enabled construction, and defining the future research agenda. The event was held on 13–14th September, 2004. The workshop brought together 45 participants from 27 organizations and representing 12 countries: Australia, Canada, Chile, China, Denmark, Finland, The Netherlands, Norway, Malta, Qatar, Singapore and the United Kingdom.

The results of all five workshops were combined to produce a roadmap of the future. The wide range of participants from over 15 different countries has enabled the roadmap to be of a global outlook. The following sections present the nD modelling roadmap for the future, which is composed of:

- an nD modelling framework, and
- an nD modelling technology framework

The nD roadmap is illustrated in Figure 12.2. It highlights areas of research needed over the coming years to enable widespread nD implementation. The diagram helps to illustrate and summarize the future topics that were suggested during the workshops. The findings of the workshops were clustered into themes and are presented as repertory grids in the final roadmap. Whilst this report highlights the barriers and opportunities of implementing nD modelling in the construction industry, the roadmap presents a research agenda for global nD uptake.

The roadmap attempts to capture all the discussions that were generated during the workshops and shows that, in order to reach the final nD destination landing place, notably the roadmap demonstrates that there are more obstacles that are human and organizational in nature than technological. Barriers of this kind were cited in all the workshops as most problematic in nature due to the unpredictability of the behaviour

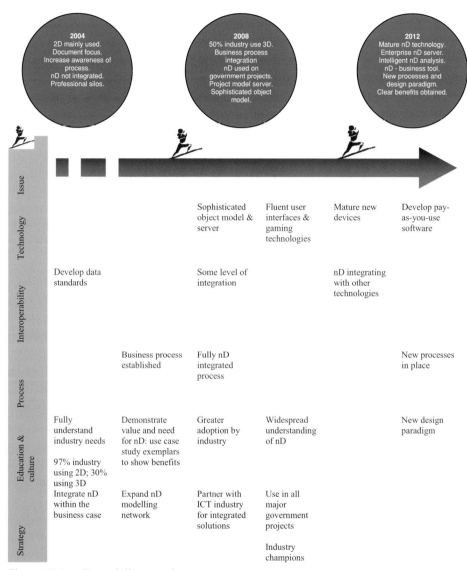

Figure 12.2   nD modelling roadmap.

of groups and individuals, and the multiple effects that that behaviour can have. Whilst the development and uptake of technology is also unpredictable, the specific issues do not persist in the same way as for the social aspects. More specifically, whilst we now have increasingly more sophisticated ways of remaining in touch as technical barriers continue to be broken down – telephones, fax machines, mobile phones, videoconferencing, wireless laptops, multimedia tools – the social problems of their usage in the workplace for collaboration and decision-making remain. Education/culture and performance measurement, and business case and process were established as being the biggest challenges.

Despite this though, there were a number of contradictions during the workshop discussions, and these are likely to result in slightly different pathways to, hopefully, the same goal. These contradictions had a more technical bias: some countries argued the lack of investment in technology is a problem, whilst others thought there was too much focus upon technology. Likewise, some argued for the strong belief in the need for change, others stated that there was a lack of awareness of the importance of this. Whilst it was accepted that the barriers and enablers of nD are global and not regional, there are still important differences in the impacts that regional culture, politics and government can have.

In focusing upon the implementation problems and practical actions for change, it is hoped that the message is driven home of the requirement to understand and change the social and cultural underpinnings before technical implementation. Of course, a significant part of this involves making explicit what it is we need to know, for example, establishing good practice exemplars. As the ideas and work related to nD modelling are developed in different contexts and on different scales, it may be that the practical benefits and added value of the concept becomes clear.

12.4. The future of the nD modelling research

The nD modelling prototype so far is a 'what-if' analysis tool that enables the impact of various design perspectives to be highlighted at a single building scale. However, the development of the nD prototype identified a number of issues that need to be accommodated, such as scalability, different actors/users, etc., and should truly mimic the design process. Architects unconsciously think about buildings in several ways while they are designing (adapted from Lawson, 2004):

- A collection of spaces – which may be indoors, outdoors or hybrids such as courtyards and atria
- A collection of building elements – such as walls, windows, doors and roofs
- A collection of systems – such as circulation, structure, skin, service
- A collection of voids and solids – as from an architectural perspective
- A series of layers – such as floor levels

When designing, architects oscillate without noticing between these descriptions of the building, thus adopting parallel lines of thought. In a similar way, if we are to fully adopt the nD concept we must first understand the human processes of designing so that it can be mimicked in the technology that we develop. We must simultaneously look at (see Figure 12.3):

- *Embrained knowledge*: Encompassing the viewpoints of different stakeholders/users of nD such as the client, architect, access auditor, etc., in terms of both feed forward and feedback of design information. Thus, it is actor configurable.
- *Process knowledge*: So that it can harness and be harnessed within various operating schematics such as the business process, operation process, etc. Thus, process configurable.
- *Encoded knowledge*: Ensuring the design conforms to the respective design standards. Thus, it is code configurable.

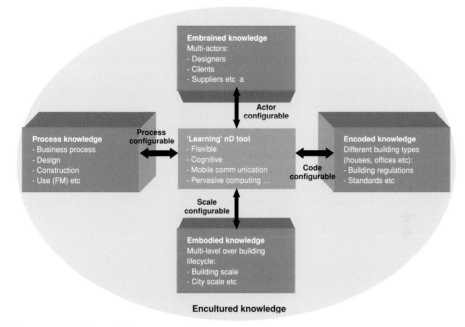

Figure 12.3   nD modelling research framework.

- *Embodied knowledge*: Enabling the scalability of use, covering a single building to city and urban use. Thus, it is scale configurable.

Therefore, the future of nD modelling is to build intelligence into the nD environment by exploring AI (artificial intelligence) technology so that it orchestrates the design process simultaneously, or as required. So far, the nD prototype tool encapsulates the encoded knowledge domain. Although state-of-the-art, it can be duly criticized as being somewhat tedious to use, as the process of information retrieval and analysis is manual. Therefore, with the paramount need for nD modelling in the twentyfirst century, the primary objective of future nD research is to automate this process intelligently in order to develop a holistic nD environment that incorporates all aspects of the nD modelling research framework (Figure 12.3). It is proposed that:

- Different users/actors will be able to utilize nD technology from their own discipline perspective (i.e. drawings for designers, cost spreadsheets for quantity surveyors, etc.)
- It can be code adapted for different building types (i.e. check conformity for housing, offices, etc.)
- It can be scaled accordingly to tackle buildings or urban environments
- It will address various processes (i.e. business, design, facilities management, etc.)
- It will be self-learning by using a number of technologies and techniques (i.e. mobile computing, pervasive computing, etc.).

In the future nD modelling environment, a multi-agent system (MAS) will be investigated to simulate the multi-disciplinary construction project environment, which will be intelligent and intuitive. The nD modelling environment has encapsulated encoded knowledge; the agent in this environment has to be able to apply such knowledge under different circumstances. Neural networks can be used to train the agent to achieve such understanding. There is a large amount of subjective information, such as accessibility criteria, in the nD modelling environment, which needs to be quantified for computer analysis; Fuzzy logic will be considered to perform the transformation. Finally, in this multi-perspective decision support environment, trade-off for conflict requirements is a common problem. Genetic algorithms will be reviewed and tested to optimize the solution of such a problem.

The future nD modelling environment aims to develop separate modules (i.e. actors, code, scale and process configurations that will 'learn' as they are being used), whereby each one will provide a discrete solution, and/or then be allowed to proactively co-operate, negotiate and exchange information autonomously in order to solve holistic 'what-if' analysis problems within a timeframe and level of detail that is conducive to the user. This will overcome the manual analysis that is apparent within the existing 3D to nD modelling prototype.

Moreover, to specifically combat the embrained/multi-actor perspective, simulation using game technology will be developed to provide potential users with a 'real' look and feel of the design through virtual characters in a virtual environment. Virtual characters have been an important part of computer graphics in recent years; characters have often taken forms such as synthetic humans, animals, mythological creatures and non-organic objects that exhibit life-like properties (such as walking lamps, etc.). They have advanced dramatically over the past decade, revolutionizing the motion picture, game and multimedia industries. In the built and human environment, virtual characters have largely been used for crowd simulation in emergency egress and pedestrian simulation (Musse and Thalmann, 2001; Penn and Turner, 2002). It is presently challenging to define all aspects of the behaviour of a complex virtual character: the desired behaviour may be impossible to define ahead of time if the characters of the virtual world change in unexpected or diverse ways (Dinerstein *et al.*, 2004). For these reasons, it is desirable to make virtual characters as autonomous and intelligent as possible whilst still maintaining animator control over their high-level goals. This can be accomplished with a behavioural model – an executable model defining how the character should react to stimuli from its environment – and/or a cognitive model – an executable model of the characters' thought processes (Tu and Terzopoulos, 1994). Both modelling techniques will be adopted to define the behaviours of potential users (e.g. children, elderly, disabled, etc.). Furthermore, the virtual characters will also be self-learning, acquiring 'new' knowledge to aid the designer to test the design of different scenarios, and it can be utilized through the support of artificial intelligent techniques, such as neural networks.

12.5. nD modelling technology framework

As a result of the five workshops, a new technology framework (Figures 12.4 and 12.5) emerged to support the nD modelling roadmap (see Figure 12.2).

The technology framework provides the architecture for different domain applications, which is based on a service-oriented platform supported by various technologies such as visualization, decision support and analysis:

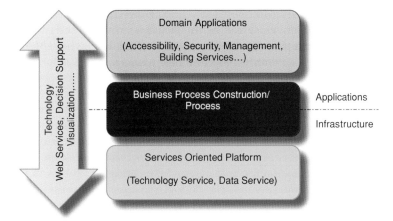

Figure 12.4   The overview of the technology framework.

- The technology framework is based on the service-oriented architecture (SOA). The participants of the workshop recommended that a common framework and interface should be provided so that applications from different domains can work together. Furthermore, common technologies such as visualization and decision support should be part of the platform. SOA is a technology that can be deployed to

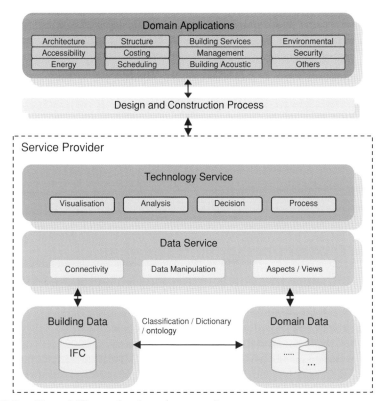

Figure 12.5   nD modelling technology framework.

support this platform. According to the World Wide Web Consortium (W3C), SOA is 'a set of components which can be invoked, and whose interface descriptions can be published and discovered'. The nD modelling framework should provide the common technology components and data access as services for each domain application.

- It was also recommended that the framework needs to be driven by the business process. Different applications will be used at different stages of the process and the data requirements will be different. The process control mechanism has to be in place to ensure the right data can be served to the right application.
- The services include technology service and data service. The technology service provides common technologies, such as visualizing the building data, multi-aspects' decision support and data analysis for thermal, structure and process control mechanisms. The data service provides the data access to two types of data sources: building data and domain data. Building data can be described as data definition and representation of the building model and it has to be supported by interoperable data standards such as IFCs. Domain data is specific data related to each domain, such as regulations for building accessibility, weather data for energy simulation, etc. Currently, these two data sources are often not linked. It is suggested that research has to be done to integrate the two data sources through developing common concepts (ontology/classification/dictionary).

Acknowledgements and contributions

The roadmap could not have been produced without the invaluable contribution from the attendees of the two academic, one national (64 participants from 39 organizations) and two international workshops (1st workshop attended by 52 participants from 28 organizations, representing 10 countries; the 2nd workshop was attended by 45 participants from 27 organizations, representing 12 countries).

References

Cyon Research (2003) *Building Information Model: A Look at Graphisoft's Virtual Building Concept*. Bethesda, MD, Cyon Research Corporation.

Dawood, N., Sriprasert, E., Mallasi, Z. and Hobbs, B. (2002) Development of an integrated information resource base for 4D/VR construction processes simulation. *Automation in Construction* **12**: 123–131.

Dinerstein, J., Egbert, P. K., Garis, H. and Dinerstein, N. (2004) Fast and learnable behavioral and cognitive modelling for virtual character animation. *Computer Animation and Virtual World* **15**: 95–108.

Graphisoft (2003) *The Graphisoft Virtual Building: Bridging the Building Information Model from Concept into Reality. Graphisoft Whitepaper*. Budapest, Graphisoft.

Kunz, J., Fischer, M., Haymaker, J. and Levitt, R. (2002) Integrated and automated project processes in civil engineering: Experiences of the Center for Integrated Facility Engineering at Stanford University. *Computing in Civil Engineering Proceedings, ASCE*, January 2002, Reston, VA, pp. 96–105.

Lawson, B. (2004) Oracles, draughtsman and agents: The nature of knowledge and creativity in design and the role of IT. *Proceedings of the International Conference on Construction Information Technology (INCITE) Conference*, 18th–21st February, Malaysia, pp. 9–16.

Lee, A., Marshall-Ponting, A. J., Aouad, G., Wu, S., Koh, I., Fu, C., Cooper, R., Betts, M., Kagioglou, M. and Fischer, M. (2003) *Developing a Vision of nD-Enabled Construction.* University of Salford, Construct IT Report.

Musse, S. R. and Thalmann, D. (2001) Hierarchical model for real time simulation of virtual human crowds. *IEEE Transactions on Visualization and Computer Graphics* **7**: 152–164.

Penn, A. and Turner, A. (2002) Space syntax based agent simulation. In: *Pedestrian and Evacuation Dynamics* (ed. M. Schreckenberg and S. D. Sharma). Berlin, Springer-Verlag, pp. 99–114.

Rischmoller, L., Fisher, M., Fox, R. and Alarcon, L. (2000) 4D planning and scheduling (4D-PS): Grounding construction IT research in industry practice. *Proceedings of the CIB W78 Conference on Construction Information Technology: Taking the construction industry into the 21st century*, June, Iceland.

Tu, X. and Terzopoulos, D. (1994) Artificial fishes: Physics, locomotion, perception, behavior. *Proceedings of the 21st Annual Conference on Computer Graphics and Interactive Techniques, ACM SIGGRAPH (Association for Computing Machinery's Special Interest Group on Graphics and Interactive Techniques)*, 24–29th July, pp. 43–50.

Further reading

Atkins, E. M., Durfee, E. H. and Shin, K. G. (1999) Autonomous flight with CIRCA-II. *Proceedings of the Workshop on Autonomy Control Software, held in association with the Autonomous Agents 99 Conference*, 1–5 May, Seattle.

Azevedo, C. P., Feiju, B. and Costa, M. (2000) Control centers evolve with agent technology. *IEEE Computer Applications in Power* **13** (3): 48–53.

Cockburn, D. and Jennings, N. R. (1996) A distributed artificial intelligence system for industrial applications. In: *Foundations of Distributed Artificial Intelligence* (ed G. M. P. O'Hare and N. R. Jennings). Hoboken, NJ, Wiley InterScience, pp. 319–344.

Drogemuller, R. (2002) CSIRO and CRC-CI FC development projects. *International Technical Management (ITM) Meeting*, 30 October, Tokyo.

Ferber, J. (1999) *Multi-Agent Systems: An Introduction to Distributed Artificial Intelligence*. Reading, MA, Addison-Wesley.

Gallimore, R. J., Jennings, N. R., Lamba, H. S., Mason, C. L. and Orestein, B. J. (1999) Cooperating agents for 3D scientific data interpretation. *IEEE Transactions on Systems, Man and Cybernetics – Part C: Applications and Reviews* **29** (1).

Heesom, D. and Mahdjoubi, L. (2004) Trends of 4D CAD applications for construction planning. *Construction, Management and Economics* **22** (2): 171–182.

Jennings, N. R., Faratin, P. J., Norman, T., O'Brien, P., Wiegand, M. E., Voudouris, C., Alty, J. L., Miah, T. and Mamdani, E. H. (1996) ADEPT: Managing business processes using intelligent agents. *Proceedings of BCS Expert Systems 96 Conference*, 16–18 December, Cambridge, UK, pp. 5–23.

Lee, A., Wu, S., Marshall-Ponting, A. J., Aouad, G., Abbott, C. A., Cooper, R. and Tah, J. (2005) *nD Modelling Roadmap: A Vision of the Future of Construction*. University of Salford, Centre for Construction Innovation.

Liston, K., Fischer, M. and Winograd, T. (2001) Focused sharing of information for multi-disciplinary decision making by project teams. *ITcon* (*Journal of Information Technology in Construction*) **6**: 69–82. Available at http://www.itcon.org/2001/6

Newnham, L. N., Anumba C. J. and Ugwu, O. O. (1999) *Negotiation in a Multi-Agent System for the Collaborative Design of Light Industrial Buildings.* Loughborough University, Technical Report No. ADLIB/02, October.

Parunak, H. D. (1999) Experiences and issues in the development and deployment of industrial agent-based systems. *Proceedings of the 4th International Conference on the Practical Application of Intelligent Agents and Multi-Agent Technology (PAAM99)*, 19–21 April, Commonwealth Institute, London, pp. 3–9.

13 Interoperable knowledge: Achievements and future challenges

Yacine Rezgui and Simona Barresi

This chapter explores the use of ICT (information and communication technologies) in the construction industry, with a focus on knowledge management (KM). The research argues that KM is facilitated by addressing software application interoperability constraints. In fact, much of the knowledge created and used on projects is embedded within these software applications. This knowledge can be mined, made sense of and exploited if made interoperable and accessible to other applications. A generic system architecture is proposed, which aims at facilitating KM while providing effective support for distributed knowledge-driven business processes. The use of a service-oriented architecture promotes the support for e-processes, implemented through a coordinated composition and invocation of web-enabled, service-oriented, corporate enterprise information systems' (EIS) applications.

13.1. Introduction

Construction is an increasingly heterogeneous and highly fragmented knowledge-intensive industry, dependent on a large number of very different professions, with a dominant number of small and medium-sized enterprises (SMEs), working and collaborating together from disparate locations on various building projects. Heterogeneity and geographical dispersion in construction lead to numerous challenges in terms of communication, data exchange and application interoperability (Rezgui and Zarli, 2006), including:

- Non co-location of individuals and teams collaborating on projects
- Lack of dominant actors to enforce ICT
- Unregulated information exchange and lack of legal admissibility of ICT-based contractual practices
- Difficulties in communication and data exchange between partners during a building project, or between clients and suppliers of construction products

In order to remain competitive, the construction industry needs to exploit the advances in ICT to achieve substantial innovation and process improvements. This chapter will, first, provide the methodological framework that underpins the research. A synthesis of the literature then follows, with the objective of identifying the various schools of thought in knowledge management. The chapter then identifies the ICT limitations related to knowledge management in construction, and provides a brief enumeration of the identified areas requiring improvements. The chapter argues the need for a process-driven, as opposed to data-oriented, approach to integration, and presents the vision and related knowledge and interoperability roadmaps developed within the context of the ROADCON (Strategic Roadmap towards Knowledge-Driven 'Sustainable' Construction) project (Hannus *et al.*, 2003; Rezgui and Zarli, 2006). The FUNSIEC (Feasibility Study for a Unified Semantic Infrastructure in the European Construction Sector) methodology, revolving around the OSIECS (Open Semantic Infrastructure for the European Construction Sector) kernel (Barresi *et al.*, 2006), is described, followed by a proposed system architecture aimed at addressing the main research question. The conclusion provides a synthesis of key findings and provides directions for future work.

13.2. Methodology

The chapter makes use of research funded by the European Commission (ROADCON IST-2001-37278 and FUNSIEC eContent 42059Y3C3FPAL2). The latter aims at enabling both effective application interoperability and the development of a new generation of services for the sector to support distributed business processes. While ROADCON (Rezgui and Zarli, 2006) provides the overall context and problem situation for KM in construction, FUNSIEC (Barresi *et al.*, 2005; Barresi *et al.*, 2006) focuses on application interoperability issues with a view to facilitating the processes underpinning knowledge activities. This is illustrated in Figure 13.1.

The research presented in this chapter adopts a positivist stance and is based on the following hypothesis:

- Information and communication technologies are advanced enough to offer suitable solutions to address key limitations faced by the industry
- SMEs are capable – through an adapted programme of training and continuous professional development, fostering maturity and capability building – of engaging into value-added collaborative activities using advanced ICTs

The research addresses the following main question:

Can the fragmented nature of the industry, which currently hinders effective collaboration, be addressed through an adapted software architecture that promotes e-processes, underpinned by knowledge-intensive activities, through seamless support for enterprise information systems interoperability?

An approach based on action research has been adopted to formulate and reach consensus around the construction requirements, ICT vision and resulting roadmap. Different data-gathering methods and instruments have been used to produce a wide scope of coverage and provide a fuller picture of the phenomena under study. These are detailed in Rezgui and Zarli (2006).

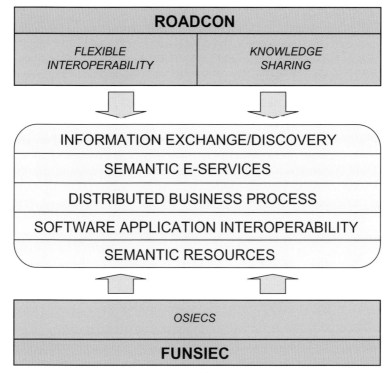

Figure 13.1   Methodological framework for the research.

13.3. Existing perspectives to knowledge management

There have been many attempts to produce a definition and there are many different views as to what exactly is meant by the term 'knowledge management' (Raub and Ruling, 2001). However, within the ongoing debate, a number of authors have proposed categorizations of knowledge management approaches. Of these, two perspectives have particular relevance to this chapter (Venters, 2001):

- *The functionalist perspective*, which attempts to 'objectify' knowledge' as a commodity that can be defined and stored, and
- *The interpretivist perspective*, which argues that knowledge cannot exist outside human experience and the social practices and structures which shape it.

Within the functionalist perspective, a number of 'models' have been proposed. Of these the most often discussed are the intellectual capital models and the knowledge category models. The intellectual capital models take the scientific view that knowledge is a commodity that has an intrinsic (and measurable) value (Roos *et al.*, 1997; Bontis, 2001). While some argue that this is a simplistic view which ignores the social activities that surround knowledge creation and sharing (McAdam and McCreedy, 1999), others such as Davenport and Prusak (1998) argue that this view correctly allows knowledge to be seen as an asset that can be marketed and traded.

Knowledge category models attempt to categorize knowledge into different forms, which may be stored, transmitted and reused through a process of 'knowledge conversion' between tacit (personal, context-specific, hard to formalize) and explicit (codified and formalized) (Nonaka and Takeuchi 1995). This conversion can occur between any forms (i.e. tacit–explicit, tacit–tacit, explicit–tacit, etc.). However, interpretive perspectives on knowledge management take the view that knowledge is socially constructed and must be seen as a part of the social structures and processes of the organization (McAdam and McCreedy, 1999).

This perspective emphasizes the ephemeral, messy nature of knowledge and the difficulties of capturing and mapping the processes of knowledge creation and use (Venters, 2001). Within this perspective there are many concepts such as 'communities of practice', as discussed by Wenger and Lave (1998) amongst others. These are seen as a 'structure' in which knowledge is created and shared and should be supported and fostered.

In the interpretivist approach, the focus shifts from capturing, sharing and disseminating knowledge 'objects' towards the support of the activities of individuals and groups. It accepts that knowledge is inherent in these social interactions and that they are as valuable in terms of knowledge management (although less easily managed) as codified knowledge 'objects'.

The research in this chapter adopts the functionalist perspective to knowledge. While this can be seen as a limitation, the authors argue that effective knowledge management can not be achieved if basic requirements, such as those related to software application interoperability, are not met. While the application of KM in construction has been reported and detailed in Rezgui (2001) and Kazi and Hannus (2002), the following section identifies the ICT requirements related to KM in the sector.

13.4. Understanding ICT requirements and needs in construction

The consultations carried out during the ROADCON project (Rezgui and Zarli, 2006), as well as the study of the results that emerged from related projects and initiatives, provided a clear picture of the current situation in relation to the use and limitations of ICT in the construction domain, including:

- Use of proprietary formats with low semantics to support software application interoperability and data exchange between different actors and companies.
- Application-centric nature of ICT. These tend to be dedicated to engineering functions within specific life cycle stages, with little support for the project life cycle.
- Importance of tacit knowledge, in that project best practices and know-how reside mainly in the mind of individuals and are available, at best, in the form of documents stored/archived on company intranets.
- Lack of knowledge reuse, in that there is a tendency to re-invent the wheel by not capitalizing on past experiences.
- Lack of scalability of the knowledge-driven systems in place, in that most available solutions offer a limited growth path in terms of hardware and software, and tend to become obsolete following upgrades or the introduction of new technologies.
- Lack of integration between the different applications in use within the organization or across organizations when collaborating on projects.

The above issues have often resulted in information inconsistencies, business process inefficiencies and difficulties in effective and efficient communication among partners during a building project. This is reflected in the strong data focus approach to integration in construction. A total of five priority areas (Rezgui and Zarli, 2006) have emerged from the requirement analysis stage with the potential of addressing the identified limitations of the industry, namely: knowledge management; legal and contractual aspects management; quality and performance management; total life cycle management; and human aspects management. This chapter specifically addresses the area related to knowledge management.

13.5. Prevailing data focus to integration in construction

Application and data integration has been a primary focus in construction. Much of this work has, for the last two decades, been centred on the definition and exploitation of product data models for the construction industry, most particularly that part of the industry producing buildings. Numerous projects have been undertaken in pursuit of this goal, both academic and industrial, some of which are mentioned here to give a flavour of the field.

Amongst the first efforts at integration were those born of the increasing use of computer-aided design (CAD) in design offices from the mid-1980s onwards. Here the necessity of transferring CAD data from one system to another resulted in *de facto* standards that persist to this day, such as the Drawing (or Data) Exchange Format (DXF) and the Initial Graphics Exchange Specification (IGES). More coordinated standards defining efforts came in the form of the STEP (Standard for the Exchange of Product Model Data) application protocols for construction (ISO, 1994). This work, inspired by previous work primarily in the aerospace and automotive fields, resulted in ISO 10303, part of the International Standard for the Exchange of Product Model Data. Latterly, the International Alliance for Interoperability defined the Industry Foundation Classes (IFCs): a set of model constructs for the description of building elements. Preceding and, in some cases, concurrent with this work, the academic research community produced several integrated model definitions, including: ATLAS (Bohms *et al.*, 1994); the COMBINE Integrated Data Model (Dubois *et al.*, 1995); the RATAS model (Björk, 1994); OPIS (Froese and Paulson, 1994); and the AEC Building Systems Model (Turner, 1988).

These research efforts were generally predicated on the use of either an integrated tool set also furnished by the respective projects, or on a central database holding all model data for access by any application used in the construction project process via some form of adapter (Björk, 1989; Björk and Penttilä, 1989). One of the most recent incarnations of the central database idea can be seen in the IFC Model Server from VTT of Finland (VTT, 2002) designed to host entire building models described in the IAI (International Alliance for Interoperability) IFC format.

Since 2003, commercial application developers in the construction domain have started to market tools manipulating fully parametric building models. This new wave of applications, termed 'building information modelling' (BIM) applications, embody much of the vision of previous academic research such as ATLAS and COMBINE, whilst still relying on data exchange standards or API (Application Programming Interface) level customization for interoperability/integration. Recently, the American National Institute of Building Sciences inaugurated a committee to look into creating a standard for life cycle data modelling under the BIM banner (NIBS, 2006). The idea here is to have a standard that identifies data requirements at different life cycle stages in order to allow a more intelligent exchange of data between BIM-enabled applications. This data has been expressed in different ways, using various formalisms and domain applications.

Among the multitude of semantic resources (SRs) developed in the construction sector, ranging from domain dictionaries to specialized taxonomies, some of the most notable efforts are listed below:

- The BS6100 (Glossary of Building and Civil Engineering terms), produced by the British Standards Institution. This is a rich and complete glossary that provides a comprehensive number of synonyms per term that can contribute towards any ontology development effort in the sector.
- The bcXML is an XML vocabulary developed by the eConstruct IST project for the construction industry. The bcXML provides the foundation for the development of the bcBuildingDefinitions taxonomy, which can be instantiated to create catalogue contents. Through bcXML, eConstruct has enabled the creation of 'requirements messages' that can be interpreted by computer applications, which can then find suitable products and services that meet those requirements.
- The ISO 12006-3 defines a schema for a generic taxonomy model, which enables one to define concepts by means of their properties, to group together concepts and to define relationships between concepts.
- The IFC model, developed by the IAI, has produced a specification of data structures with the aim of supporting the development of the BIM, where all the information about the whole-life cycle of a construction project would be stored and shared among the actors involved.
- The OmniClass Construction Classification System (OCCS) developed in Canada by the Construction Specifications Institute, addresses the construction industry's information management needs through a coordinated classification system.

All of the above resources, although different in terms of formalism, scope, details and applicability, can be used in a complementary manner. Providing an infrastructure to map these resources helps to overcome problems related to SRs' different formalisms and inconsistencies, and enables the effective reuse of existing construction-related SRs. This, in turn, facilitates the efficient use of knowledge within the sector and supports the implementation of e-services for the construction domain.

13.6. The process-driven vision

Data-centred approaches to integration have suffered from a number of problems, including the lack of context. Whilst very structured data, such as a building model, has a well-defined internal context, its place in the project is less certain in terms of its relationship to other information and the people who use it. An integration that places information in a context that is easily understood by those working on the project, i.e. one abstracted away from file stores and computers and geared more towards processes and project context, is a superior arrangement. Both STEP and IFC could be argued to incorporate some aspects of context, modelling as they do some project scheduling and organizational elements of the domain. This, however, still fails to incorporate information that exists outside the model and is not executable as a process in its own right.

Finally, for large international standards' efforts, agility is something of a problem. Once the standard is agreed, changing it can take a considerable amount of time, which in an age of rapidly evolving business needs can turn a formerly helpful system into a hindrance. Given the problems outlined with data level integration, basic ingredients of an environment better suited to both organizational and end-user needs are given hereafter. Such an environment should be based on business and/or project processes as the natural level at which people interact with their work in an organizational or project context. There is already a trend towards business process integration in the wider commercial context, with many vendors offering 'solutions' in the domain. These systems normally tend to be focused on integrating legacy applications into a process workflow in a very organizationally introspective way. In an environment typified by short-term partnering arrangements between multiple organizations, a more eclectic approach to integration is required to support the creation of short-lived, project-specific business systems integrated and operated at the process level.

Here we draw on object-oriented principles, in that we envisage a number of components, each furnishing a small piece of functionality in a fully encapsulated independent fashion. These components would be published as web services and composed into higher level business process components, again self-contained. Each business process would service some pertinent part of the construction project life cycle. It may be the case that the same process component is published in several different guises to suit different organizational or disciplinary perspectives on the project process it supports. For example, the view of a component supporting commissioning of an HVAC (heating, ventilation and air conditioning) installation from the engineer's point of view would be different to the view required by the client who ultimately needs to approve it.

The components, therefore, will also be required to adapt to the context from which they are executed. This model of arbitrary combinations of process components, or 'e-processes' as one might call them, allows for greater flexibility in the definition and construction of business systems to support construction projects. Using this model it should be possible to select 'best-of-breed' process components on a project-by-project basis, the systems effectively configuring themselves based on a description of process needs. Even reconfiguring the project business system during the life of the project to reflect changes to the description of needs would be feasible.

The question of information in context becomes less of an issue with process-based systems as information is firmly rooted as the fuel for the process. A clear and thorough analysis of information needs at each stage of a project process will serve both to identify data needs and contextualize that data within the framework of the process. This leaves scope for continuing research into information structuring and management, primarily as a means to better understand and situate it in the process, rather than as an end in its own right as has often been the case to date. Further, such research could move to a more knowledge-based mode, where a focus on integrating unstructured data through semantic analysis or indexing, perhaps employing techniques from the information extraction field, would help to locate data automatically in the frame of the process. Benefits may also be derived from uncovering previously unseen linkages between various elements of project data using such analysis methods. Kosovac *et al.* (2000) and Scherer and Schapke (2005) have undertaken research in this or closely related areas.

This vision does not necessarily mandate that the whole project process be mapped and serviced by process components from the outset. It is an integral part of the vision that as much or as little of the project process as is feasible or desirable should be supported. Thus it is possible to integrate processes incrementally as the underlying functionality becomes available and is proven to be effective by the early adopters.

The technology to do most of this already exists or is being developed. For example, in order to have process components (essentially a composite web service) integrate seamlessly as we have envisioned, it will be necessary to describe them at a semantic level in addition to the basic input and output messages defined by the standard WSDL (Web Service Description Language). For this purpose Ontology Web Language for Web Services (OWL-S Coalition 2005)) could be employed. OWL-S allows for the ontologically grounded description of web services, both 'atomic' and composite, making the description computer-interpretable. The language supports the notion of profiles that describe both what the service (and process component) offers and the requirements of a searching agent that are met in the service, with each service potentially having multiple profiles. Thus searching for a service that meets the requirements of part of a model of process needs can be automated.

Further work is underway to design a method for describing and reasoning about the dynamic behaviour of a service prior to its invocation to further ensure compatibility between services (Solanki *et al.*, 2004). OWL-S also allows for other ontologies, taxonomies, etc. to be used to describe various sub-parts of the service, for example in order to classify the service into a construction-specific scheme for our purposes. It is here that an element of discord may be re-introduced by the use of disparate ontologies in the underlying service descriptions. This is where FUNSIEC (Barresi *et al.*, 2005; Barresi *et al.*, 2006) offers interesting solutions.

13.7. Nature and type of e-services in construction

While electronic data interchange (EDI) techniques, which have been around for nearly two decades, require dedicated data links between involved parties, the internet provides cheap and ubiquitous modes of interaction and collaboration. It has thus accelerated the emergence of new ways of working and conducting business, leading to the development of electronic marketplaces where suppliers and potential customers are brought together. Over the last few years, traditional e-commerce is giving way to a new paradigm known as 'e-service'. e-Services include all interactive services that are delivered on the internet using advanced telecommunications, information and multimedia technologies (Roth, 2000). This emerging paradigm represents a coherent point of view that challenges many of the traditional assumptions about how to use the online environment to raise profits (Rust and Lemon, 2001). It is based less on reducing costs through automation and increased efficiency and more on expanding revenues through enhancing service and building profitable customer relationships (Ruyter *et al.*, 2001; Rust and Kannan, 2003).

As argued by several authors, including Van Riel and Ouwersloot (2005), e-services can be classified according to two dimensions: their origin and the nature/type of goods supplied (Table 13.1). The first dimension relates to how well an e-service provider is rooted in the physical world. e-Service providers having their origins as a dot.com can

Table 13.1 e-Services' classification in construction.

Classification Origin of e-services in construction		
	Virtual	Physical
Virtual	Specialized portals aiding in selecting the right products, taking into account cost, quality and performance parameters	Software manufacturers that have developed service-based solutions for their software products
Product	Access to semantic resources in the construction domain: thesauri, dictionaries, etc.	Access to construction regulations online Access to construction organization details and yellow pages
Physical	Virtual team work and enterprise solutions, e.g. Buildonline, Bricsnet, Buzzsaw, Citadon	Product/equipment manufacturers and suppliers Selling construction products online

be thus distinguished from those having traditional roots. For the latter, the portal site itself is an extension, whereas dot.com starters have always operated in the e-world, with the portal as the original brand. The second dimension concerns the type of products that are supplied. Some e-service providers supply tangible goods or traditional services: the websites mainly function as a distribution channel, an on-line store or an interface with customers. In contrast, other e-service providers offer virtual products, remaining within the borders of the e-world: search functions, communication tools, information, downloadable software, etc.

Based on the framework depicted in Table 13.1, there are several variants of the e-Services models in construction characterized by their intended use. These are categorized below.

13.7.1. e-Services as enhanced customer-driven e-commerce

This is where traditional e-commerce approaches that tended to focus on automating 'product selling' practices have been extended to provide a service dimension to the customer using customer relationship management (CRM) techniques. Moreover, there is a difference between conducting basic e-commerce purchases and adopting e-services. In comparison to one-time, e-commerce-based purchases, the e-service adoption decision is typically more complex, as e-services initiate a long term relationship between the consumer and service provider (Featherman and Pavlou, 2003). e-Services can take the form of consumer and trading portals providing advanced market-maker e-services, such as reverse auctions and collective bidding. A great many services in the construction sector falling under this category have emerged over the last few years. These have mainly been developed in the area of product manufacturers and suppliers.

13.7.2. e-Services as a new software licensing paradigm

Increasingly complex business processes require the use of a variety of software packages but only a few packages are used on a daily basis. This infrequent usage pattern often does not justify purchasing full licences and, therefore, motivates the need for a more flexible way to use and pay for software usage (Fenicle and Wahls, 2003). Although, in the late 1990s, a host of start-ups began offering applications delivered over the web using a pay-per-use cost model, most of them went out of business, mainly because large corporate customers baulked at the idea of allowing untested outsiders to run their most important applications. But, recently, both corporations and smaller businesses have become more comfortable with this way of buying software (Kerstetter and Greene, 2004).

Many software publishers are, today, reconsidering their software distribution methods. They deliver application software and services (software maintenance, upgrades, staff training, etc.) over the web on a lease or subscription basis instead of the traditional perpetual licensing system (Zhang and Seidmann, 2003). Several leading software vendors have started adopting this new business model, emphasizing cost saving factors. Analysts now estimate that by 2010 as much as half the software sold to corporations will be paid for on a monthly basis, as part of a long term contract or a monthly rental fee, or even on a pay-per-use basis (Kerstetter and Greene, 2004).

This new form of software licensing provides a software service that includes configuration, maintenance, training and access to a help-desk. It enables organizations to rent, as opposed to purchase, software. This involves using only the functionality directly needed by the user, therefore reducing cost and increasing work efficiency. In the construction sector, such services include solutions from Bricsnet (http://www.bricsnet.com), Buzzsaw (http://www.buzzsaw.com), BuildOnline (http://www.build-online.com) and Citadon (http://www.citadon.com).

13.7.3. e-Services as a means to enable e-processes

Businesses now have a new opportunity to rethink fundamental aspects of information technology (IT). One of the key capabilities within e-services can also be applied to elements of a business IT infrastructure to facilitate the execution of distributed business processes, implemented through a coordinated composition and invocation of web-enabled, service-oriented, corporate enterprise information systems applications. For example, messaging/email can now be provided to the employees of an enterprise using a service-based approach. This frees time and money for the IT organization to invest elsewhere – into areas that can better serve their customers and their partners. The e-services model has been used by some leading organizations to integrate their in-house legacy as well as commercial software applications better, thus becoming ubiquitous and providing better support to mobile users. Some leading construction organizations have started investigating and investing into these technologies (Rezgui and Meziane, 2005).

13.8. Towards seamless interoperability and knowledge sharing

The ROADCON vision statement for the future use of ICT in the sector has been defined as follows (Rezgui and Zarli, 2006):

> Construction in the Knowledge Society uses Information and Communication Technologies as key enablers to meet main societal and industrial challenges. The Construction sector is driven by total life cycle performance through knowledge intensive and functional integration of products and processes using a model-based approach to pave the way towards a knowledge driven industry.

This vision and the resulting proposed roadmap are detailed in Hannus *et al.* (2003) and Rezgui and Zarli (2006). This section focuses on aspects of the roadmap relevant to this research, namely 'flexible interoperability' and 'knowledge sharing'.

Interoperability between companies' ICT applications represents a current major issue in construction, posing numerous limitations on the actors' ability to efficiently and effectively cooperate in construction life cycles and supply-chains. Typically ICT applications' communication is characterized by point-to-point access, based on incomplete, proprietary (vendor or user-dependent) non-adaptable interfaces at a low semantic level, in which frequently only the syntax aspects are agreed. Generally, as revealed in ROADCON surveys, 90% of all data exchange is done manually, requiring a great degree of human (interpretation) effort and introducing numerous errors in the process. The situation is even worse when considering the reuse of 'functionality' or services.

The *flexible interoperability* vision is based on the general assumption that a complete ICT system will involve multiple services provided by various service providers. However, these services mainly address the same kind of information; hence, they have to communicate about the same objects. Agreement on communication between software application/components is therefore vital, particularly in the construction domain where projects always involve external communication and collaboration. Depending on the specific scenario, different types of agreements can be envisaged, including: agreement on the form of syntax, semantic for information and/or functionality.

Currently, various standardization initiatives are pushing this problem forward and investigating the use of neutral definitions to facilitate systems' interoperability. In this context, ontology approaches, especially web-based ones, are seen as a key opportunity. The next generation web will be characterized by a higher level of semantics (model-based semantic web) and tight integration with functionality (web services or aggregations of them). As such, it will allow rigid monolithic ICT applications and interface schema specifications to be replaced by flexible distributed components, which will be processing distributed content based on common ontologies. This is illustrated in Figure 13.2, which provides a staged roadmap spanning a period of 2–15 years aimed at implementing the proposed flexible interoperability roadmap.

In the knowledge sharing vision (Figure 13.3), experiences should be shared. Access to the right information (both internal and external to the organization) should be timely and accurate. However, currently a great deal of information is stored in personal and departmental archives in the form of paper-based and digital documents. Experiences from projects are generally not captured or retained efficiently and in most cases only reside in the minds of the people involved in the project. Sharing or disseminating knowledge is not common practice and solutions are regularly re-invented in every project.

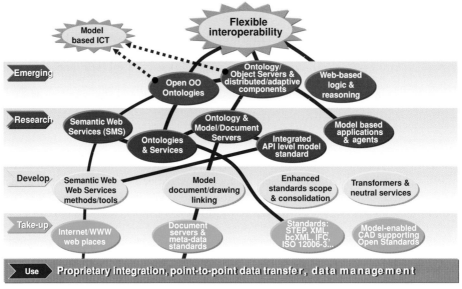

Figure 13.2   Roadmap for flexible interoperability.

To enable effective knowledge sharing within and between organizations, the following factors and technologies should be explored and adapted to construction:

- *Knowledge and practice pattern recognition*: Mechanisms and tools to identify, for example, usage patterns in the execution of a particular task using some application. These patterns may then be used by application wizards to help users in the

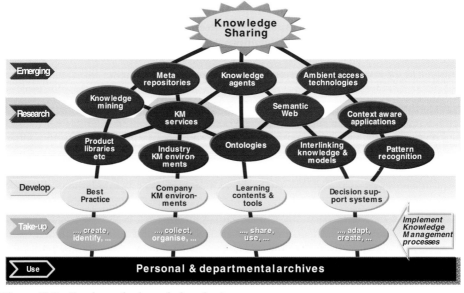

Figure 13.3   Roadmap for knowledge sharing.

execution of such tasks (e.g. provision of the first ten steps) without the user having to navigate through complex menus.

- *Context- and knowledge-aware applications*: Applications that are context-sensitive and can recognize what the user is aiming to do. They should thence be able to provide appropriate guidance, menus and make available the relevant information.
- *Comprehensive and intelligent product libraries*: Intelligent digital catalogues of building products. They should contain substantial product information (much more than simple geometry) in parametric form. As an example, they could contain built-in support for engineering analysis and product configuration, and guidelines for the work implementation of the product.
- *Knowledge portals (KM environments)*: KM environments at an industry level are needed to enable individuals retrieve shared best practices and experiences. These should ideally be transparent to the users and be accessible by different applications and search services. Furthermore, they should provide relevant groupware functionality at an industry (e.g. network of experts) level.
- *High level semantic representations through ontology*: These will help with the identification of key concepts and their interrelationships. Ontologies should not be too generic or too large. Rather, life cycle phases, or specific topics (e.g. facilities management) should be developed in detail. A meta-ontology should be built on top of these to allow for interoperability and mapping between these ontologies when and where needed.
- *Knowledge-related services*: These services should facilitate inter-enterprise knowledge management through the provision of simple services such as searching, and sophisticated services such as e-tendering. These services may be subscribed to on an as-need basis.

Also, as illustrated in Figure 13.3, advances in ICT are expected to bring major contributions in the following areas:

- *Knowledge mining*: Tools for the retrieval of knowledge (including business logic and rules) from different information sources and applications. This should be automated, with the captured business logic and rules made reusable in the form of application components.
- *Meta repositories*: These will provide definitions of, and relationships and mappings between, different information repositories, knowledge sources and ontologies. As an example, through their support, when a search is made for a particular item (e.g. a standard), then only one instance of that standard will be retrieved with a note that the same is also available in given locations. In simple terms, instead of hundreds of links/pointers to the same information, only one direct link to the source of information will be provided.
- *Semantic web*: This will enable a paradigm shift in the way individuals, and particularly applications, solicit information from the internet. As opposed to human interpretable web content, annotations and intelligence will be added to content for ease of retrieval and interpretation by different applications.
- *Knowledge agents*: Intelligent knowledge agents will act as a transient entity between individuals and/or applications and knowledge sources. They will (if necessary through automation) be able to modify and adjust queries so as to retrieve the required information from the relevant sources. Furthermore, the results may

be ranked and categorized (automatically) based on the typical preferences of the user/application. Many more such applications of knowledge agents can be envisaged.

- *Ambient access technologies*: Ubiquitous, personalized and context-dependent access to knowledge is necessary and will be provided through ambient access technologies. These technologies will be based on an integrated use of ontologies, semantic web, context-aware applications, knowledge processes, personal usage patterns, mobility, etc.

While the roadmaps shown in Figure 13.2 and 13.3 provide directions for future knowledge management in construction, the following section provides a methodology for achieving this vision and enabling seamless integration of services through semantic resource mapping.

13.9. Managing semantic incompatibility between services: The FUNSIEC approach

The key idea behind FUNSIEC is to be found in the generally accepted assumption that in order to improve communication and information sharing, within and between organizations during a construction project, and to enable the development of a new generation of services for the construction domain, accessing single, construction-related semantic resources is no longer sufficient. Despite the growth of initiatives targeting the development of domain-specific ontologies and SRs, most of these tend to be country-specific and not adapted to the multinational nature of the sector. Also, the developed resources tend to be specialized for dedicated applications or engineering functions, e.g. product libraries.

The FUNSIEC project aimed to overcome the above problems by investigating the feasibility of building an Open Semantic Infrastructure for the European Construction Sector (OSIECS), which supports integration and interoperability between various domain-specific semantic resources. Schema-matching represents a fundamental approach in the development of OSIECS. In fact, a semantic infrastructure supporting integration must inevitably deal with the problems inherent to heterogeneous SRs, which may differ in both structure and terminology. Through schema-matching, two schemas are compared and the mapping between elements that correspond semantically to each other is produced (Li and Clifton, 1994; Milo and Zohar, 1998; Rahm and Bernstein, 2001). However, schema-matching is considered to be a time-consuming and error prone process, due to the fact that it is still predominantly performed manually.

The methodology proposed as part of the FUNSIEC project suggests a semi-automatic approach to the matching process (Barresi *et al.*, 2006), which involves the development of a semi-automatic software tool (OSIEC Kernel), responsible for the generation of the OSIEC meta-model and model. The FUNSIEC methodology (Figure 13.4) consists of the six-phase approach summarized below:

- *Domain scoping*: Characterization of the domain to be covered by the infrastructure (e.g. e-procurement). During this phase the specification of various scenarios can be used to facilitate the selection of domains to be covered, by describing how the

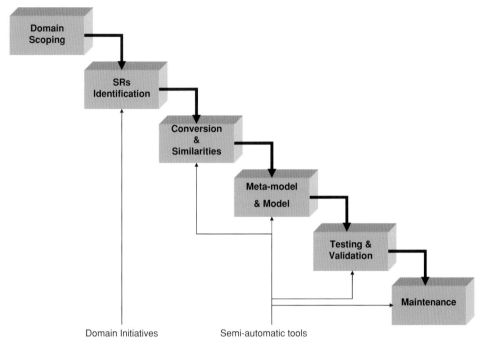

Figure 13.4   FUNSIEC methodology.

infrastructure is expected to be used and specifying the type of information it will provide.

- *Semantic resources identification*: Identification of SRs to be included in OSIECS. During this phase the following SRs have been selected: IFC (part of the IFC kernel), the LexiCon (supported the ISO DIS 12006), the e-COGNOS ontology and the bcBuildingDefinitions taxonomy.
- *Conversion and similarity detection*: Solving syntax-related problems by converting each SR meta-schema and schema into a neutral language (OWL). Detection and validation of the similarities existing between the SRs meta-schemas and subsequently the ones existing between the different SRs schemas.
- *Meta-model and model construction*: The outputs of the previous phase lead to the creation of the OSIECS meta-model and model. These are mapping tables that identify and establish the semantic correspondence between the entities forming the SRs.
- *Testing and validation*: This phase aims at testing the infrastructure, assessing the relevance of concepts and relationships as well as verifying the consistency and coherency of concepts.
- *Maintenance*: During the maintenance phase, the results obtained during the project will be extended by incorporating new SRs into the existing group of resources. In order to achieve this, new mappings and methods will have to be considered.

By following the methodology depicted in Figure 13.4, experts were able to develop the OSIECS kernel, map four SRs between each other and produce the OSIECS meta-model and the OSIECS model. In addition to providing an infrastructure for mapping SRs in the construction sector, OSIECS provides a base for the development of a new generation of e-services for the domain. In the future, new SRs will have to be selected to integrate the OSIECS meta-model/model (Barresi *et al.*, 2006).

13.10. A proposed construction generic system architecture

Intensive intra- as well as inter-enterprise collaboration is a fundamental characteristic of the construction sector. In order to support the intra-enterprise processes as well as the formation and operation of construction projects, a layered service-oriented architecture that makes use of established work, initiatives and standards in the web services domain has been specified (Curbera, 2002; Sutherland and Van Den Heuvel, 2002; Ferris and Farrell, 2003; Van Den Heuvel and Maamar, 2003). Each layer represents the main building blocks supporting intra-enterprise business processes while enabling the formation of virtual enterprises that best illustrate the dynamics of a construction project.

As illustrated in Figure 13.5, the upper layer of the architecture represents the intelligent multi-modal interfaces (including personal computers, personal digital assistants, wearable and head-mounted displays) that are used to interact with other actors and access a wide range of specialized services as well as related information and knowledge repositories. The 'Service infrastructure management' layer provides access to API functions necessary for all aspects of invocation, registration and de-registration of services from third-party service providers, as well as their publication in a dedicated registry.

The 'Business process specification' layer (BPS) includes the API functions that enable service composition in order to implement a given business process. This supports the concept of e-process resulting from the semantic composition of two or several services, and is based on a set of core services concerned with service coordination, transaction and security ('Secure service transactions and coordination' layer). The 'Semantic integration' layer (supporting OSIECS) includes all the interoperability support services described earlier in the chapter, and at the heart of the ROADCON roadmap. The 'Web-serviced enterprise information systems' layer represents all web-serviced commercial and corporate applications that are ready for invocation and use as part of a service composition exercise in order to implement a business process. Any existing enterprise information system (EIS) or legacy application has the potential to be promoted to become a web service. Actors can use adapted software to collaborate in a project. They also have the possibility to extend their existing portal (or EIS application) by implementing relevant API functions and/or providing transparent access to relevant core services.

13.11. Conclusions

This chapter has presented the results of research aimed at facilitating knowledge management while supporting the vision of electronic business processes, enabled by the use of interoperable enterprise information systems. While ROADCON provided the overall context and problem identification for knowledge management in construction, FUNSIEC focused on application interoperability with a view to facilitating the processes underpinning knowledge activities. In particular, it suggested a complementary

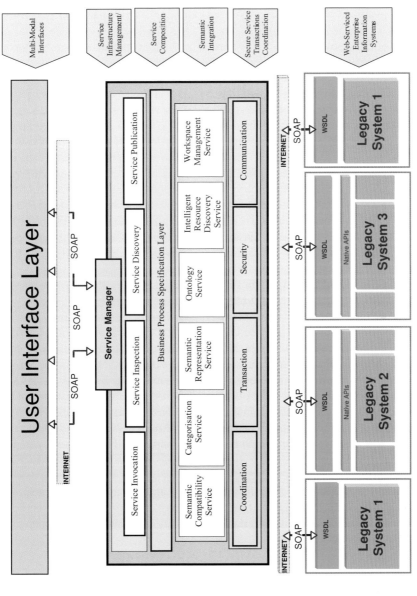

Figure 13.5 e-Process generic system architecture.

use of existing construction-related semantic resources, enabled by the development of an open semantic infrastructure for the sector (OSIECS). Such infrastructure helps to overcome problems related to SR heterogeneity, facilitating the efficient use and reuse of knowledge within the sector, and providing support for the deployment of e-services in the targeted domain. A detailed roadmap providing a pragmatic short, medium and long term implementation of the overall vision that resulted from ROADCON can be found in Hannus *et al.* (2003). Likewise, further details on semantic interoperability using OSIECS can be found in Barresi *et al.* (2005, 2006).

As was initially shown from the literature and subsequently elaborated upon in the empirical research, the construction industry has been operating in a mode that is inefficient due to various causes, including inadequate technological support. A technological solution, it has been shown, had to support the central business processes; allow integration of systems and interoperability between disparate applications used in non-collocated teams whilst, at the same time, taking into account the fact that the industry runs on thin financial margins. While ICT is definitely advanced enough to offer adapted solutions to the industry, the construction end-users' needs and aspirations go beyond what is achievable by automation and digital support. Furthermore, the human and economic elements still prohibit full immediate advancement, as identified in the requirement capture work and confirmed in the interviews/questionnaire results' analysis. What is not clear, however, is the extent to which the majority of the industry would react to such developments as are proposed in the ROADCON roadmap, concomitant with the possible process re-engineering and potential change management involved.

Moreover, increasingly complex business processes require the use of a variety of software packages, only a few of which are used on a daily basis. This infrequent usage pattern often does not justify purchasing full licences, motivating a need for a more flexible way to use and pay for specialized ICT applications. A solution in this respect, as promoted by ROADCON, may be made available by federating services from various non-collocated organizations, and making the tools and services they offer available via ubiquitous web browsers. A potential new business model would be that of 'rental' ICT services via a platform designed to support construction intra- and inter-enterprise processes. This new form of licensing provides a software service that includes configuration, maintenance, training and access to a help-desk. It enables organizations to rent software as opposed to purchasing it. This involves using only the functionality directly needed by the user, therefore reducing cost and increasing work efficiency.

The authors have pursued their research aimed at implementing the models and concepts presented in this chapter (Rezgui, 2007a, b). Future directions include the support for dynamic, as opposed to static, composition of services to implement distributed processes.

Acknowledgements

Special thanks go to all the partners involved in the ROADCON and the FUNSIEC consortiums, together with other organizations that collaborated in various ways in this research. These include: GTM Construction, Nemetschek AG, CSTB, VTT, TNO, BBRI, University of Loughborough, University of Salford, Uninova and Fraunhofer

IRB. The authors would also like to thank the 250 organizations, research and academic institutions that participated in one form or another to the requirement capture, vision and roadmap formulation. The whole list can be found at www.roadcon.org and www.funsiec.org.

References

Barresi, S., Rezgui, Y., Lima, C. and Meziane, F. (2005) Architecture to support semantic resources interoperability. *Proceedings of the ACM Workshop on Interoperability of Heterogeneous Information Systems (IHIS05)*, 4 November, Bremen, Germany. ACM Press, pp. 79–82.

Barresi, S., Rezgui, Y., Meziane, F. and Lima, C. (2006) Methodology to support semantic resources integration in the construction sector. *Proceedings of the Eighth International Conference on Enterprise Information Systems (ICEIS 2006)*, 23–27 May, Paphos, Cyprus. INSTICC Press, pp. 94–101.

Björk, B. C. (1989) Basic structure of a proposed building product model. *Computer-Aided Design* **21** (2): 71–78.

Björk, B. C. (1994) RATAS project — developing an infrastructure for computer-integrated construction. *Journal of Computing in Civil Engineering* **8** (4): 400–419.

Björk, B. C. and Penttilä, H. (1989) A scenario for the development and implementation of a building product model standard. *Advances in Engineering Software* **11** (4): 176–187.

Bohms, M., Tolman, F. and Storer, G. (1994) ATLAS, a STEP towards computer integrated large scale engineering. *Révue Internationale de CFAO* **9** (3): 325–337.

Bontis, N. (2001) Assessing knowledge assets: a review of the models used to measure intellectual capital. *International Journal of Management Reviews* **3** (1): 41–60.

Curbera, F. (2002) Unravelling the web services web: An introduction to SOAP, WSDL, UDDI. *IEEE Internet Computing* **2** (6): 86–93.

Davenport, T. and Prusak, L. (1998) *Working Knowledge: How Organizations Manage What They Know*. Boston, Harvard Business School Press.

Dubois, A. M., Flynn, J., Verhoef, M. H. G. and Augenbroe, F. (1995) Conceptual modelling approaches in the COMBINE project. *Presented at the COMBINE Final Meeting*, Dublin, http://erg.ucd.ie/combine/papers.html [Accessed 3 December 2006.]

Featherman, M. S. and Pavlou P. A. (2003) Predicting e-services adoption: A perceived risk facets perspective. *International Journal of Human–Computer Studies* **59**: 451–474.

Fenicle, B. and Wahls, T. (2003) A methodology to provide and use interchangeable services. *Proceedings of the 2003 ACM Symposium on Applied Computing (SAC)*, 9–12 March, Melbourne, FL, ACM Press, pp. 1140–1146.

Ferris, C. and Farrell, J. (2003) What are web services. *Communications of the ACM* **6** (46): 31.

Froese, T. and Paulson, B. (1994) OPIS, an object-model-based project information system. *Microcomputers in Civil Engineering* **9**: 13–28.

Hannus, M., Blasco, M., Bourdeau, M., Bohms, M., Cooper, G., Garas, F., Hassan, T., Kazi, A. S., Leinoinen, J., Rezgui, Y., Soubra, S. and Zarli, A. (2003) Construction ICT roadmap. *ROADCON Project Deliverable Report D52, IST-2001-37278*. Available at http://www.roadcon.org [Accessed 2 December 2006].

ISO (1994). *ISO 10303-1:1994 Industrial Automation Systems and Integration – Product Data Representation and Exchange – Part 1: Overview and Fundamental Principles*. Geneva, International Standards Organization, TC 184/SC 4.

Kazi, A. S. and Hannus, M. (Guest eds) (2002) Editorial: ICT support for knowledge management in construction. *Electronic Journal of Information Technology in Construction (ITcon)* [Special Issue on ICT for Knowledge Management in Construction] **7**: 67–61.

Kerstetter, J. and Greene J. (2004) Pay-as-you-go is up and running. *Business Week* **January 12th**: 93–94.

Kosovac, B., Froese, T. M. and Vanier, D. J. (2000) Integrating heterogeneous data representations in model-based AEC/FM systems. *Proceedings of CIT 2000 – The CIB-W78, IABSE, EG-SEA-AI International Conference on Construction Information Technology*, 28–30 June, Reykjavik, Iceland; G. Gudnason (ed.), Icelandic Building Research Institute, **2**: 556–567.

Li, W. and Clifton, C. (1994) Semantic integration in heterogeneous databases using neural network. *Proceedings of the 20th International Conference on Very Large Data Bases*, 12–15 September, Santiago de Chile. San Francisco, Morgan Kaufmann Publishers, pp. 1–12.

McAdam, R. and McCreedy, S. (1999) The process of knowledge management within organisations: A critical assessment of both theory and practice. *Knowledge and Process Management* **6** (2): 101–112.

Milo, T. and Zohar, S. (1998) Using schema matching to simplify heterogeneous data transaction. *Proceedings of the 24th International Conference On Very Large Data Bases*, 24–27 August, New York City. San Francisco, Morgan Kaufmann Publishers, pp. 122–133.

NIBS (2006) National Institute of Building Sciences, http://www.nibs.org [Accessed 2 December 2006].

Nonaka, I. and Takeuchi, H. (1995) *The Knowledge-Creating Company: How Japanese Companies Create the Dynamics of Innovation*. New York, Oxford University Press.

OWL-S Coalition (2005) OWL-S: Semantic markup for web services. Available at http://www.daml.org/services/owl-s/1.1/overview/

Rahm, E. and Bernstein, P. A. (2001) A survey of approaches to automatic schema matching. *VLDB Journal* **10** (4): Dec.

Raub, S. and Ruling, C. C. (2001) The knowledge management tussle – speech communities and rhetorical strategies in the development of knowledge management. *Journal of Information Technology* **16** (2): 113–130.

Rezgui, Y. (2001) Review of information and knowledge management practices state of the art in the construction industry. *The Knowledge Engineering Review Journal* **3** (16): 241–254.

Rezgui, Y. (2007a) Role-based implementation of a virtual enterprise: A case study in the construction sector. *Computers in Industry* **58** (1): 74–86.

Rezgui, Y. (2007b) Exploring virtual team-working effectiveness in the construction sector. *Interacting with Computers (Elsevier)* **19** (2): 96–112.

Rezgui, Y. and Meziane, F. (2005) A web services implementation of a user centred knowledge management platform. *Journal of Intelligent Information Technology* **1** (4): 1–19.

Rezgui Y. and Zarli, A. (2006) Paving the way to the vision of digital construction: A strategic roadmap. *Journal of Construction Engineering and Management* **132** (7): 767–776.

Roos, J., Roos, G., Dragonetti, N. C. and Edvinsson, L. (1997) *Intellectual Capital: Navigating the New Business Landscape*. London, Macmillan Press.

Roth, A. V. (2000) Service strategy and the technological revolution. In: *POM Facing the New Millennium: Evaluating the Past, Leading With the Present and Planning the Future of Operations* (ed. J. A. D. Machuca, T. Mandakovic). Production and Operations Management Society and the University of Sevilla, pp. 159–168.

Rust, R. T. and Kannan, P. K. (2003) E-service: A new paradigm for business in the electronic environment. *Communication of the ACM* **46** (6): 37–42.

Rust, R. T. and Lemon, K. N. (2001) E-service and the consumer. *International Journal of Electronic Commerce* **5** (3) (Spring): 85–102.

Ruyter, K. de, Wetzels, M. and Kleijnen, M. (2001) Customer adoption of e-service: An experimental study. *International Journal of Service Industry Management* **12** (2): 184–206.

Scherer, R. J. and Schapke, S.-E. (2005) Constructing building information networks from proprietary documents and product model data. *Proceedings of cib-w78 2005 22nd Conference on Information Technology in Construction* (ed. R. J. Scherer, P. Katranuschkov and S.-E. Schapke), 19–21 July, Dresden, Germany. CIB publication, pp. 343–348.

Solanki, M., Cau, A. and Zedan, H. (2004) Augmenting semantic web service descriptions with compositional specification. *Proceedings of the 13th International Conference on World Wide Web*, 19–21 May, New York City. New York, ACM Press, pp. 544–552.

Sutherland, J. and Van Den Heuvel, W. J. (2002) Enterprise application integration and complex adaptive systems. *Communications of the ACM* **10** (45): 59–64.

Turner, J. (1988) *AEC Building Systems Model.* Working paper ISO/TC/184/SC4/WG1, October 1988. Geneva, ISO.

Van Den Heuvel, W. J. and Maamar, Z. (2003) Moving towards a framework to compose intelligent web services. *Communications of the ACM* **10** (46): 103–109.

Van Riel, A. C. R. and Ouwersloot, H. (2005) Extending electronic portals with new services: Exploring the usefulness of brand extension models. *Journal of Retailing and Consumer Services* **12** (3): 245–254.

Venters, W. (2001) *Review of the Literature on Knowledge Management. C-Sand working paper.* Available at http://www.C-Sand.org.uk [Accessed 3 December 2006].

VTT (2002) *IFC Model Server.* Available at http://cic.vtt.fi/projects/ifcsvr/ [Accessed 3 December 2006].

Wenger, E. and Lave, J. (1998) *Communities of Practice: Learning, Meaning, and Identity.* Cambridge University Press, UK.

Zhang J. and Seidmann, A. (2003) The optimal software licensing policy under quality uncertainty. *ACM Proceedings of the International Conference on Electronic Commerce (ICEC2003),* 1–3 October, Pittsburgh, PA. New York, ACM Press, pp. 276–286.

14   Knowledge management systems in the future

Matthew Bacon

A new form of 'disruptive' technology that has yet to be discovered may well shape the knowledge management systems of the future. Emerging technologies that are only now entering the public domain might well be such a disruptive technology. However, because it is people that create knowledge, a knowledge management system must be more than a technological solution. This means that such a system is also likely to be shaped from working practices that have evolved from those of today, but are yet to become mainstream.

In this chapter the relationship between knowledge management systems and virtual prototyping (VP) is considered. The author suggests that current VP technology lacks the ability to harness knowledge and, on the contrary, can only reflect knowledge-based decisions by the users of such technology. He argues that a new form of enterprise system, which combines both technologies, could transform businesses' decision-making in the future.

In order to consider what form knowledge management systems of the future might take, a useful starting point is to reflect on both current practices as well as emerging technologies. The author speculates how both practice and emerging technologies could shape the knowledge management system of the future. This chapter reflects on both aspects and seeks to build a picture of what might be possible in a timeframe of 15 years.

14.1. Introduction

The disciplines of knowledge management and virtual prototyping have been converging gradually for some years. However, I am not aware of, and neither have I been able to find, examples of where they have actually converged to any useful degree. This is surprising, because in architecture, engineering and construction if virtual prototyping is the means by which information and building modelling is integrated into a virtual environment then surely it is knowledge that has been created from this integration? After all, what is the purpose of the virtual prototype if it is not meant to inform?

That knowledge is created, I do not imagine for one moment that anyone would dispute. However, the newly created knowledge is not persistent, in that it is not stored by the team in a form that is reusable and in a form that others could benefit from at a later time.

This scenario leads the author to the following proposition:

That it must be possible to create a knowledge management system with which collaborating teams could store, retrieve and exploit design, engineering and construction knowledge. Furthermore, that such a system could be used in concert with a virtual prototype in order to inform those teams.

This raises questions such as how should 'knowledge' be stored and made retrievable in such a way that it could be exploitable? In what form should such design and construction knowledge be in, so that it is useful and readily accessible to designers working with the virtual prototype?

The objective of this chapter is to explore the concepts surrounding this proposition and to discuss how emerging as well as future technologies might enable this proposition to be realized. In doing so I wish to look ahead to the next 15 years, and envisage a future world where the disciplines of knowledge management and virtual prototyping have converged.

In setting out to answer these questions we need to understand what we mean by design and construction knowledge when used in conjunction with virtual prototyping. To look forward we also need to look back.

14.2. Design and construction knowledge

An example of the convergence of knowledge management and virtual prototyping disciplines is on the Terminal 5 development for BAA plc at Heathrow airport. Laing O'Rourke, with the support of BAA plc and design partner Connell Mott MacDonald, found that they could bring greater efficiency to the design process, which in turn enabled production to commence earlier with a corresponding earlier handover of the facility to the occupier. They achieved this by revolutionizing the design process by encouraging cross-functional collaboration. The team amalgamated the reinforcement detailing, manufacture and installation processes within a 3D digital prototype. The primary aim of this work was to replicate the pre-installation phase of the construction process, eliminating errors and omissions before work began on site.

In the example cited above, the availability of design information as well as the design and construction knowledge of the collaborating professionals would have been essential for the successful prototyping and evaluation of the pre-installation phase. Indeed, BAA has previously found that repeatability (of processes) is the key to a predictable delivery of facilities in terms of time and cost. BAA has practised repeatability on numerous occasions: The production of the BAA Lynton office buildings was repeated at Heathrow, Gatwick and Stansted airports. On each occasion the design and construction was reviewed and subsequently improved on the building that followed. Design and construction data was reused and, by challenging specific construction sequences, the construction process was optimized. At the Gatwick building an example of this concerned the construction of edge beams that needed to be constructed prior to the cladding being erected. The following knowledge was required:

- The structural implications of different edge beam configurations
- The effect of different edge beam design on the cladding erection process, in terms of commencement of work and the subsequent duration of the work
- The implication on subsequent construction operations with the earlier completion of the building cladding

The successful re-engineering of the design and the subsequent construction sequencing was brought about by a team-based approach (this team was referred to as the 'Office Product team'), where design information was managed on a central server, and the design and construction team members openly shared their construction knowledge.

The inspiration behind this desire to share knowledge came from Sir John Egan, CEO of BAA plc, who demanded that the project teams 'virtually construct' new buildings prior to them being actually constructed on site. He argued that it was unacceptable that the first time that the design was tested was on a building site, and insisted that construction sites were not to be used for this purpose.

It was with this background that the re-engineered design for the Gatwick Lynton offices was run through a simulation based on the information derived from the central server. Construction sequences demonstrated that substantial time savings would be possible when compared to the previous contracts. However, the design and construction knowledge that was elicited in this collaborative process was not in any kind of office product-knowledge management system. On the contrary, the knowledge remained with the team members.

The lack of a knowledge management system means that teams tend to go through a continual cycle of 'learn–forget–relearn'. This is both wasteful and inefficient. For example, construction sequencing knowledge relating to cladding sequencing on the Europier project (1992–1995), completed at BAA some years prior to the office buildings at BAA (1997–1999), would only have been available to the project team if they had been aware of an internal review for that project. As an aside, it is interesting to note that the report also emphasized the need for properly structured data from both the design and construction teams, because without this it would be impossible to derive the potential benefits from the model (i.e. knowledge) such as constructing sequencing management and clash-detection checking.

Some years later the Office Product team, for which I was an expert adviser, was experiencing exactly the same problems as the Europier team. Not only was a lack of structured data an issue, but so too was compatible software. Design work was carried out on different systems, which required the managing contractor to re-key significant amounts of the design information to create a coordinated building model. Re-keying was needed not just because of data loss in the data conversion process but also because of poor 'drafting' (poor layer management, lack of consistent origin, disconnected entities and so on). The management contractor made up for these deficiencies through the re-keying process. Just as with the Europier project some years earlier, a coordinated model was the outcome of an activity that followed the design process, not as an integrated part of it. The coordinated model embedded the knowledge of the participants in that it was implicit in the design. However, because the knowledge was not explicit, in that it was neither visible nor retrievable (through a search engine for example), teams that studied the model some time later would find it very difficult, if not impossible, to elicit the latent knowledge from the coordinated model.

To summarize, we have seen that construction knowledge is embedded in the minds of the collaborating teams. We have seen that when different disciplines from different stages of the design and construction process collaborate then it is out of this process that knowledge has the potential for being created. We have seen that if the right environment is created, they can use this knowledge to inform their design and construction decisions and that they can reflect these decisions in the digital design information within the building model.

However, what we have not seen is the ability to harness that knowledge in any form of knowledge management system. But what do we mean by a knowledge management system? Is it a technology-based solution, a sort of sophisticated information management system, or is it something quite different?

14.3. The industry context for a knowledge management system

In 1996 I contributed to the Construct IT Implementation Plan 1996–2001 (Department of the Environment, 1996). In discussing the rationale for action the report stated:

> Industry electronic captive libraries and information databases will play a valuable role in helping the industry to capture, manage and exploit this knowledge better ... Knowledge of design solutions and best practice allows organisations to learn from and improve their performance.

Contrast the above-mentioned assertion with the following statement (Deloitte Touche, 2006):

> Pouring billions of dollars into information technology does not help business if CFO's and CIO's do not have access to the right information quality (IQ), according to a new global research report ... A majority of business decision makers say they don't have ready access to high quality, reliable, and useful information on operating and financial performance at their companies. As a result of this IQ shortcoming, decision makers are forced to spend time building special reports and analyses, or reconciling the 'multiple versions of the truth' provided by their disparate, non-integrated business processes and IT systems.

This report published in May 2006, some ten years later than the Construct IT report, describes a situation that has changed little over the intervening period. Information is still held largely on paper. Knowledge is still, for the greater part, held in people's heads. Technology and process integration may have improved, but there are still significant challenges in this regard. In gazing forward during the next 15 years to knowledge management systems in the future, is it realistic to expect the situation to improve? Before I explore that question it is important to note that the Deloitte report (Deloitte Touche, 2006) emphasizes three fundamental challenges:

(1) Multiple instances of the truth [I would suggest that this arises where information is replicated and subsequently modified]
(2) Disparate and non-integrated processes [I would suggest that this is where information is not shared from one process to another – such as a feedback review process not informing a subsequent design process]
(3) Disparate and non-integrated IT systems [I would suggest that this is where data is unable to be shared seamlessly between different systems, meaning that it has to be re-keyed and thus creating 'another version of the truth']

In spite of this, the reader will no doubt be aware of examples of good practice where collaborating teams have addressed some or all of the above-mentioned issues, and I would expect that the Terminal 5 example quoted earlier is one of those. However, it is not unreasonable to assume that much of the construction industry is still struggling with the issues highlighted in the Deloitte report. From a knowledge management perspective, one of the key problems for the industry is that it is substantially document-oriented as distinct from data-oriented. It is this (almost slavish) dependence on unstructured information in documents (as a means of creating and storing information) which I contend is one of the major reasons for the complexities we face in attempting to harness knowledge. In reality, an individual is more likely to seek out a body of information, and from this they might then harvest knowledge. However, they have to know where to look for it and certainly they will be unable to query any form of industry-wide knowledge management system, such as that conceived in the Construct IT Implementation Plan referred to earlier.

My contention is that it is information, which is the 'DNA' of knowledge, which must be managed first before we can attempt to manage knowledge. The collection and storage of information is therefore an essential part of the foundation of a knowledge management system. So too is the classification of the information in a form that is relevant to the users of the system. One important means of classifying information is to put it into context. In doing so, it provides information with meaning, and it is from this meaning that the user is able to derive knowledge. Many industries outside of the construction industry are developing ontological frameworks to provide context, in what is often referred to as 'meta-data'. The most prominent examples of this can be found from the Dublin Core (see http://dublincore.org). This framework provides the means for defining meta-data in industry specific contexts. Used in conjunction with another framework called RDF (Resource Definition Framework), which is in effect an open standard for defining meta-data on the web, information can be given meaning by storing it in the context in which it has been created. I believe that the Dublin Core framework offers the potential for a standard classification system for construction information and knowledge.

In practical terms, we could design web-based applications where information has been classified according to a construction industry standard and use RDF to manage the storage of that data according to that standard. We could then use RDF-compliant web browsers to search for information in context. This would be an important step towards the building of a knowledge management system, and in the short term could be akin to an industry electronic library mentioned earlier.

The fact remains that some ten years after the Construct IT report was published we still have no captive industry electronic libraries. But neither has the construction industry shed the shackles of a document-oriented approach to information management, and neither do we have any kind of widely accepted information or knowledge classification system (or even a framework) in the industry.

14.4. A concept design for a knowledge management system

My own reflection on this situation is that if were to set out to create a knowledge management system then four critical perspectives would need to be addressed:

(1) *A process centric perspective.* Issues: Processes for knowledge capture – a reflective process for analysis and synthesis and a process for knowledge delivery. Control of the process.

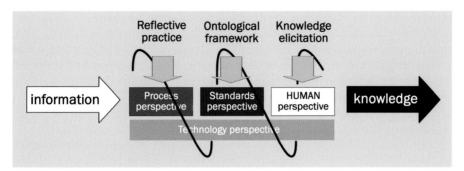

Figure 14.1   Concept for a knowledge management system.

(2)   *A standards centric perspective*. Issues: Classification of information and knowledge (Dublin Core). Data standards (RDF).
(3)   *A human centric perspective*. Issues: Human interaction with the system. Human–machine interfaces. Knowledge elicitation.
(4)   *A technology centric perspective (which underpins all other perspectives)*. Issues: Data storage, data retrieval, knowledge elicitation and knowledge delivery.

Collectively, these perspectives could form what we would understand to be a 'knowledge management system'. We can think of the system as akin to a pipeline where information flows in at one end and, as it passes through it, it is processed according to a user's requirement. Knowledge flows out at the other end. A scenario might be where a user would conduct a search for knowledge appropriate to their need. The system would first process the query, and in doing so would synthesize relevant information, such as its process relevance, and its Dublin Core meta-information. Then, according to the type of technology the user was deploying, the system would deliver the knowledge in a form appropriate to their need.

Alternatively, users could use such a system to download their knowledge (a process of knowledge elicitation). In this case it would be knowledge that would be classified and put into context. This would then be used to deliver knowledge to another user when searching for knowledge in that context. Figure 14.1fig 1 illustrates these concepts.

14.5. A process centric approach to knowledge management

How might experts impart their knowledge to a knowledge management system? The act of writing this chapter is one means of me imparting my knowledge – but the knowledge is not structured and neither is it stored in a database-driven system, able to be searched for and retrieved. Context in this case (virtual prototyping) gives meaning to the text. However, in writing in another context this knowledge might be expressed in quite different terms and therefore the reader would derive quite different knowledge. In BAA I experimented with knowledge-based software, and whilst this demonstrated that it was possible to programmatically influence the design through such a system, the challenge that remained was how to capture expert knowledge within it. The importance of context was realized. But without a formalized (even standardized) notion of context, the capturing of relevant knowledge became impossible.

Another challenge in eliciting knowledge is that knowledge creation relies on post-information production processing. It relies on expansive patterns of thinking, where people are continually learning how to learn together (Liker, 2004). People and teams need time to reflect on their experiences and having done so they can elicit knowledge from them. This was demonstrated in the Terminal 5 example (Concrete, 2002), and from the author's own experiences.

In quality processes there is usually a post-project completion period where the team carry out a post-contract review. The purpose of such a review is to reflect on the project, asking such questions as: How well did we meet the project objectives? What could we have done differently? What do we know now that we did not know then, and how could we use this knowledge better on the next project?

An example of post-information production processing is the work of the Gatwick Office Product Team example cited earlier, where there was a policy of holding post-project reviews with the team. On one such occasion the team was confronted with the following question:

By how much can we reduce the superstructure construction time from 38 weeks?

If one were to use Toyota's famous 'five-why analysis', which requires detailed problem solving (Liker, 2004), the team would be asked:

(1) **Q**: Why does construction take 38 weeks?
 A: Because it takes 20 weeks for the frame and cladding construction and 18 weeks for internal fit-out. The latter cannot commence until the cladding has been completed.
(2) **Q**: Why cannot the cladding be completed earlier?
 A: Because the edge beam construction, which follows on from the floor plate construction, has to be completed before the frame is stiff enough to receive the cladding loads.
(3) **Q**: Why cannot the edge beam construction be changed so that the floor plate construction can incorporate a stiffening detail?
 A: No reason why not.
(4) **Q**: If we can incorporate a construction like this, why cannot we commence cladding earlier?
 A: We could.
(5) **Q**: Why don't we phase construction from one end of the building to the other, in order that frame construction and cladding take place concurrently rather sequentially, and how much time would we save if we did so?
 A: 6 weeks. We could reduce the programme from 38 weeks to 32 weeks.

The team established that in order to speed up construction, the detailing of the structure would have a major effect on construction sequencing. The project team knew that the longer construction took, the more expensive the final account was likely to be. Through this process, knowledge was being elicited from the team.

The team accumulated this knowledge and this was then embedded in the design. However, as explained earlier, the knowledge was not explicit in the design and consequently if a different team were to inherit the project, the building model would be 'silent' as to why the detailing was configured in the way that it was.

Even if the knowledge were not explicit, would the information that led to the creation of that knowledge be accessible to that team in a form that would be useful, as described earlier? This is the challenge for the storage and retrieval of information and knowledge. Without a knowledge management system where information is stored in context then the answer to this question is likely to be 'no'. This is because documents which are used to create and store the information do not embrace the concept of context, with the consequence that it is difficult to search for it and retrieve it. Perhaps the solution is to use something like the system 1 described earlier to categorize information at the point of creation. Such a system would mean that knowledge elicitation could at least be made easier and ideally much more efficient. This would obviously be highly desirable.

In my own work in developing the IFM Smart Information System I have found that process context is of critical importance in this regard. I reasoned that whenever users of the system create information it should be automatically categorized with the process context (and other meta-data) in which it was created. Process context means the business process in which the information was created, as well as the role of the person in the process that created the information. This is very useful, because a person working in a design process will use the same information quite differently to another person working in procurement or a construction process, for example. However, when both disciplines work together using the same information, then knowledge has the potential for being created. The example of the Terminal 5 team cited earlier, where a cross-functional team worked collaboratively and amalgamated relevant information from across different processes, was able to create useful knowledge to inform both the design process and the subsequent pre-installation phase of the process. Likewise, the Gatwick cross-functional team collaborated and amalgamated information to create knowledge concerned with speeding up the construction process. The team learnt from collaborating with the cladding panel manufacturer that cladding panel dimensions have a profound effect on the cladding panel manufacturing process. The consequences of cladding panel dimensions ripple through financial, cost and time control decision-making. This knowledge led to the standardization of cladding panel dimensions.

To summarize, in order to achieve knowledge elicitation there needs to be a pool of structured information available to business users (teams) from which they are able to create knowledge. They require access to a comprehensive body of information (data) from which they are able to elicit knowledge (knowledge elicitation) to inform their processes. The business users would need to synthesize the information (data) to create knowledge that could be programmed into the knowledge management system. To create knowledge, experts need to be analysing the information and establishing patterns and relationships in it. They would then contextualize this and embed it into the system. This is the process centric approach to knowledge management.

A knowledge management system is clearly not just a storing and retrieving solution (the standards dimension). It must also way of doing business (the process dimension), and one where people set out to learn together (the human dimension).

14.6.　A standards centric approach to knowledge management

However, the challenge remains: that of classifying knowledge for the benefit of other teams on subsequent projects. Without standards such as the Dublin Core discussed earlier, a knowledge management system could not function. Again, my own experience of working in BAA is relevant. BAA considered that design and engineering standards were an essential part of the elicitation of knowledge, and that through these standards such knowledge could be passed onto other integrated teams. They reasoned that if expert knowledge could be embedded in design and engineering standards then this could be the means by which knowledge could be managed. The problem with this approach is that the knowledge that shaped the design standard was not explicit (just as was the case with the coordinated building model referred to earlier), and briefing teams could not always remember why the standard had been specified in the way that it had. This resulted in the constant challenging (usually because cost and/or time needed to be driven out of the design) by project teams and the 'learn–forget–relearn' syndrome.

The standards driven approach is further complicated because one design or engineering standard could contradict another standard – what was appropriate for one context might be inappropriate for another. Furthermore, such standards were created using a document paradigm and the document was the information storage mechanism. With so many design and engineering standards in place there arose another challenge, which was concerned with compliance. The challenge was how to ensure that the design output and the contract specifications accurately reflected the requirements of these standards. Design teams would complain of information being too difficult to find, and too difficult to understand at what phase of the project process the standards should be incorporated into the design. This required the authors of these standards having to provide an internal consultancy role to the project team. In doing so, they would advise the project team as to the application of a specific standard. In other words they were being required to impart their knowledge of the standard in the design process.

My own experience in this situation is that from such collaboration the team will determine key design criteria, which in software design we would refer to as the 'business rules'. Essentially, the business rules are the representations of knowledge. These rules can be codified and scripted and can be stored in a database for subsequent reuse. However, in BAA, because the standards information was unstructured, there was no opportunity to create such a rule-base. To explain this, in my recent work in Integrated FM Limited (IFM) we created a rule-base for enabling a national contractor to work with their suppliers using an automated transaction process. In doing so, the author worked with the relevant business disciplines in the contractor's business and elicited their knowledge and thus the business rules concerning the transactional relationships with the contractor's suppliers. The author found that it was possible to codify these rules and, more importantly, apply them at specific stages of a transactional process. In doing so, IFM configured their Smart Information System to automatically process information according to these rules, which would have otherwise been carried out by a knowledgeable manager. The rules were codified in an XML (human readable) database. However, the reason for the existence of these rules had to be documented separately, and thus the knowledge of the transactional process was stored in the contractor's corporate domain.

In operating the system we are in fact operating a virtual prototype of a business process. Each rule was 'commented in' to the code, in that the reason for the rule and the logic relating it to a business process were made explicit.

Stored in a database in the form of standards, the business rules could become the means by which the virtual prototype is configured to meet specific business requirements. In essence, this is how manufacturing systems are configured. An article from the University of Reading confirms the value of this approach, 'Expert system fuses human and computer knowledge' (University of Reading, 2006). It states:

> Mass manufacturing of products, such as aluminium plate, requires millions of pounds worth of investment every year, and aluminium plate producers naturally seek to maximise the productivity and profitability of rolling mills. Now, scientists at the University of Reading, in conjunction with the University of Leicester and Alcoa, have developed a new state-of-the-art fused expert system that has shown through plate rolling trials how mills could work at optimum performance levels...

> ... The EPSRC-funded research (following on from groundbreaking work by University of Leicester) involved the combining of knowledge-elicitation and data-mining techniques to develop the fused expert system...

> ... Knowledge elicitation involves establishing important facts and heuristics (rules of thumb) from plant experts, whereas data mining is the process of analysing data, often using advanced artificial intelligence techniques, in order to identify patterns or relationships.

> 'The fusion of these two techniques produced an expert system that successfully rolled aluminium plate without significant shape defects', said Dr Browne.

> The methodology is transferable to all the other plate alloys and it is applicable to many other industrial problems.

In both this example and that from my own business in IFM, a common factor was the desire to elicit knowledge to inform a specific objective. This could be described as, 'what is it that we are aiming to achieve and what do we know about it?' Again, this is another example of where context is important. The business rules in themselves do not mean anything. However, in the context of a transactional process, which is required to deliver specific outcomes, they are given meaning. It is what the Reading University team (University of Reading, 2006) referred to as the 'patterns and relationships' in the process. It should now be appreciated why a knowledge elicitation process has therefore to be so much more than simply the storage and management of information. It has to be a foundation stone of a knowledge management system as much as searchable information repositories have to be.

Having captured the knowledge through a knowledge elicitation process, a business needs a means of retrieving it that is appropriate to the context in which it is required. A rule-based system such as that described above is clearly one starting point. Applied to a repeatable process it is possible to envisage how knowledge might be delivered to the point of need relevant to a specific context.

However, in a business environment where processes tend to be much more abstract, and to an extent chaotic (by this I mean that they are not systematic), the application of a knowledge management system becomes a significant challenge. It is a particular challenge because the management of information, as we have seen from the Deloitte report (Deloitte Touche, 2006) still remains a significant problem for the enterprise. If a business is unable to manage information, how can it effectively deploy a knowledge management system?

14.7. Data management standards

Disparate and non-integrated processes and IT systems are a significant management challenge as we have already seen. But what does this mean in the context of a knowledge management system?

We have seen how all too often data has to be re-keyed in order to assemble it from different sources and different systems. This in itself creates 'multiple instances of the truth'. By this I mean that we now have information pertaining to the same subject on two independent systems and this leads to 'multiple instances of the truth', which erode the integrity of the information. If the base information or data is wrong then so too may be the knowledge that is derived from it. Consequently, standards have to apply not just to the classification of information, but to the data that provides the information as well. If the data is in a proprietary form, that cannot be searched, analysed and categorized, then it will be useless to the enterprise. If it is in a form that constantly requires re-keying then the integrity of the information will be undermined.

This is where open standards offer much value to the enterprise. Open standards are the means by which data can be managed independent of the application that created it. Open standards enable data to be shared seamlessly with other systems. Open standards enable relationships (both *ad hoc* and structured) to be applied to data to enable it to be assembled for different uses in different contexts. It is by this means that different collaborating teams can build their own knowledge from the same information used by other teams, who themselves are creating their own accumulated knowledge.

The obvious open standard that could be deployed for a knowledge management system is XML. Its extensible capability means that data can be tagged with attributes appropriate to different contexts. Because it is platform-independent it can be referenced to any other data objects, and by this means a network of information relationships can be constructed that become the 'DNA of knowledge' for the enterprise.

14.8. A human centric approach to knowledge management

In the author's experience, the age-old saying 'not invented here' is as true in business today as it was in the past. Perhaps the 'learn–forget–relearn syndrome' is an inexorable part of the culture of present day businesses. Some critical proponents of innovation argue that a knowledge-based approach is irrelevant – that past knowledge which has been derived from previous working practices and business constraints is unnecessary, and indeed constrains creative thinking. As Toyota discovered (Liker, 2004):

> The critical task when implementing standardisation is to find that balance between providing employees with rigid procedures to follow and providing the freedom to innovate and be creative to meet challenging targets consistently for cost, quality and delivery. The key to achieving this balance lies in the way that people write standards as well as who contributes to them.

Yet creative thinking devoid of past learning can also be highly dangerous. Established working practices have evolved for very good reasons. Corporate compliance, fiscal control and risk management are just three reasons why a knowledge management system informed by lessons learnt in these disciplines needs to be stored for the benefit of the 'learning organization'. This is what Toyota refers to as the 'Toyota Way' of learning – standardization punctuated by innovation, which gets translated into new standards (Liker, 2004).

Anecdotal evidence, as we shall see in a moment, suggests that organizational culture results in people not wanting to share their knowledge even if the infrastructure and processes exist to do so. It is often remarked that of all the professions, lawyers have most to benefit from the sharing of their knowledge resources (Rusanow, 2006):

> A large firm may find that there is little sharing of knowledge across practice groups and offices. There are a number of cultural reasons for this. Where the partner compensation model rewards the individual rather than the firm, practice groups tend to operate as separate business units, focused only on growing their own practices. There is no incentive to share work with others, since there may be no reward for referring work to colleagues. Indeed, there may be overlap in areas of practice between lawyers in different practice groups. These groups may be competing with each other in the market. Lawyers may also believe that their knowledge base is their power base, and that sharing that knowledge would dilute their value.

From these comments it is clear that integrated working practices and incentivization should be important facets of a knowledge management system. Toyota asserts that a culture needs to exist where people learn how to learn together.

Learning together was a central tenant of the BAA product teams, as much as it is in Toyota. It is in such teams that trust is created. But can a culture of knowledge transfer operate between teams? This is the challenge for the construction professions as much as it is for legal professions. People are unlikely to make this happen. Pressures of work and their own team-focused objectives tend to prevent this from happening. Without a knowledge management system in place, knowledge transfer can only happen as an *ad hoc* process. The system therefore has to be able to prompt the user, guide the user and inform the user where knowledge exists as they are creating their own information and knowledge. This where technology has the potential to perform the 'glue' that makes the system serve the needs of the enterprise.

14.9. Envisioning the future

Human understanding of business needs associated with technology that is becoming ever more adaptable and user-oriented, should provide us with the confidence that knowledge management in the next 15 years will be revolutionized. We are moving inexorably from a paper-centric world to a data-centric world, where more and more of the information that we access is through the digital medium. 'Joined-up' collaborative processes as well as 'joined-up' integrated technologies are the essential catalysts for change. Innovative technologies that challenge the machine–computer interface will provide much greater access for all. Smart systems that recognize our needs and deliver information to the point of need will become the norm. Computer systems integrated with the human-form, and which enable us to dynamically share information with the world around us, will become as ubiquitous as the mobile phone.

Is this just fanciful? Where is the reality in the virtual world that we are exploring? What does this mean for knowledge management systems in the next 15 years?

Earlier in the last decade (1996) virtual reality technologies had been investigated as a means of integrating computer knowledge with human knowledge. Faraj suggested that, 'users can transfer their thoughts to the integrated environment in a more efficient manner' (Faraj, 1996). I am not sure that the writer had this in mind when the following article was published (Laurance, 2006):

> He is unable to move or breathe on his own after his spinal cord was severed in a knife attack five years ago. But thanks to a dramatic scientific advance Matthew Nagle, 25, can now pick up objects, open e-mails, change the channel on the television, and play computer games. And, incredibly, he does it all using the power of thought alone.

> The quadriplegic – paralysed from the neck down – is part of an experiment at the cutting edge of neural implants research that enables him to operate a computer and a robotic arm with his brain ...

> The idea of using thoughts to control the world around us was once the realm of science fiction. With the development of the latest generation of brain implants, it has moved an important step closer to reality.

It would seem that there is now a distinct possibility that information synthesized in the human brain could be used to create knowledge, which could then be stored on a computer system. Work carried out at the University Reading provides an insight into the 'art of the possible' (Warwick, 2003):

> Dr. Kevin Warwick, professor of Cybernetics at the University of Reading, UK, undertakes research into artificial intelligence, control and robotics. A long-term advocate of the benefits of enhancing human operations with technology, the Cyborg scientific research projects being conducted at Reading University demonstrate the potential for a range of medical and commercial applications.

Whether or not we will have the capability to achieve knowledge elicitation within 5, 10 or 15 years through some form of personal interaction with a network, the question still remains as to how we contextualize it so that others can retrieve it? Does such technology enable us to make a step change in the way in which we manage knowledge?

Over the past few years Autonomy has led the market in the contextualization of information. From the front page of their web site (Autonomy, 2006) the reader learns that:

> Autonomy is the acknowledged leader in the rapidly growing area of Meaning Based Computing (MBC) ...

> ... The last few years have seen explosive growth in the use of unstructured information, which includes documents, emails, telephone conversations and multimedia. More than 85% of all information inside an enterprise is now unstructured and this 'human-friendly' information has traditionally been difficult for computers to understand and use. Meaning Based Computing solves this problem.

> Meaning Based Computing enables computers to understand the relationships that exist between disparate pieces of information and perform sophisticated analysis operations with real business value, automatically and in real-time.

'Meaning based computing' is an emerging discipline. It is now being exploited in television content. BT launched a television service in December 2006 called 'BT Vision'. One aspect of their service is that it is able to automatically 'cherry-pick' the most exciting parts of a football match and create football highlights (Fildes, 2006).

> The football highlights service selects clips based on the level of motion on the pitch, the number of camera angles and the audio levels in the game. In this way it can recognise when a goal is scored, or a player sent off, based on the replays, the excitement of the commentator and close ups of players and the crowd.

Like unstructured text-based documents, television footage is also unstructured. Meaning-based computing offers a rich information resource from this unstructured data to infer meaning and apply meta-data to it, enabling it to be searched and amalgamated with other information. With the possibility of a number of sophisticated knowledge elicitation tools becoming available in the future, meaning-based computing may be the start of a revolution in the way in which we manage information and knowledge in a knowledge management system. It will bring convergence between knowledge elicitation and knowledge classification – the holy grail of a knowledge management system.

There still remains the not insignificant issue of encouraging business users to exploit the value of this resource. FT Digital Business described it in these terms (*Financial Times*, 14th December 2005):

> 'You can look at technologies in the lab and see a number of things happening', says Rick Rashid senior vice president for Microsoft Research. 'But the caveat is that it's hard to predict how people will use technology, how it will get adapted into their lives.'

Maybe the vision of the future (2054) displayed in the film *Minority Report*, where Tom Cruise walks through a shopping mall and is 'bombarded' with advertising based on his known spending habits, will become the norm. Perhaps the boundaries between how people use technology and how technology uses people will become blurred? Perhaps how well technology will adapt to people will be as important as how well people will adapt to the technology and so cause further blurring of the boundaries at the human–technology interface?

In 'current world' we are expected to search for information. Perhaps taking a clue from *Minority Report*, our present focus is wrong and that in 'future world' we should expect knowledge management systems to deliver information (and knowledge) to the user based on their known needs in the process (role, responsibilities, etc.) within which they are working at that time?

This is actually not that far removed from what can happen in 'current world'. We are quite familiar with intelligent building management systems that proactively manage a building's key systems, in that they warn the facility manager when the monitored systems are functioning outside of tolerance. Indeed, in IT systems there are similar monitoring technologies at work. In my own work, Integrated FM technologies monitors services on their hosted systems and warns the system administration function when the systems services are nearing critical or indeed have failed.

A building management system uses managed processes, business logic (rules) and system monitoring to deliver its functions. Why should not a business system function in the same way? Why shouldn't a virtual prototype connected to a knowledge management system be configured to receive real-time business data to simulate facility usage, dynamically informing business managers of system events that require management action? In my own work in Integrated FM, our systems notify business managers when transactions in the business process require their management attention. In BAA we connected a virtual model of a security comb with a simulation engine in order to analyse the effect of different security regimes on passenger queue lengths. It is not too far beyond the present to connect such data to a real-time business management system.

We must expand our thinking to consider that a virtual prototype connected to a knowledge management system can be so much more than a virtual prototype of a facility, which is just one view of the world we inhabit. We need to recognize that it can also be a prototype of any business system.

It is not beyond the realms of possibility to envisage that a virtual facility could embody a virtual business and so become a real-time replication of an existing facility. As users entered and left the building, accumulated in groups, carried out various functions, the building systems would adapt automatically to the received data. The virtual facility would simulate scenarios based on trends of usage and prepare systems for rapid response should the environmental demands dictate the need to do so, based on the events taking place. I am reminded of a study tour to Japan in 1996, where I witnessed Toyota modelling the production line for a new car. There were virtual people, virtual processes, all functioning according to the business rules that they had been programmed with. And the value of this? Well they carried out time and motion studies in order to optimize the process. Is it not that big a step to take to model a business in the same way? Will 'Six Sigma' (a business performance management system) become the performance model in the knowledge management system that drives the simulation engine?

However, we have learnt in the intervening years that business processes must be adaptable. They cannot be 'hard-wired' into the business model. Consequently, whatever technology is developed it must be adaptable. Systems designed for specific functions are systems of today's software paradigm. Systems that adapt themselves to the working practices of the user, or the needs of the business, are more likely to be the business systems of the year 2023.

Contrast the above vision with today's world. There are business intelligence reporting systems and business process management tools, but they are for the greater part static systems that report on what has happened, or at best provide trend reporting on discrete business processes. There are event-driven systems that provide control in manufacturing processes but few event-driven systems in business processes. There are wireless systems and public area networks over which they operate. We are seeing increasing convergence between different technologies. Each of these technologies point to the future of a need for a knowledge management system that is fully interactive. They point to a system that is responsive and provides expert advice for process optimization, just as the aluminium rolling plant system conceived by Reading University and others, referred to earlier, does for a manufacturing process. I believe that they point to what is called 'pervasive computing', which may be the infrastructure technology for the knowledge management system of the future.

IBM and Cambridge MIT (Massachusetts Institute of Technology) are just two examples of where 'pervasive computing' is being actively researched. The following examples explain the concepts:

From MIT (2004) we learn:

The Cambridge–MIT Institute has launched the Pervasive Computing Community to explore the challenges of a world where computers are burgeoning in numbers and influence even as they shrink in size.

Researchers from Cambridge University and MIT will collaborate with students, industrial partners and other organizations to explore the challenges of a networked wireless world. The Pervasive Computing Community will work on issues involved in letting computer users be genuinely 'nomadic' and to be able to access information everywhere. Researchers will also work on developing new computer vision and speech processing technologies that will make it easier for people to interact with computers.

And from IBM (*Web Sphere Journal*, 2004):

People typically think of wireless solutions in terms of mobile phones and PDA's, but it will be an integrated infrastructure – technology, hardware, software, and services – and enabling the communication between many devices, including chips, refrigerators, and cars – that will emerge as the key to the growth of wireless solutions. Handhelds are just the tip of the iceberg. Wireless technology is rapidly growing to support telematics, home gateways, consumer electronics, appliances, and Web-enabled self-service kiosks.

However, I foresee a world where the virtual world and the real world interact. It requires pervasive computing, but it also requires simulation. After all, how can effective decisions be made if we are not able to understand the consequences of the decisions that we are inclined to make? It follows, that as a business manager I don't just need information, but I need foresight: a knowledge management system must deliver foresight.

The simulation tools that are the 'engine room' of the virtual prototype need to be programmed. Different simulation tools will be required to simulate many different needs. They will provide the user with foresight. Each simulation tool will be driven from the knowledge management system. Business rules, business logic, business constraints, business processes, business standards and business data will all be synthesized to drive the simulation engine.

But why stop at the virtual prototype? Why shouldn't this evolve into the virtual facility and from there into the virtual business? Indeed why shouldn't it become an essential part of business systems as much as the ERP (Enterprise Resource Planning) system is to the corporate organizations of today? Back in today's world, the business intelligence community is actively investigating this potential (Swarby, 2006):

How can the insights be distributed and used by the right people, at the right time? Increasingly, the focus is on delivering targeted information at the right time – not trying to make sense of great warehouses of data.

The virtual business would interact with business managers or vice versa. Event-driven business systems would prompt the manager when business processes were reaching a critical stage. Performance trends would be analysed 'on the fly' and the manager would be directed to critical tasks that would need to be performed in order to achieve a predictable delivery of service or product.

As managers create information, a knowledge management system of the future might be scanning databases and returning relevant information to the user – not of assumed relevance but of real relevance based on the multitude of thought processes passing through their brain. Business intelligence delivered in real-time serving the needs of the user is an obvious need.

As business managers need to make decisions the system would enable them to run multiple simulations, such as cause and effect scenarios. They would not just require insight into the business. They would require foresight too in order to understand the potential consequences of those decisions – the consequences on the P&L account, the consequences on shareholder value, or the more intangible consequences on staff and customer satisfaction.

But to get anywhere near that vision we need to be effective in the management of information. We need to be able to elicit knowledge and we need to capture that knowledge in a form that is useful, and that can be reused. We need to be able to configure simulation engines with that knowledge and to transform it into configurable business rules that drive dynamic business processes. Most importantly, we need our knowledge management system to be adaptable in order that it can be adapted to the needs of the business. It needs the ability to simulate and predict and deliver new knowledge, derived from what it has already learnt. It probably needs to acquire some of the attributes of the human brain. Maybe it can be programmed like a neural network, where knowledge from you and me is downloaded into it as easily as we currently download information from the internet?

As The FT Digital Business article speculated (*Financial Times*, 14th December 2005):

In the very near future a small group of technologists will set out on the greatest adventure of them all . . . to search for the next big thing. Can they find it?

This is the challenge for the technology road map creators of today . . . can they, can we, really create a knowledge management system that is responsive to the needs of us mere mortals?

References

Autonomy (2006) *Autonomy: Understanding What Matters*. Available at http://www. autonomy.com/content/Autonomy/index.en.html [Accessed 13 August 2006].

Concrete, (2002), June

Deloitte Touche (2006) *Do Executives have the right IQ?*

Department of the Environment (1996) *Construct IT (Bridging the Gap) – Implementation Plan 1996–2001*, p. 23.

Faraj, I. (1996) The Organisation and Management of Construction, Vol. 3: Managing Construction Information (ed. D. A. Langford and A. Retik). London, E&FN Spon, p. 144.

Fildes, N. (2006) BT computers calculate football highlights. *The Independent* 18 July, p. 34.

Financial Times (2005) A technology odyssey. *The Financial Times* 14 December, p. 1. [Editorial]

Laurance, J. (2006) Introducing the first bionic man. *The Independent* 13 July, p. 18.

Liker, J. (2004) *The Toyota Way*. New York, McGraw-Hill, p. 250.

MIT (2004) *CMI Launches Pervasive Computing Initiative*. Available at http://web.mit.edu/newsoffice/2004/pervasive-0331.html [Accessed 13 August 2006].

Rusanow, G. (2006) *Knowledge Management is a Business Imperative*. Available at http://www.llrx.com [Accessed 13 August 2006].

Swarby, P. (2006) Intelligent processes – the information management imperative. *Information Age – Business Briefing* No. 54 [Magazine supplement]. London, Infoconomy Ltd.

University of Reading (2006) *Expert System Fuses Human and Computer Knowledge*. http://www.engineeringtalk.com/news/unf/unf100.html [Accessed 13 August 2006].

Warwick, K. (2003) *Are Chip Implants Getting Under Your Skin?* [Editorial] Available at http://www.synopsys.com/news/pubs/compiler/art3_chipimplan-mar03.html?NLC-insight&Link=Mar03_Issue_Art3 [Accessed 13 August 2006].

Web Sphere Journal (2004) *Pervasive Computing*, 7 February 2004. Available at http://websphere.sys-con.com/read/43850.htm [Accessed 13 August 2006].

Part 5
Other challenges: Agent technologies, security, regulations and management control

15 Future agent-driven virtual prototyping environments in construction

Joseph H. M. Tah

The current generation of virtual prototyping (VP) systems used in construction is still not smart enough to cope with the increasingly complex and dynamic nature of construction projects, and with the problems facing the construction industry and society at large. This chapter argues that the next generation of VP systems will need to draw heavily on agent-based technology to handle the complexity that characterizes construction projects. The nature of this complexity is briefly examined, and leads us to characterize construction projects as complex adaptive systems that lend themselves well to multi-agent systems' solution techniques. An overview of agent technologies, followed by a review of the state-of-the-art applications in construction is presented. The typical areas in which agent technologies have been deployed in other industry sectors are discussed. The challenges facing the deployment of agent-based systems and emerging technologies that can be used in combination with agents to overcome some of these challenges are briefly discussed. It is concluded that future agent-driven environments for virtual prototyping will allow us to model and simulate complex construction projects in ways that are not currently possible.

15.1. Introduction

The construction industry is experiencing unprecedented change and dynamic conditions resulting from societal demands for building and infrastructure assets with ever-increasing standards of performance, constantly diminishing environmental impacts and steadily reducing costs of construction, operation and decommissioning. This makes the job of predicting the future needs of inhabitants very problematic. The situation is exacerbated by the global quest for sustainability, which encompasses numerous contemporary issues such as the effects of climate change, natural resource over-consumption, an impending energy shortage, greenhouse gas emissions, transport, pollution, crime, conservation, economic regeneration, ageing population, disaster management, etc.

This is fuelling the development of an increasing number of new methods, materials, technologies, processes and innovative practices aimed at improving buildings, infrastructure, spaces and communities with respect to a multitude of sustainability performance considerations and indicators. As the number of methods and technological options increases, so does the complexity and associated cost of choosing amongst alternative combinations for a given situation. Informed decisions require the management of vast amounts of information and knowledge about the combinations of available options and the simulation of their performance. It is almost impossible to apply manual methods and physical prototypes comprehensively and it has long been recognized that the use of computer-based virtual prototyping offers a solution. Significant research and development efforts have been made to develop various generations of virtual prototyping systems. These systems have evolved from early 3D CAD (computer-aided design) software, through 4D to nD modelling (Lee *et al.*, 2006) more recently. Advanced virtual and augmented reality techniques have also been developed and used with varying degrees of success. Although significant advances have been made in the development of virtual prototyping systems, further advances are required to allow us to deal adequately with the increasing complexity of construction projects.

15.2. The complex nature of construction

In addressing the contemporary problems facing society and the construction industry discussed above, it has been widely acknowledged that it is not sufficient just to be concerned with the physical built assets, but that the interplay between assets and the social, economic and environmental consequences arising out of human activity also needs to be addressed. However, this presents major challenges due to inherent complexities within and between the elements that constitute the physical built assets and the organizational systems used in their procurement, delivery, use and disposal.

Complex interactions exist between buildings, open spaces, bridges, transport networks, utility networks, communication networks and other infrastructural assets. These assets are characterized by complexity, from their inception, through design, construction and operational life, to their disposal or recycling. For example, in buildings alone, there are multiple interacting subsystems such as building materials and components, the structure, the building's fabric, building services, utilities, communication networks and the external environment. The procurement of these assets involves complex interactions between multiple stakeholders and organizations operating in complex value and supply networks, often with conflicting perspectives on their creation, configuration and use. These organizational networks are characterized by social, economic, legal, regulatory and cultural interactions that strongly influence the management of the planning, design and production processes involving varying degrees of cooperation, competition and conflict.

The host of contemporary problems, together with inherent complexities in construction, presents considerable challenges when searching for appropriate solutions. Existing conventional problem-solving methods are unable to deal adequately with this complexity, given that built assets need to constantly evolve and adapt in response to rapidly changing societal demands. Furthermore, construction is a complex multi-disciplinary field that requires integrating expertise and input from various constituencies, with many people processing and exchanging complex heterogeneous information over complex human and communication networks, in the context of many changing constraints. Thus, construction projects can be considered as exhibiting the characteristics of complex adaptive systems (Holland, 1995).

Complex systems are open, non-linear dynamic systems, with numerous internal and external connections which are subject to turbulence. Such systems evolve and adapt over time. A common feature of these systems is that organized behaviour emerges from the interactions of many simple parts. This feature has been exploited extensively through the application of multi-agent systems' techniques in the analysis of complex systems. In agent-based modelling, the model consists of a set of agents that encapsulate the behaviours of the different elements or entities within the complex system. The interaction between multiple agents in the system results in an emergent behaviour of the overall system. It has been widely acknowledged that the agent-based approach provides a metaphor and framework for channelling problem- solving approaches from diverse disciplines into the design of software systems capable of handling complexity. The ensuing section presents a brief review of the state-of-the-art in the development and application of agent technology.

15.3. An overview of agent technology

Currently, there is no general consensus on the definition of an agent, but the description provided by Wooldridge and Jennings (1995) is probably the most widely accepted. They define a software agent as 'a self-contained program capable of controlling its own decision making and acting, based on its perception of its environment, in pursuit of one or more objectives'. Typically, an agent works for and/or on behalf of a user, and acts to support the user in achieving his/her objectives. In general, agent-based solutions involve several such agents (i.e. a multi-agent system (MAS)) interacting to provide a solution for a given problem. Wooldridge and Jennings (1995) specify four main attributes that determine agenthood:

- *Autonomy*: The ability to function largely independent of human intervention
- *Social ability*: The ability to interact 'intelligently' and constructively with other agents and/or humans
- *Responsiveness*: The ability to perceive the environment and respond in a timely fashion to events occurring in it
- *Pro-activeness*: The ability to take the initiative whenever the situation demands

Nwana and Ndumu (1999) present a complementary view of agenthood that subsumes the above definition but includes learning ability, i.e. the ability to improve performance over time. It is not expected that a system should have all these attributes to be called an agent. However, most agent systems will posses a substantial subset of these attributes. Agents are well suited to a variety of applications (Jennings, 1999; Luck *et al.*, 2005), although they are particularly appropriate for applications that involve open, complex and distributed computation and communication across networked computers (Jennings, 2001).

In the main, agent technologies are autonomous software systems that can decide what they need to do for themselves. Agents are capable of operating in dynamic and open environments and often interact with other agents, which may include both people and software. Luck *et al.* (2005) group the agent technologies into three categories, according to the scale at which they apply: organizational level; interaction level; and agent level.

15.3.1. Organizational level

Technologies at the organizational level relate to agent societies as a whole, and deal with issues of organizational structure, trust, norms and obligations as well as self-organization in open agent societies. Research in this area draws on other disciplines, such as sociology, anthropology and biology, and focuses on technologies for designing, evolving and managing complex agent societies. Work is continuing on building dynamic agent organizations to support virtual organizations, dealing with methods for teamwork, coalition formation, management, assessment, coordination and dissolution. This exploits aspects of the emerging visions of the web, the Grid, ambient intelligence and ubiquitous computing (all presented later in this chapter) in which agents come together to deliver composite services. The need to properly assign roles, powers, rights and obligations to agents, in order to handle security and trust aspects in an open online environment, is becoming increasingly important.

There is also a lot of interest in using the theory of self-organization to handle dynamism in contemporary open computing environments. A self-organizing system functions through contextual local interactions, without central control, and often results in emergent behaviour that may or may not be desirable. Components aim to achieve individually simple tasks, but a complex collective behaviour emerges from their mutual interactions. Such a system modifies its structure and functionality to adapt to changes in requirements and the environment based on previous experience. In multi-agent systems, emergent phenomena are the global system behaviours that are collective results originating from the local agent interactions and individual agent behaviours. To achieve the desired global emergent system behaviour, local agent behaviours and interactions should comply with some behavioural framework dictated by a suitable theory of emergence. Unfortunately, too few theories of emergence are currently available and existing ones still require improvement (Luck *et al.*, 2005). As a consequence, new theories of emergence are being developed based on inspiration from natural or social systems.

Much work is also being undertaken in the area of trust and reputation, as interacting agents need to reflect the relationships between their human counterparts in real-life organizational settings in order to make appropriate decisions. The field of trust, reputation and social structure seeks to capture human notions such as trust, reputation, dependence, obligations, permissions, norms, institutions and other social structures in electronic form (Ashri *et al.*, 2006; Garcia-Camino, 2006).

15.3.2. Interaction

Technologies and techniques at the interaction level are concerned with communications between agents. These include technologies related to communication languages, interaction protocols and resource allocation mechanisms. Here research draws on work in disciplines such as economics, political science, philosophy and linguistics to develop computational theories and technologies for agent interaction, communication and decision-making. There is a need for mechanisms that allow agents to coordinate their actions automatically without the need for human supervision in many applications. Much work has been done to develop a wide range of different types of coordination and cooperation mechanisms, such as: coordination protocols (which structure interactions to reach decisions); emergent cooperation (which can arise without any explicit communication between agents); coordination media (or distributed data stores that enable asynchronous communication of goals, objectives or other useful data); and distributed planning (which takes into account possible and likely actions of agents in the domain) (Luck *et al.*, 2005).

15.3.3. Agent level

In a multi-agent society, conflicts are likely to arise from agents trying to satisfy their respective goals simultaneously. In such circumstances, agents will need to enter into negotiations with each other to resolve conflicts. Negotiation is a key form of interaction that enables agents to arrive at a mutual agreement regarding beliefs, goals or plans. Although a considerable amount of effort has been devoted to developing negotiation protocols (Smith, 1980; Cammarata *et al.*, 1983; Sycara, 1988; Klein, 1991; Muller, 1996; Beer *et al.*, 1999), resource-allocation methods (Alder *et al.*, 1989; Sathi and Fox, 1989; Conry *et al.*, 1992) and task allocation (Davis and Smith, 1983; Durfee and Montgomery, 1990; Zlotkin and Rosenschein, 1994), a lot of work still needs to be undertaken to create robust computational mechanisms and strategies for handling negotiations.

Technologies and techniques at the agent level are solely concerned with individual agents and include procedures for agent reasoning and learning. This is the main premise of artificial intelligence work, which aims to build systems that can reason and operate autonomously in the world. Most agent research has drawn heavily on this prior work, and the current focus on agent-based systems is mostly on the organization and interaction levels. Much recent work focuses on the representation of computational concepts for the norms, legislation, authorities, enforcement, etc. which can underpin the development and deployment of dynamic electronic institutions (Vázquez-Salceda, 2004) or other open multi-agent system environments. There is also ongoing work on the automation of coalition formation for virtual organizations in complex settings such as in Grid applications. The ability of agents to learn as they perform their tasks is still limited. Although there has been progress in many areas, such as evolutionary approaches and reinforcement learning, these are yet to be used in real-world applications.

In addition to technologies at these three levels, consideration also needs to be given to technologies that provide infrastructure and supporting tools for agent systems, such as agent programming languages and software engineering methodologies. These supporting technologies and techniques provide the basis for both the theoretical understanding and the practical implementation of agent systems.

15.4. The state-of-the-art of application of agent technology in construction

Several attempts have been made to explore the application of agent technology to address construction problems. These efforts have largely focused on design- and management-related areas and a brief review of selected work is presented here.

15.4.1. Agent-based design

A great deal of work in agent-based design has been undertaken to support design generation and performance assessment. Much focus has been on coordinating the activities and information flow between the multiple specialists that characterize the construction industry. Fenves *et al.* (1994) developed an integrated building design environment (IBDE) which consisted of several agents for design generation and critique. The design generators are knowledge-based systems that contribute to the emerging design. The critiques evaluate the design, such as in constructability assessment, and make recommendations for re-design. Chiou and Logcher (1996) developed an agent-based system for design in which agents represent areas of specialist knowledge and are able to perform design and checking tasks.

Heckel *et al.* (1996) developed an agent collaboration environment (ACE) which supports collaboration amongst members of the design team by providing infrastructure for a community of cooperating design agents that assist users. Agents are organized to reflect tasks and workflows between different design teams in a typical organization setting. The agents provide design assistance in checking designs against code compliance and automating routine design tasks, thus enhancing productivity. Ndumu and Tah (1998) demonstrated the use of MAS in collaborative design to explore the interaction and negotiation between four software agents, representing architects, structural engineers, quantity surveyors and building services engineers, in the re-design of a three-storey building in response to a change of use order from a client. Ugwu *et al.* (2000) developed the ADLIB system to explore issues involved in agent-based collaborative design and the extent to which certain design tasks can be automated. The design domain used was light industrial buildings with a focus on portal frames. Agents were developed to reflect the different participants in the design process and coordinated their requirements through negotiations by adopting the monotonic concession protocol.

Maher *et al.* (2003) have proposed a situated agent model to virtual world platforms in which an avatar, a visual representation of a user, has agency. As a rational agent, the avatar can respond to events in the world, either through the human control of the avatar or via the avatar's agent. The agent interacts with the world on the user's behalf and directly in response to the user's activities in the world. Rational agents are used here to reason about the use and design of the virtual world.

A lot of work has also been undertaken on the application of agents to design in related engineering disciplines, and a brief overview of selected efforts is presented here. Cutkosky *et al.* (1993) described the Palo Alto Collaborative Testbed (PACT), which integrated four legacy concurrent engineering systems into a common framework. It involved 31 agent-based systems arranged into a hierarchy around facilitators. The agents cooperated in the design and simulation of a robotics device, reasoning about its behaviour from the standpoint of four engineering disciplines: software, digital, analog electronics and dynamics. Balasubramanian *et al.* (1996) developed a multi-agent architecture for the integration of design, manufacturing and shop floor control activities. In the RAPPID (Responsible Agents for Product-Process Integrated Design) project, agents were used to resolve conflicts among designers (Parunak, 1996). Petrie (1997) developed ProcessLink, aimed at providing a technical infrastructure and methodology for integrating geographically distributed engineers, designers and their heterogeneous tools. Other researchers (for example, Khedro and Genesereth, 1994; Wellman,

1995; Case and Lu, 1996; Shen and Barthes, 1996; Bento and Feijó, 1997; Roseman and Wang, 1999; Shen *et al.*, 2001; Bilek and Harman, 2002)) have also developed various agent-based systems to facilitate collaborative engineering design.

More recently, Lawson (2005) describes the use of existing CAD architectural systems as using the computer as a 'draughtsman', since these systems are incapable of aiding creative design. He envisions a future for the computer as 'agent', which will be web-based, intelligent and capable of learning and being proactive. He acknowledges that elements of this either already exist or have at least undergone feasibility tests on rough development models. The previous agent-based design work presented above provides the evidence to support this view.

15.4.2. Agent-based project and enterprise management

There has been a lot of interest on modelling projects and organizations as complex systems using multi-agent systems' techniques. One of the early demonstrators in this area was the Advanced Decision Environment for Process Tasks (ADEPT), developed by Jennings *et al.* (1996) to demonstrate how multiple autonomous agents could manage the nearly 100 business tasks which make up a typical British Telecommunications plc business process. In the construction domain, Oliveira *et al.* (1997) developed a multi-agent system for resource management in civil construction (MACIV), aimed at exploring the use of MAS to enable decentralized management of the different resources in a construction company. In MACIV, negotiation follows a five-step process as follows: announcing; task evaluation; selection; market manipulation; and price adjustment. The last two steps are repeated until all but one agent have been eliminated from the process, or the system reaches a set timeout. Each agent makes decisions according to its estimate of the cost for the execution of each task, in terms of the travel cost, depreciation cost, operation cost and profits.

Pena-Mora and Wang (1998) developed a CONVINCER model to facilitate the negotiation of conflict resolution in large-scale civil engineering projects. Kim *et al.* (2000) developed a project schedule coordination model based on a compensatory negotiation framework, in which a project can be re-scheduled dynamically by project participants based on their resource profiles. Ren *et al.* (2001) developed MASCOT for construction claim negotiations, where agents are used to negotiate construction claims on behalf of their individual owners. Tah (2005) developed a prototype system to explore the potential for use of multi-agent systems' techniques to model and simulate the preplanning of collaborative procurement in construction supply networks. Udeaja and Tah (2005) developed a prototype multi-agent based system for modelling and simulating the procurement of materials in the construction supply chain. Xue *et al.* (2005) designed an agent-based framework for construction supply chain coordination and developed a prototype multi-agent system using multi-attribute negotiation and multi-attribute utility theory. Schnellenbach-Held *et al.* (2005) developed an agent-based marketplace for AEC (architecture, engineering and construction) -bidding in Germany. The focus was on providing adequate security for the agent-based marketplace to meet the requirements of the (German) Digital Signature Act, and to satisfy the official contracting terms for the award of construction contracts.

Although a good number of agent-based applications have been developed, prototyped and evaluated, no one has yet built a robust and fully functional multi-agent system that has been successfully deployed in the construction industry. However, there have been some successful deployments in other industry sectors, a selection of which is presented in the ensuing section.

15.5. Deployment of agent technology in industry

Currently, agent technologies are in use in a few industry sectors and focus on applications such as (Luck *et al.*, 2005):

- Automated trading in online marketplaces for financial products and commodities
- Simulation and training in defence
- Network management in utilities
- User interface and local interaction management in telecommunication networks
- Scheduling and optimization in logistics and supply chain management
- Control systems' management in industrial plants
- Simulation modelling to guide decision-makers in public policy domains, such as transport and medicine

The notion of agent-based computing has been adopted enthusiastically in the financial trading community, where autonomous market trading agents are said to outperform human commodity traders. One example is the Zero Intelligence Plus (Zip) autonomous adaptive trading agent algorithm developed by Dave Cliff at HP Labs (Cliff, 2003). It works by calculating the best trading strategy for continuous double auctions, the trading basis of most financial markets. Zip traders have the ability to 'learn' from their actions, using simple machine learning rules.

In the manufacturing sector, Daimler–Chrysler implemented an agent-based system on one factory floor to allow individual work pieces to be directed dynamically around the production area. The intention was to implement flexible manufacturing to meet rapidly changing operational targets, and it is claimed that this resulted in a significant increase in productivity (Luck *et al.*, 2005). Magenta Technology (Magenta, 2007), a software company specializing in the commercial use of multi-agent technology, claims to have developed many applications for several clients in scheduling supply chains, semantic search, text understanding and document classification as well as pattern recognition.

Although some of these applications can be implemented as closed systems inside a single company or organization, agent technologies are most suited to domains that involve interactions of multiple organizations. A good domain is that which involves purchasing decisions along a supply chain and which requires the participation of companies along that supply chain. Here, the application of agent technologies requires agreement and coordination of multiple companies. Domains characterized by a multitude of variables, with complex interdependencies, that change dynamically and thus cannot all be known in advance, are also suitable candidates for agent technologies. This requires agent solutions because of their ability to dynamically adapt to changes in the environment and thus to offer real-time optimization. Good examples can be found in network planning and transportation optimization domains, where agent-based solutions have successfully replaced conventional systems because these were limited in their ability to cope with the increasing complexity, and especially with the dynamics of globalized transportation business (Dorer and Calisti, 2005).

15.6. Challenges

The practical application of agent technology in an industry context still faces a number of challenges, which require addressing through further research. Currently, there is a lack of sophisticated software tools, techniques and methodologies to support the specification, development, integration and management of agent systems compared to more mature technologies such as object-oriented programming. This is probably one of the main reasons for the lack of widespread development and use of agent technologies in practice.

The ability of multi-agent systems to adapt to changing environments and to cope with autonomous components means that they often exhibit properties that were not predicted or desired. Striking the right balance between adaptability and predictability is a major challenge that has yet to be resolved both theoretically and practically. The development of tools for the verification of multi-agent systems that are likely to exhibit emergent behaviour is still a major challenge.

In order for users to gain confidence in using applications that draw on multiple software agents, developed by multiple developers and running in open and dynamic environments, techniques are needed for expressing and reasoning about trust and reputation in these environments. There is a lot of ongoing research on trust and reputation with some degree of success, but robust and practically usable results are yet to emerge. Linked to this is the notion of virtual organizations which have been identified as one of the key contributions of Grid computing, but the development of procedures and methods for the automation of virtual organization creation, management and dissolution still represents a major research challenge.

Existing agent coordination mechanisms are inspired by actions in human society, such as the common auction protocols. However, these are unsuitable for problems that require the satisfaction of multiple objectives, which is typical in the construction industry, necessitating the development of new coordination mechanisms which is still proving to be a major challenge. A related major research challenge is the need for effective negotiation strategies and protocols that establish the rules of negotiation, as well as languages for expressing service agreements, and mechanisms for negotiating, enforcing and reasoning about agreements, disagreement and their justifications (Luck *et al.*, 2005).

In practice, people continuously change and adapt to their environments and learn through interactions with others. Although there has been a lot of work on learning and adaptation, this has focused more on single-agent learning contexts. Thus, there is a need for further work on learning in multi-agent contexts.

There is still a lack of mechanisms for explaining and tracing the reasoning behind results and outputs produced by agent-based applications. Mechanisms need to be developed that allow users to trace how particular outputs have been produced in order to inspire confidence and increase uptake in industry.

In multi-agent systems different entities use distinct information models. Although there has been a lot of work on developing techniques to bridge the semantic gaps between multiple systems, many challenges remain. Much work still needs to be undertaken to realize scalable and practically usable ontologies together with robust techniques for schema and semantic mediation between systems using different ontologies. In the construction industry, there is a need for a model of the computer representation of the built environment that can provide the basis of a common language that can be understood by multiple agents. The model itself needs to be defined at a level of information granularity that allows for computational interpretation and inference. The basis of such a model exists in previous and ongoing interoperability initiatives as indicated in the ensuing section, but the realization of a robust ontology in this area is still a major challenge.

15.7. Emerging related technologies

A number of emerging technologies that, when used in combination with agent technologies, present very interesting prospects in realizing sophisticated virtual prototyping systems in the future. Selected significant emerging technologies are presented, including: ontologies; the semantic web, web services and services-oriented computing; and pervasive computing.

15.7.1. Ontology

Previous developments in artificial intelligence (AI) and emerging developments in the semantic web demonstrate how the use of formal knowledge representation, typically in the form of an ontology, leads to machine-interpretable descriptions that can be used for automated reasoning. These also demonstrate how the adoption of an ontology that provides a common vocabulary and shared knowledge leads to improved semantic interoperability. The creation of semantic models offers three key advantages:

(1) It promotes reuse and interoperability among independently created software components
(2) It facilitates the use of ontology-supported representations based on formal and explicit representation leading to improved automation
(3) The explicit modelling of entities and the relationships between them allows deep and insightful analysis to be performed

A virtual model of a construction project can only be used to support deeper decision-making if knowledge about its constituent components, their interrelationship and their behaviour is available to the user. We need a way of describing and representing information and knowledge about construction projects in a manner that allows us to handle its inherent complexity, share a common understanding and develop software systems that interoperate with each other. The problems arising out of the lack of common language formats for exchanging and sharing information in computer systems have been well documented. There have been significant efforts and progress towards developing international standards for sharing information related to construction projects and the built environment, such as those of the International Alliance for Interoperability for building information in IFCs (Industry Foundation Classes) (Froese, 2003), and the Open Geospatial Consortium (OGC) for GIS data (OGC, 2007). However, there are still major challenges in establishing a comprehensive common language for application in practice (Tanyer *et al.*, 2005).

15.7.2. Semantic web

The web has rapidly become a powerful medium for communication, commerce and culture. However, its power is limited by the ability of humans to sieve through the vast amount of information it represents. The semantic web is based on the idea that the data on the web can be represented in such a way that it can be automatically processed by machines and integrated across different applications (Berners-Lee *et al.*, 2001). The development of a common language for the exchange of semantically rich information between software agents is a key aspect of realizing the semantic web. This presents an opportunity for developing a range of agent applications that exploit the semantic web (Benjamins *et al.*, 2003).

15.7.3. Web services and service-oriented computing

Web services and service-oriented architectures (SOA) provide a means of reducing or eliminating impediments to the interoperable integration of applications, regardless of their operating system platform or implementation language. This is an approach to systems' development in which software is encapsulated into components called 'services'. Services interact through the exchange of messages that conform to published interfaces. The interface supported by a service is all that concerns any prospective consumer, as implementation details of the service itself are hidden from all consumers of the service. Thus, distributed systems can be characterized as collections of service provider and service consumer software components that can be loosely coupled to provide specific application functionally. Just as agents may perform tasks on behalf of a user, a web service provides this functionality on behalf of its owner. Many researchers see web services as providing an infrastructure that is almost ideal for use in supporting agent interactions in a multi-agent system (Luck *et al.*, 2005). Thus, the increasing adoption of SOA will likely lead to an increase in the uptake of agent technologies, since provider and consumer web services environments are seen as a form of agent-based system.

15.7.4. Pervasive computing technologies

The emerging notion of pervasive computing has been widely acknowledged as providing the basis for the next generation of computing environments. Pervasive computing involves the creation of physical environments saturated with embedded computing and wireless communication, yet gracefully integrated with human users. We can envision and create a future in which user-friendly ICT (information and communication technologies) becomes an integral part of the built environment, with integrated nano- and micro-processors, sensors and actuators connected via high-speed wireless networks and accessible through a variety of new ergonomically appropriate multi-modal interfaces and visualization devices. Thus, we can develop agents in virtual environments that exhibit behaviours in response to stimuli received from devices embedded in their real-world equivalents in real-time.

15.7.5. Ambient intelligence

The notion of ambient intelligence has largely arisen through the efforts of the European Commission in identifying challenges for European research in the Information Society Technology programme. Aimed at a seamless delivery of services and applications, it relies on the areas of pervasive or ubiquitous computing, ubiquitous communication and intelligent user interfaces. The vision describes an environment of potentially thousands of embedded and mobile devices (or software components) interacting to support user-centred goals and activity, and suggests a component-oriented view of the world in which the components are independent and distributed. The consensus is that autonomy, distribution, adaptation, responsiveness and so on, are key characteristics of these components, and in this sense they share the same characteristics as agents.

Ambient intelligence requires these agents to be able to interact with numerous other agents in the environment around them in order to achieve their goals. Such interactions take place between pairs of agents (in one-to-one collaboration or competition), between groups (in reaching consensus decisions or acting as a team) and between agents and the infrastructure resources that comprise their environments (such as large scale information repositories). Interactions like these enable the establishment of virtual organizations, in which groups of agents come together to form coherent groups able to achieve overarching objectives. The environment provides the infrastructure that enables ambient intelligence scenarios to be realized.

15.7.6. Intelligent virtual agents

Major advances have been made over the last couple of decades in computer graphic modelling for image synthesis, animation and virtual reality, revolutionizing the motion picture, interactive game and multimedia industries. First-generation purely geometric models have advanced to more elaborate physics-based models for animating particles, rigid bodies, deformable solids, fluids and gases. We can now simulate and animate a variety of real-world, physical objects with stunning realism. In recent years, research in behavioural modelling has progressed towards self-animating characters that react appropriately to perceived environmental stimuli. Cognitive modelling using AI techniques, including knowledge-representation, reasoning and planning, is applied to produce characters with some level of intelligence that mimic real-world phenomena. These characters have been termed 'intelligent virtual agents' (IVAs). They are autonomous, graphically embodied agents in an interactive, 2D or 3D virtual environment. They are able to interact intelligently with the environment, other IVAs and especially with human users. This has generated a lot of research interest on the use of AI techniques in virtual environments to create the so-called 'intelligent virtual environments' (Luck and Aylett, 2000). Although there are commercial game development environments that claim to be AI-based, they do not really implement true agent techniques due to the challenges presented above, and therefore further research is needed.

15.8. Discussion and conclusions

Although existing demonstrations of the application of multi-agent systems in the construction domain are still largely research prototypes and scalable only under controlled and simulated conditions, this is likely to change in the future given the increasing rate of innovation and the use of emerging related technologies discussed above. The construction industry should be able to draw on the experiences of other industry sectors, where agents have been successfully used for automated trading, logistic scheduling and supply chain coordination, to develop similar applications in the near future for use in closed corporate environments at the very least. The lack of standard methodologies, tools and languages has hampered the development and demonstration of a wide range of practical demonstrators in practice, but this is expected to change in the near future as work is being undertaken to establish industry-strength development techniques and development platforms. The development of open multi-agent systems is still a long way off, as obtaining agreement amongst multiple stakeholders on the objectives of an open system will be difficult, including, of course, security concerns.

The emergence of networked environments such as those of the Grid will facilitate the development of agents to support cross-organizational collaborative working and business processes in the near future. It is anticipated that development methodologies, languages and tools will be much more mature and systems will be designed on top of standard web services and services-oriented infrastructures, and the Grid. Such systems will be developed to enable automated scheduling and coordination between different departments within a company, or a closed group of customers and suppliers engaged in electronic procurement in a supply network. In their paper on 'Construction IT and the "tipping point"', Brandon *et al.* (2005) observe that the use of virtual prototyping and intelligent knowledge Grids has the potential to transform the construction industry beyond recognition in the future.

In the future, agents will be designed by different designers to collaborate in multi-agent systems by adopting published requirements and open standards. This will be supported by emerging, commonly agreed, modelling languages and much more mature ontologies and integrated development environments which will facilitate the development process. Such open systems will have to deal with issues such as robustness against malicious agents. Examples of such systems may include corporate business-to-business electronic procurement systems permitting participation by any supplier rather than a closed business network, using heterogeneous agents.

In the long-term future, multi-agent systems will be developed by diverse teams to operate in open computing environments that involve diverse participants and address multiple application domains that cut across multi-organization boundaries. In such environments, agents will have the ability to learn the rules and regulations for participation through interaction with other agents, rather than being pre-encoded as these cannot be determined in advance. These agents will have the ability to self-select appropriate communication protocols, and mechanisms and strategies for participation will be undertaken automatically with little human intervention. Such agents will be able to establish dynamic *ad hoc* coalitions and virtual organizations that will be managed and dissolved automatically when collaborating to pursue shared goals, as is typical in real-life construction projects.

In conclusion, the combination of multi-agent systems techniques with the emerging technologies presented in this chapter holds a promising future. We will be able to produce true intelligent virtual prototyping software environments capable of mimicking real-world phenomena and behaviours. This will allow us to model and simulate complex construction projects in the future in ways that are not currently possible.

References

Alder, M. R., Davies, A.B., Weihmayer, R. and Worrest, R. W. (1989) Conflict resolution strategies for non-hierarchical distributed agents. In: *Distributed Artificial Intelligence*, Vol. II (ed. L. Gasser and M. Huhns). London, Pitman Publishing, pp. 139–161.

Ashri, R., Payne, T. R., Luck, M., Surridge, M., Sierra, C., Aguilar, J. A. R. and Noriega, P. (2006) Using electronic institutions to secure grid environments. *Proceedings of the Tenth International Workshop CIA 2006 on Cooperative Information Agents*, 11–13 September, University of Edinburgh, Scotland, pp. 461–475.

Balasubramanian, S., Maturana, F. and Norrie, D. H. (1996) Multi-agent planning and coordination for concurrent engineering functionality. *International Journal of Cooperative Information Systems* [Special Issue: *Agent Based Information Management*], 5 (2–3): 153–179.

Beer, M., d'Inverno, M., Jennings, N. R., Luck, M., Preist, C. and Schroeder, M. (1999) Negotiation in multi-agent systems. *Knowledge Engineering Review* 14 (3): 285–289.

Benjamins, V. R., Contreras, J. and Prieto, J. A. (2003) Agents and the Semantic Web, *AgentLink Newsletter* Issue 13, pp. 10–11. An AgentLink Publication, available at http://www.agentlink.org/newsletter/ [Accessed October 2007].

Bento, J. and Feijó, B. (1997) A post-object paradigm for building intelligent CAD systems. *Artificial Intelligence in Engineering* 11 (3): 231–244.

Berners-Lee, T., Hendler, J. and Lassila, O. (2001) The semantic web. *Scientific American* May. Available at http://www.sciam.com [Accessed 25th April 2007].

Bilek, J. and Harman, D. (2002) Collaborative structural engineering based on multi-agent systems. *Proceedings of the 9th International EG-ICE Workshop on Advances in Intelligent Computing in Engineering*, 1–2 August, Darmstadt, Germany, pp. 49–60.

Brandon, P., Li, H. and Shen, Q. (2005) Construction IT and the 'tipping point'. *Automation in Construction* 14 (3): 281–286.

Cammarata, S., McArthur, D. and Streeb, R. (1983) Strategies of distributed problem solving. *Proceedings of the Eighth International Joint Conference on Artificial Intelligence*, 8–12 August, Karlsruhe, Germany, pp. 767–770.

Case, M. P. and Lu, Y. S. (1996) Discourse model for collaborative design. *Computer-Aided Design* [Special Issue: *Computer-Aided Concurrent Design*] 28 (5): 333–345.

Chiou, J. D. and Logcher, R. D (1996) *Testing a Federation Architecture in Collaborative Design Process, Final Report*. Champaign, IL, CERL (Construction Engineering Research Laboratory of the US Army Corps of Engineers), Report no. R96-01.

Cliff, D. (2003) Explorations in evolutionary design of online auction market mechanisms. *Electronic Commerce Research and Applications* 2 (2): 162–175.

Conry, S. E., Kuwabara, K., Lester, V. and Meyer, R. (1992) Multistage negotiation in distributed constraint satisfaction. *IEEE Transactions on Systems, Man and Cybernetics* 21 (6): 1462–1477.

Cutkosky, M. R., Englemore, R. S., Fikes, R. E., Genesereth, M. R., Gruber, T. R., Mark, W. S., Tenenbaum, J. M. and Weber, J. C.: PACT (1993) An experiment in integrating concurrent engineering systems. *IEEE Computer* **26** (1): 28–37.

Davis, R. and Smith, R. G. (1983) Negotiation as a metaphor for distributed problem solving. *Artificial Intelligence* **20**: 63–109.

Dorer, K. and Calisti, M. (2005) An adaptive solution to dynamic transport optimization. In: *Proceedings of the Fourth International Joint Conference on Autonomous Agents and Multi-Agent Systems: Industry Track* (ed. M. Pechoucek, D. Steiner and S. Thompson), 25–29 July, Utrecht University, The Netherlands, New York, ACM Press, pp. 45–51.

Durfee, E. H. and Montgomery, T. A. (1990) A hierarchical protocol for coordinating multi-agent behaviours. *Proceedings of the 8th National Conference on Artificial Intelligence (AAAI-90)*, 29 July–3 August, Boston, MA, pp. 86–93. [Association for the Advancement of Artificial Intelligence]

Fenves, S., Flemming, U., Hendrickson, C., Maher, M. L., Quadrel, R., Terk, M. and Woodbury, R. (1994) *Concurrent Computer-Integrated Building Design*. New Jersey, Prentice Hall PTR.

Froese, T. (2003) Future directions for IFC-based interoperability. *ITcon* [Special Issue: *IFC-Product Models for the AEC Arena*] **8**: 231–246.

Garcia-Camino, A., Rodriguez-Aguillar J., Sierra, C. and Vasconcelos, W. (2006) A rule-based approach to norm-oriented programming of electronic institutions. *ACM SIGecom Exchanges* **5** (5): 33–40.

Holland, J. H. (1995) *Hidden Order: How Adaptation Builds Complexity*. Reading, MA, Addison-Wesley.

Jennings, N. R. (1999) Agent-oriented software engineering. *Proceedings of the 9th European Workshop on Modelling Autonomous Agents in a Multi-Agent World: Multi-Agent System Engineering (MAAMAW-99)* (ed. F. J. Garijo and M. Boman), 30 June–2 July, Valencia, Spain. London, Springer-Verlag, pp. 1–7.

Jennings, N. R. (2001) An agent-based approach for building complex software systems. *Communications of the ACM* **44** (4): 35–41.

Jennings, N. R., Faratin, P., Norman, T. J., O'Brien, P., Wiegand, M. E., Voudouris, C., Alty, J. L., Miah, T. and Mamdani, E. H. (1996) ADEPT: Managing business processes using intelligent agents. *Proceeding of the BCS Expert Systems '96 Conference*, 16–18 December, Cambridge, UK, pp. 5–23.

Khedro, T. and Genesereth, M. R. (1994) Concurrent engineering through interoperable software agent. *Proceeding of the First International Conference on Concurrent Engineering, Research and Applications*, 29–31 August, Pittsburgh, PA.

Kim, K., Paulson, B. C., Petrie, C. J. and Lesser, V. R. (2000) *Compensatory Negotiation for Agent-Based Project Schedule Coordination*. Stanford University, CIFE Working Paper #55.

Klein, M. (1991) Supporting conflict resolution in cooperative design systems. *IEEE Transactions on Systems, Man and Cybernetics* [Special Issue: *Distributed Artificial Intelligence*] **21** (6): 1379–1390.

Lawson, B. (2005) Oracles, draughtsmen and agents: The nature of knowledge and creativity in design and the role of IT. *Automation in Construction* **14** (3): 383–391.

Lee, A., Wu, S., Marshall-Ponting, A., Aouad, G., Tah, J. H. M. and Cooper, R. (2006) nD modelling – a driver or enabler for construction improvement? *RICS Research Paper Series* **5** (6), September: 45.

Luck, M. and Aylett, R. S. (2000) Applying artificial intelligence to virtual reality: Intelligent virtual environments. *Applied Artificial Intelligence* **14** (1): 3–32.

Luck, M., McBurney, P., Shehory, O. and Willmott, S. (2005) *Agent Technology: Computing as Interaction. A Roadmap for Agent Based Computing*. University of Southampton on behalf of AgentLink III.

Magenta (2007) Magenta Technology. Available at http://www.magenta-technology.com/index.php [Accessed May 2007).

Maher, M. L., Smith, G. and Gero, J. S. (2003) Design agents in 3D virtual worlds. *International Joint Conference on Artificial Intelligence*: IJCAI03 Workshop on Cognitive Modeling of Agents and Multi-Agent Interaction (ed. R. Sun) ,. 9–11 August, Acapulco, Mexico, pp. 92–100.

McGraw, K. D., Lawrence, P. W., Morton, J. D. and Heckel, J. (1996) The agent collaboration environment: an assistant for architects and engineers. *Proceedings of the Third Congress on Computing in Civil Engineering*, 17–19 June, Anaheim, CA. New York, ASCE Press, pp. 739–745. Muller, H. J. (1996) Negotiation principles. In: *Foundations of Distributed Artificial Intelligence* (ed. G. M. P. O'Hare and N. R. Jennings). New York, Wiley, pp. 139–164.

Ndumu, D. T. and Tah, J. H. M. (1998) Agents in computer-assisted collaborative design. In: *AI in Structural Engineering* (ed. I. Smith). *Lecture Notes in Artificial Intelligence No. 1454*. Berlin, Springer, pp. 249–270.

Nwana, H. S. and Ndumu, D. T. (1999) A perspective on software agents research. *The Knowledge Engineering Review* **14** (2): 125–142.

OGC (2007) *The Open Geospatial Consortium, Inc.* Available at http://www.opengeospatial.org/ [Accessed October 2007). Oliveira, E. D., Fonseca, J. M. and Steiger-Garcao, A. (1997) MACIV: a DAI based resource management system. *Journal of Applied Artificial Intelligence* **11** (6): 525–550.

Parunak, H. V. D. (1996) Application of distributed artificial intelligence in industry. In: *Foundations of Distributed Artificial Intelligence* (ed. G. M. P. O'Hare and N. R. Jennings). New York, Wiley, pp. 139–164.

Pena-Mora, F. and Wang, C. (1998) Computer-supported collaborative negotiation methodology. *Journal of Computing in Civil Engineering* **12** (2): 64–81.

Petrie, C. (1997) *Process Link Coordination of Distributed Engineering, Technical Report*. Center for Design Research, Stanford University, CA.

Ren, Z., Anumba, C. J. and Ugwu, O. O. (2001) Development of a multi-agent system for construction claims negotiation. *Journal of Advances in Engineering Software* **16** (5): 359–394.

Roseman, M. and Wang, F. (1999) A component agent-based design-oriented model for collaborative design. *Research in Engineering Design* **11**: 193–205.

Sathi, A. and Fox, M. (1989) Constraints directed negation of resource reallocations. In: *Distributed Artificial Intelligence*, Volume II (ed. N. H. Michael and L. Gasser). Los Altos, CA, Morgan Kaufmann, pp. 163–193.

Schnellenbach-Held, M., Denk, H. and Albert, A. (2005) Agent-based virtual marketplace for AEC-bidding. In: *Agents & Multi-Agent Systems in Construction* (ed. C. J. Anumba, O. O. Ugwu and Z. Ren). London, Taylor and Francis, pp. 209–232.

Shen, W. and Barthes, J. P. (1996) An experimental multi-agent environment for engineering design. *International Journal of Cooperative Information Systems* **5** (2–3): 131–151.

Shen, W., Norrie, H. D. and Barthes, J. (2001) *Multi-Agent Systems for Concurrent Intelligent Design and Manufacturing*. London, Taylor and Francis.

Smith, R. G. (1980) The contract net protocol: high-level communication and control in a distributed problem solver. *IEEE Transactions on Computing* **29** (12): 1104–1113.

Sycara, K. P. (1988) Resolving goal conflicts via negotiation. *Proceedings of the Seventh National Conference on Artificial Intelligence (AAAI-88)*, 21–26 August, St Paul, MN.

Tah, J. H. M. (2005) Towards an agent-based construction supply network modelling and simulation platform. *Automation in Construction* **14** (3): 353–359.

Tanyer, A. M., Tah, J. H. M. and Aouad, G. (2005) An integrated database to support collaborative urban planning: The n-dimensional modelling approach. *Proceedings of the ASCE International Conference on Computing in Civil Engineering*, 12–15 July 2005, Cancun, Mexico.

Udeaja, C. and Tah, J. H. M. (2005) Multi-agent systems based procurement in the construction material supply chain. In: *Agents & Multi-agent Systems in Construction* (ed. C. J. Anumba, O. O. Ugwu and Z. Ren). London, Taylor and Francis, pp. 272–309.

Ugwu, O. O., Anumba, C. J., Newnham, L. and Thorpe, A. (2000). Agent-oriented collaborative design of industrial buildings. *Proceedings of the 8th International Conference of Computing in Civil & Building Engineering, ICCCBE-VIII* (ed. R. Frucher, F. Pena-Mora and W. M. K. Roddis), 14–17 August, Stanford University, CA, pp. 333–340.

Vázquez-Salceda, J. (2004) *Role of Norms and Electronic Institutions in Multi-Agent Systems: The HARMONIA Framework*. Berlin, Birkhäuser Varlag.

Wellman, M. P. (1995) A computational market model for distributed configuration design. *Artificial Intelligence for Engineering Design, Analysis and Manufacturing* **9**: 125–133.

Wooldridge, M. and Jennings, N. R. (1995) Intelligent agents: Theory and practice. *The Knowledge Engineering Review* **10** (2): 115–152.

Xue, X., Li, X., Shen, Q. and Wang, Y. (2005) An agent-based framework for supply chain coordination in construction. *Automation in Construction* **14** (3): 413–430.

Zlotkin, G. and Rosenschein, J. (1994) Cooperation and conflict resolution via negotiation among autonomous agents in non-cooperative domains. *IEEE Transactions on Systems, Man and Cybernetics* [Special Issue: *Distributed Artificial Intelligence*] **21** (6): 1317–1324.

16 The nature of virtuality and the need for enhanced security in the virtual world

Grahame S. Cooper

16.1. Introduction

Throughout history, and in many areas of endeavour, humans have sought to free themselves from the constraints of their physical existence and capabilities. Physical examples include the creation of powered vehicles and flying machines, which reduced the dependence on geographical constraints. Much of modern architecture attempts to free the spatial and aesthetic design from constraints imposed by structural requirements, a principle implicitly espoused by Le Corbusier's *Five Points of Architecture* published in 1926 (Korzilius, 1999).

Often, the desired separation is brought about by the creation of some form of abstraction layer that acts as a buffer to decouple the physical world from the particular human activity. A particularly ubiquitous example of this is the concept of money, where some form of currency acts as a proxy for goods and services that are to be traded. This approach frees humans from the constraints that would exist in a system of direct bartering, eliminating the need for a would-be trader to find a trading partner with complementary needs. In this context, money, supported by the associated processes of trading, accounting and auditing, may be thought of as 'virtual goods and services'. It is interesting that, whilst the use of physical tokens (cash) to represent currency has created an illusion of reality for money, most money nowadays is stored in the form of bits of data within financial systems, which automate much of the trading and accounting, again emphasizing its virtual nature.

In each case, one of the key consequences has been to enable human activity to be carried out in a less rigidly planned, more flexible manner, with greater scope for *ad-hoc* decisions to be made in the light of previously unknown or unanticipated changes in the environment or needs. Greater freedom is made possible by allowing us to change our minds in more and more arbitrary ways. However, one potential cost of this freedom is a reduction in the reliability and stability of the environment, on which much of our decision-making is traditionally based.

This chapter explores the way in which virtualization, developed and applied in many areas of information and communication technology, is leading humankind towards a world that will be characterized by high levels of flexibility and agility. It goes on to raise some concerns regarding the implications of such a change in relation to security, which, due to our need to apply only limited forms of rationality, has traditionally relied on familiarity, predictability and stability of the environment in which we exist.

16.2. Recent virtualization trends in IT – The nature of virtuality

Nowhere is the desire to relieve constraints through 'separation' more apparent than in the development of computer and communication technologies. In turn, these technologies have allowed us to achieve separation in ways that were previously impossible. For decades, the mantra of the software engineering discipline has been 'separating the concerns' (see, for example, Sommerville, 2004), a concept that encourages software architectures to be developed to support, at least in principle, the reuse and adaptation of software artefacts. This has been the driving force behind the whole development of languages for describing software, leading to what we call 'object-oriented programming' and more recently 'aspect-oriented programming' (Elrad *et al.*, 2006), which attempts to disentangle orthogonal areas of concern completely and allow them to be pursued independently of one another.

Separation is emerging in many areas of technology as a key imperative, which is usually achieved through some form of 'virtualization'. In this context, virtualization is the act of making some aspect of the world appear to be different than it really is, and is supported strongly by the related concept of 'abstraction' – hiding the details of how something is implemented and characterizing it purely by its outward properties and behaviour. In some cases, this may extend to the complete fabrication of a 'world' with highly complex relationships, processes and parallel activities.

It is useful to look briefly at some examples of how virtualization in IT is helping to create greater freedom for organizations, removing constraints that previously existed because of the physical design of the technological systems.

16.2.1. Virtual LANs (Howald, 2003; Sarbagyshov and Munshey, 2003)

A traditional, ethernet-type local area network is typically divided into subnets for ease of management of the network and the traffic it carries. A subnet consists of a number of computers linked together by a single part of the network (the subnet), which is linked to the rest of the network by a router. Typically, a subnet will be created for a particular department or workgroup, so it is likely that the machines on a particular subnet will have similar requirements, will make use of similar services, be appropriate recipients of broadcast messages within the subnet and that much of their communication will be contained within the one subnet. In this case, they can often be managed collectively and be provided locally with servers that are accessed primarily from within the same subnet. However, these advantages require that the people working in the department or workgroup are geographically co-located, which may not always be the best option for the organization. The Virtual LAN (or VLAN) is an abstraction layer that has been introduced, over the last 10 years or so, to separate the concept of a subnet from the physical structure of the network. A VLAN provides a logical grouping of computers that are managed together, have closely related IP addresses and appear to be on the same subnet despite their being located in physically different parts of the network. This allows the logical structure of the network to be altered in any way required when staff move, when buildings change their purpose and even when an organization is restructured, thereby freeing the organization from physical constraints imposed by networking hardware.

16.2.2. Storage virtualization

In corporate servers, a typical issue facing system administrators has been the choice of how best to partition the hard discs to utilize disc storage in the most efficient manner, whilst still leaving scope to accommodate growth and anticipating changing requirements in the future. In cases where multiple data storage devices are employed, the use of the storage is constrained even further as decisions have to be made regarding the placing of particular parts of the data on particular physical devices, which may then be difficult to change in the future. Increasing storage capacity by the addition of new devices may require significant restructuring of the data. In recent years, approaches have been developed to relieve organizations from the physical constraints of their data storage devices by the introduction of an abstraction layer allowing the creation of 'logical volumes'. These appear to the user and to the software applications to be individual storage devices but may, in fact, be distributed over a number of separate 'physical volumes' (the actual installed discs and other devices on which the data is stored). A very good example of this approach is provided by the Linux Logical Volume Manager, which is rapidly being adopted as a standard feature by a number of Linux distributions (Teigland and Mauelshagen, 2001). In this way, new storage may be added or the use of existing storage reconfigured even whilst the system is running, allowing an organization quite quickly to make significant changes to the way it operates without the high cost of restructuring critical systems. Storage area network technologies are able to take this concept to an even higher level, in which a logical volume may be distributed across several devices connected across a communications network (Brinkmann *et al.*, 2003).

16.2.3. Virtual machines

The concept of a 'virtual machine' has been around for a considerable time in the mainframe computer world, whereby several users could make use of a single computer as though each were the sole user of the machine. However, developments in machine virtualization in the PC end of the market are now advancing rapidly, resulting in what has been termed a 'virtualization renaissance' (Figueiredo *et al.*, 2005). Machine virtualization systems such as VMWare (http://www.vmware.com/) and Xen (Barham *et al.*, 2003 and the University of Cambridge Computer Laboratory (the Xen™ virtual machine monitor; http://www.cl.cam.ac.uk/Research/SRG/netos/xen/)) allow a single physical computer to host a number of virtual computers, each possibly running a different operating system, in such a way that they are indistinguishable on the network from separate physical machines. The latest microprocessors from Intel and AMD provide hardware support for machine virtualization, further increasing both the performance and flexibility that may be achieved (Uhlig *et al.*, 2005). It is even possible to move a virtual machine, complete with its current state, from one physical machine to another across a network without stopping that virtual machine or interrupting the service it provides (Clark *et al.*, 2005; Nelson *et al.*, 2005). The virtual server approach to computing makes it possible to set up and take down networked services very quickly, even for very short periods, with only a fraction of the effort and cost that would be incurred if a physical computer had to be procured, configured and integrated into the infrastructure. Technologies such as LivePC (see http://www.moka5.com/) are making the creation and management of virtual machines extremely easy, even for people who are not experts in managing computers. With this technology, virtual machines for specific purposes may be made available to users, allowing IT providers to manage large-scale IT resources with the effectiveness and control that has previously been associated with thin-client models, whilst still providing the power and freedom enjoyed by users under a thick-client model (Kozuch *et al.*, 2004).

16.2.4. Grid computing

The Grid (Foster *et al.*, 2001) promises to revolutionize access to computing resources in much the same way that the world-wide web has completely changed people's access to information. Grid technology and infrastructures allow computing resources to be pooled and accessed independently of physical location, allowing virtual organizations to be created with shared computing resources. Grid technologies make use of a number of abstraction layers (Amin *et al.*, 2004) to separate the use of computing resources from the details of where those resources reside or how they are accessed. Grid infrastructure makes use of service-oriented architectures (Foster, 2006–07-28), including, most recently, web services (Heinis *et al.*, 2005), to provide access to abstract resources in standard ways. This allows computing resources to be allocated to tasks as needed, pooling disparate resources when necessary and allowing optimal usage to be maintained. Services (including, nowadays, web services) provide abstractions of functionality that may be used without knowledge of where or how they are provided. It is easy to see that Grid computing provides an extremely powerful platform for virtual machine technologies to operate on (Figueiredo *et al.*, 2003), and one can envisage a whole infrastructure in which virtual machines are able to move around freely across the Grid as requirements and resource constraints dictate.

16.2.5. Virtual reality and virtual environments

These attempt to create apparent environments for humans to occupy, which do not exist in reality. This is done by creating appropriate sensory inputs to reflect the inputs that would have been available (or a subset of them) if the human occupied an equivalent real environment. In this way, humans may be made to experience situations that are not easily reproduced in the physical world for purposes such as training, communication, fun or design activities.

16.2.6. Virtual prototyping

Many definitions of virtual prototyping have been proposed (Wang, 2002). In the most general terms, virtual prototyping combines computer modelling of phenomena, such as mechanical systems, with visualization and other technologies supporting a proximal interaction with users, to allow design options to be explored without the costs and constraints of having to build, implement and transport physical prototypes. Since computer models may be altered and reconfigured relatively easily, this approach allows many more options to be explored than would be possible with physical models.

It also serves to improve the potential for design collaboration, not only between co-located workers (Smith, 2001) but also at a distance (Kim *et al.*, 2001), thereby reducing the geographical constraints to design collaboration. The ease of modification of virtual prototypes introduces the possibility to make changes to designs and investigate the implications much closer to, or even during, the time of manufacture, again allowing for greater flexibility in planning and allowing more agile, *ad-hoc* approaches to be adopted.

16.3. From virtual prototypes to virtual products

Currently, the majority of computer modelling is performed at the design stage, with real, physical artefacts being created on the basis of simulations often carried out in virtual prototypes. However, the existence of reliable models, along with cheap, powerful computing resources, brings the possibility of creating efficiencies and flexibility in the products themselves. If it is possible to use virtual prototypes based on computer modelling to optimize the design of a product to operate effectively across a range of operating conditions, then why not build the computer model into the product itself and allow the properties of the product to adapt to the conditions prevailing at any particular time. This provides the possibility to optimize the performance of the product across all operating conditions.

This approach has also been evident for many years in traditionally high-tech areas such as military aircraft. Traditionally, aircraft designs achieved stability by creating an inherently stable geometry, for example making use of the dihedral wing formation to achieve natural roll stability. However, it is well known (see for example Birkenstock's *Unstable Aircraft Design: The Computer at the Controls*, 1999) that unstable designs provide greater manoeuvrability and agility, which is particularly important for military aircraft. In order to allow a pilot to control an unstable aircraft, fly-by-wire systems involving active control provide a layer of abstraction that effectively turns the highly unstable aircraft of the physical world into a virtual, stable aircraft having high manoeuvrability with which the pilot may interact more effectively. Such technological approaches are also becoming more evident even in domestic motor vehicles, where quite sophisticated engine management systems, braking systems, steering systems and traction control systems are placed between the driver and the physical characteristics of the vehicle.

As well as real-time control applications, various items of software may now be provided with a product to extend its functionality, ease its management and provide ancillary services (such as maintenance). Such products have been referred to as 'extended products' (Jansson *et al.*, 2003), and are likely to appear more frequently in the context of construction, particularly as the emphasis shifts towards whole-life cycle approaches to construction, resulting in a need to integrate products with design software and particularly software supporting facilities management.

16.4. When the virtual world spills over

It is clear from the previous sections that the technologies are coming together to create a world in which much of our time will be spent interacting with the physical world through layers of 'virtuality' in order to carry out our activities. However, one of the earliest practical manifestations of 'virtuality' in common use is in the gaming area, where significant numbers of people spend some of their time occupying the virtual world of one of the 'massively multi-player online role-playing games' (MMORPGs). In these communities, players compete for resources and climb a virtual social ladder, either by fair means or foul. One organization takes this to the extreme, by creating an environment in which people can almost live a second life (http://secondlife.com/). For many people, however, the separation between the virtual and real worlds is becoming ever more difficult to distinguish. A search on the ebay online trading service (http://www.ebay.co.uk/) for 'world of warcraft' will reveal a number of items advertising the sale of 'gold' within this very popular online game. However, whilst the 'gold' in question is purely virtual – just bits within the database of the game – the purchase price for the 'gold' is in real US dollars. There are many such sellers on sites like ebay, selling virtual artefacts in a variety of online games, who are presumably making a reasonable living from their businesses.

Clearly here, the virtual world is spilling over into the real world in quite dramatic ways. The virtual artefacts in the game are, in fact, no less real than any other goods or services that have the potential to improve someone's quality of life (or at least their perception of their quality of life), and are therefore just as likely to be traded in exchange for currency. Indeed, as we saw in the first section of this chapter, the currency itself is probably no more real than the gold in the game mentioned in the previous section. One online game provider, 'Project Entropia', has taken this further by providing a card (*BBC News*, 2006) that allows game players to convert virtual cash into real cash (and vice versa), and to buy virtual property in the game in exchange for real cash. Furthermore, they have defined a guaranteed fixed exchange rate with the US dollar, in which 10 PED (Project Entropia dollars) = 1 US dollar (see Project Entropia website: http://www.entropiauniverse.com/en/rich/5357.html). A very interesting analysis of the trade in game-based virtual goods is available in a Masters thesis presented at the University of Helsinki (Lehdonvirta, 2005).

16.5. The increased need for security

Whatever we think of the moral and ethical issues surrounding the phenomenon of people trading virtual goods in a game for real currency, it is beginning to reveal some of the dangers that are inherent to virtual worlds in general. The real value attached to virtual property has not failed to catch the attention of criminals. Examples appear with increasing frequency of crimes occurring in the virtual world that have an impact in the real world (Yan and Randell, 2005). Fraudulent techniques that are already known in the 'real' online world, such as 'phishing' and identity theft, are happening in the virtual world (Terdiman, 2005), with no less severe consequences. In the role-playing game arena, it is easy to think of these incidents as just cheating in a game. However, when virtual assets in the game are being traded for significant amounts of real-world currency, the picture becomes somewhat more serious. It is becoming apparent that law-makers will need to give careful consideration to the status of crime in the virtual world as it can have social consequences as severe as crime in the real world (Lastowka and Hunter, 2006).

It has been proposed (Laird and van Lent, 2000) that interactive games provide the most productive arena for artificial intelligence (AI) research. Indeed, many of the games' manufacturers themselves provide some quite convincing AI-based robotic players within their games, and this is an area where considerable investment is likely to be placed. However, there are increasing reports of third-party 'bots' – artificial intelligence robots – playing online games, ranging from poker to sophisticated role-playing games, in order to make financial gain. Gains may be realized either by selling a service to game players to carry out 'power levelling'[1], or by building up virtual assets within the game, which may then be sold for real currency. Whilst methods exist or are under development to counteract these phenomena (Golle and Ducheneaut, 2005), it will always be a constant battle as the criminals develop methods to avoid the counter-measures that are put in place.

An even more sinister trend that has been reported (Barboza, 2005) has been the exploitation of vulnerable people in developing countries by employing them as game 'farmers' (sometimes referred to as 'gold farmers') to play games for long periods for payment. Whilst this does put money into the pockets of the 'farmers', it is not clear that it is really helping to develop the economy in the long term.

Many of the techniques that are used to compromise security in the 'real world'[2] involve some form of social engineering[3] (Rusch, 1999). Essentially, this involves the use of deception to cause a person to carry out an action that they would not normally consider. These range from the very crude approaches, such as the email that invites the user to 'click here to see naked pictures of Claudia Schiffer', to quite sophisticated approaches that provide a range of visual and possibly audible cues to mislead the user into providing confidential information. Phishing is a technique that has become quite prominent, which involves the creation of a replica site that closely resembles an online bank or financial service, and misleads people into entering usernames and passwords, which may then be used to gain access to the user's bank and/or credit card accounts. Despite the fact that the risks of phishing attacks have been fairly prominent in the news during recent years, there are still significant numbers of people who fall victim to such attacks (Dhamija *et al.*, 2006). The research quoted has shown that there is a significant risk of deception even in cases where the user is aware of the risk and is expecting to be subjected to an attack.

A separate, but related risk involves the fraudulent use of a person's identity for nefarious purposes, often referred to as 'identity theft' (Berghel, 2000). This has been particularly prevalent in the area of credit card fraud outside of the internet. However, it is becoming a common problem on the internet due to the 'virtual' nature of identity, even in the current internet arena, and is certainly exacerbated greatly by the existence of the internet (Wang *et al.*, 2006).

---

[1] Power levelling services involve playing a game on behalf of a player in order to build up that player's status within the game. This may involve either 'bots' playing games automatically or teams of low-paid workers in developing countries playing games as full-time employment.

[2] 'Real world' is in quotes here because it is arguable that the internet, in which the kinds of incident to be described here occur, could be referred to as a virtual world itself.

[3] Note that the term 'social engineering', traditionally associated with macro-level manipulation of social norms and behaviour, has been adopted and adapted within the Information Systems Security community to mean the application of deception to cause an individual to take some risky action that they would, or should, normally avoid.

16.6. Identity and trust in a virtual world

Identity management is a topic that is drawing much attention, since it is intimately linked to access control and trust (Jøsang *et al.*, 2005). As virtuality extends into many areas of life, the ways in which humans have instinctively managed the verification of identity and the management of trust are increasingly becoming challenged.

Identification in electronic environments is generally managed through some form of digital credentials. Whilst many of the technical problems associated with digital credentials are solved (see, for example Brands, 2002, for an in-depth technical overview), there remains a large gap between the technologies that may be applied and the understanding and attitudes of the people who need to use those technologies (Dourish *et al.*, 2004).

Much work in the field of social science makes a distinction between trust – which may be seen as a substitute for rationality in certain forms of decision-making – from the consequences of trust, such as risk-taking and relationship building (Möllering, 2001). In any case, decisions involving trust generally involve the application of some form of bounded rationality (Simon, 1965) in order that decision-making is tractable in everyday life within the constraints of human cognitive processing capability. It has been proposed that this involves the application of heuristics (Newell, 2005) to make decisions based on familiar cues. A problem with carrying out transactions in a virtual world is that many of the familiar cues may be absent, or different from those cues the inhabitant expects to see. The inhabitants may not have the knowledge, experience or capability to make a rational assessment of the risks involved in a transaction, even if sufficient time and cognitive processing power were available.

This issue has been amply illustrated by the success of social engineering (referred to earlier) as a means to compromise information system security (Dhamija *et al.*, 2006). In fact, the abstraction of technological details, which involves the creation of some symbolism or virtual objects (such as the yellow padlock in the corner of a web browser), may even exacerbate the problem. It attempts to simplify the world by creating a virtual world, which is precisely what the attacker attempts to do when carrying out a social engineering attack (such as phishing).

Research is ongoing to attempt to address the gap between security technology and users in the presence of increasing levels of virtualization. However, there is a clear need for closer interaction between social scientists, psychologists and the information security community, which is still largely technically oriented.

16.7. Conclusions

It has been argued that virtualization is a phenomenon that, though it has been around for a long time, is occurring much more rapidly across many areas of life, supported by technology. Rather than being seen as separate from the real world, virtualization needs to be seen as an intrinsic part of the nature of our interaction with the world. This brings many challenges, as many of the familiar cues that we normally use to orient ourselves and support tractable forms of decision-making through bounded rationality can no longer be relied upon.

This issue is particularly important in relation to security, which is primarily concerned with the management of risk in our interactions with our environment and with other people. Whilst many technological innovations in the security area have been brought about by the existence of technology, the key element in enforcing that security is still the human being. A security system is only as reliable as the people that use it. Much internet-crime is based on the capability of technology to make things appear to be different to the way they are in reality, and it is the human being that is the weakest link.

This, then, raises the concern that, as we introduce into our interactions with the world greater levels of virtuality, which is itself concerned with making the world appear to be different than reality, there is a great danger that security may be compromised even further. The solution will undoubtedly involve technology, but it will need to pay close attention to the people involved, their levels of understanding and attitude towards security as well as the manner in which the interactions between the people and the technologies are carried out and managed.

References

Amin, K., Hategan, M., von Laszewski, G. and Zaluzec, N. J. (2004) Abstracting the Grid. *12th Euromicro Conference on Parallel, Distributed and Network-Based Processing (PDP'04)*, IEEE Computer Society p. 250.

Barboza, D. (2005) Ogre to slay? Outsource IT to Chinese. *New York Times*. Available at http://www.nytimes.com/2005/12/09/technology/09gaming.html?ex=1291784400 &en=a723d0f8592dff2e&ei=5090&partner=rssuserland&emc=rss

Barham, P., Dragovic, B., Fraser, K., Hand, S., Harris, T., Ho, A., Neugebauer, R., Pratt, I. and Warfield, A. (2003) Xen and the art of virtualization. *SOSP '03: Proceedings of the Nineteenth ACM Symposium on Operating Systems Principles*, 19–22 October, Bolton Landing, New York. New York, ACM Press, pp. 164–177.

BBC News (2006) Cash card taps virtual game funds. *BBC News* Tuesday, 2 May 2006. Available at http://news.bbc.co.uk/1/hi/technology/4953620.stm [Accessed 9th September 2006].

Berghel, H. (2000) Identity theft, social security numbers, and the web. *Communications of the ACM* **43**: 17–21.

Birkenstock, W. (1991) Unstable aircraft design: The computer at the controls. *Flug Revue Online* **99**: 74. Available at http://www.flug-revue.rotor.com/ frheft/FRH9909/FR9909e.htm [Accessed 9th September 2006].

Brands, S. (2002) A technical overview of digital credentials 2002. Available at http://www.it.lut.fi/kurssit/03-04/010635000/luennot/26-Tmt-ek-2004_Digital-credentials-technical-overview.pdf [Accessed 1st December 2006].

Brinkmann, A., auf der Heide, F. M., Salzwedel, K., Scheideler, C., Vodisek, M. and Rückert., U. (2003) Storage management as means to cope with exponential information growth. *Proceedings of SSGRR (Scuola Superiore G. Reiss Romoli) 2003*, 28 July–3 August, L'Aquila, Italy. Available at http://wwwcs.uni-paderborn.de/cs/ag-madh/WWW/german/vodisek/ssgrr03w.pdf [Accessed 1st September 2006].

Clark, C., Fraser, K., Hand, S., Gorm Hansen, J., Jul, E., Limpach, C., Pratt, I. and Warfield., A. (2005) Live migration of virtual machines. *Proceedings of the 2nd Symposium on Networked Systems Design and Implementation, NSDI'05*, USENIX, 2 May, Boston, MA. Available at http://www.usenix.org/events/nsdi05/tech/full_papers/clark/clark.pdf [Accessed 31st August 2006].

Dhamija, R., Tygar, J. D. and Hearst, M. (2006) Why phishing works. *CHI '06: Proceedings of the SIGCHI Conference on Human Factors in Computing Systems* (ed. R. Grinter, T. Rodden,

P. Aoki, E. Cutrell, R. Jeffries and G. Olson), 22–27 April, Montreal, Canada. New York, ACM Press, pp. 581–590. Available at http://doi.acm.org/10.1145/1124772.1124861

Dourish, P., Grinter, E., de la Flor, J. D. and Joseph, M. (2004) Security in the wild: user strategies for managing security as an everyday, practical problem. *Personal Ubiquitous Computing* **8**: 391–401.

Elrad, T., Filman, R. E. and Bader, A. (2001) Aspect-oriented programming: Introduction. *Communications of the* ACM **44**: 29–32. Available at http://doi.acm.org/10.1145/383845.383853 [Accessed 8th September 2006].

Figueiredo, R., Dinda, P. and Fortes, J. (2003) A case for grid computing on virtual machines. *Proceedings of the 23rd International Conference on Distributed Computing Systems (ICDCS'03)*, 19–22 May, Providence, Rhode Island, pp. 550–559.

Figueiredo, R., Dinda, P. and Fortes, J. (2005) Guest editors' Introduction: Resource virtualization renaissance. *Computer* **38**: 28–31. Available at http://ieeexplore.ieee.org/xpls/abs_all.jsp?arnumber=1430628 [Accessed 31st August 2006].

Foster, I. (2006-07-28) Globus Toolkit Version 4: Software for service-oriented systems. *Journal of Computer Science and Technology* **21** (4): 513–520 [Boston, Springer].

Foster, I., Kesselman, C. and Tuecke, S. (2001) The anatomy of the Grid: Enabling scalable virtual organizations. *International Journal of Supercomputer Applications* **15**: 200–222.

Golle, P. and Ducheneaut, N. (2005) Preventing bots from playing online games *Computers in Entertainment* **3**: 3.

Heinis, T., Pautasso, C., Deak, O. and Alonso, G. (2005) Publishing persistent Grid computations as WS resources e-science. *First International Conference on e-Science and Grid Computing (e-Science'05)*, pp. 328–335 [IEEE Computer Society].

Howald, R. (2003) New ethernet. *Electronic Engineering Times*. Available at http://www.eetasia.com/ARTICLES/2003AUG/B/2003AUG16_NTEK_PD_ID_TA.PDF [Accessed 31st August 2006].

Jansson, K., Kalliokoski, P. and Hemilä, J. (2003) Extended products in one-of-a-kind product delivery and service networks. In: *eChallenges e-2003*, 22–24 October 2003, Palazzo Re Enzo, Bologna, Italy. Available at http://virtual.vtt.fi/cobtec/files/e2003_jansson_173_final.pdf [Accessed 30th August 2006].

Jøsang, A., Fabre, J., Hay, B., Dalziel, J. and Pope, S. (2005) Trust requirements in identity management. *ACSW Frontiers '05: Proceedings of the 2005 Australasian Workshop on Grid Computing and e-research*, 30 January–4 February, Newcastle, New South Wales, Australian Computer Society, Inc., pp. 99–108.

Kim, Y., Choi, Y. and Yoo, S. B. (2001) Brokering and 3D collaborative viewing of mechanical part models on the Web. *International Journal of Computer Integrated Manufacturing* **14**: 28–40. Available at http://dx.doi.org/10.1080/09511920150214884

Korzilius, L. (1999) Vers une architecture and Villa Savoye – A comparison of treatise and building. Available at http://www.lesterkorzilius.com/pubs/ma/pdf/vua_vs.pdf [Accessed 31st August 2006].

Kozuch, M., Satyanarayanan, M., Bressoud, T., Helfrich, C. and Sinnamohideen, S. (2004) Seamless mobile computing on fixed infrastructure. *Computer* **37**: 65–72.

Laird, J. E. and van Lent, M. (2000) Human-level AI's killer application: Interactive computer games. *Proceedings of the Seventeenth National Conference on Artificial Intelligence*, 31 July–2 August, Austin, TX. Menlo Park, CA, The AAAI Press, pp. 1171–1178

Lastowka, F. G. and Hunter, D. (2006) Virtual crime *New York Law School Law Review*. Available at SSRN (Social Science Research Network): http://ssrn.com/abstract=564801 [Accessed 9th September 2006].

Lehdonvirta, V. (2005) *Economic Integration Strategies for Virtual World Operators*. Masters thesis, Helsinki University of Technology, Department of Computer Science and Engineering.

Möllering, G. (2001) The nature of trust: From Georg Simmel to a theory of expectation, interpretation and suspension. *Sociology* **35**: 403–420.

Nelson, M., Lim, B. and Hutchins, G. (2005) Fast transparent migration for virtual machines. *Proceedings of USENIX 2005 Annual Technical Conference*, 10–15 April, Anaheim, CA. Available at http://www.usenix.org/event/usenix05/tech/general/nelson.html [Accessed 4th August 2006].

Newell, B. R. (2005) Re-visions of rationality? *Trends in Cognitive Sciences* **9**: 11–15.

Rusch, J. J. (1999) The 'social engineering' of internet fraud. *Proceedings of the 9th Annual Conference of the Internet Society, INET'99*, 22–25 June, San Jose, CA. Available at http://www.isoc.org/inet99/proceedings/3g/3g_2.htm [Accessed 9th September 2006].

Sarbagyshov, F. and Munshey, S. (2003) *INTERNETWORKING – Project VLAN Technology*. Albert-Ludwigs-Universität, Freiburg Institut für Informatik. Available at http://www.ks.uni-freiburg.de/download/inetworkSS05/practical/VLAN-report.pdf [Accessed 31st August 2006].

Simon, H. A. (1956) Rational choice and the structure of the environment. *Psychological Reviews* **63**: 129–138.

Smith, R. C. (2001) Shared vision. *Communications of the ACM* **44**: 45–48. Sommerville, I. (2004) *Software Engineering* (7th edn). Reading, MA, Addison-Wesley.

Teigland, D. and Mauelshagen, H. (2001) Volume managers. *Linux Proceedings of the FREENIX Track: 2001 USENIX Annual Technical Conference*, 25–30 June 2001, Boston, MA. Available at http://www.usenix.org/publications/library/proceedings/usenix01/freenix01/teigland.html [Accessed 31st August 2006].

Terdiman, D. (2005) Virtual goods, real scams. *c|net news service*, September 12. Available at http://news.com.com/Virtual+goods,+real+scams/2100-1043_3-5859069.html [Accessed 30th August 2006].

Uhlig, R., Neiger, G., Rodgers, D., Santoni, A., Martins, F., Anderson, A., Bennett, S., Kagi, A., Leung, F. and Smith, L. (2005) Intel virtualization technology. *Computer* **38**: 48–56. Available at http://ieeexplore.ieee.org/xpls/abs_all.jsp?arnumber=1430631 [Accessed 1st September 2006].

Wang, G. G. (2002) Definition and review of virtual prototyping. *ASME* [American Society of Mechanical Engineers]*Transactions, Journal of Computing and Information Science in Engineering* **2**: 232–236.

Wang, W., Yuan, Y. and Archer, N. (2006) A contextual framework for combating identity theft. *Security and Privacy Magazine, IEEE* **4**: 30–38.

Yan, J. and Randell, B. (2005) A systematic classification of cheating in online games. *NetGames '05: Proceedings of the 4th ACM SIGCOMM*[ACM Special Interest Group on Data Communications]*Workshop on Network and System Support for Games*, 10–11 October, Hawthorne, New York, T. J. Watson Research Center. New York, ACM Press, pp. 1–9.

17 The future of virtual construction and regulation checking

Nicholas Nisbet, Jeffrey Wix and David Conover

Virtual construction will bring together information from many sources into a building design proposal, which can then be tested to see how it will perform in the real world. Performance here means not only the visual aesthetic but also energy, environmental impact, service life, whole-life cost, fire safety, accessibility and much more. To bring all of this to bear at a virtual construction requires considerable discipline in identifying, organizing and integrating information, what it is, where it comes from, who is responsible for it and so on. If information is identified or presented differently from discrete sources, many of the potential benefits of virtual construction are lost. Currently, regulatory controls impact on the freedom to explore design options.

Consistent information results from agreements by industry to use a common expression of ideas. These are now emerging through the development of dictionaries in various countries; these will develop to provide a consistent and multi-lingual ontology that can be used in web-based virtual construction and in regulatory domains. This chapter examines the application of standards in both the domains. It shows that the major obstacles to the wider adoption of automatic building regulatory approval are being addressed so that the risk of delay and abortive costs associated with the approval gateway can be removed. The focus will be on the germination of current ideas and how these will be likely to grow to meet real business need.

17.1. Introduction

This chapter examines a particular business case for the use of virtual construction, and examines the role of information standards in creating a practical and economic solution.

One of the first, and still one of the most impressive, implementations of virtual construction has been developed in Singapore. The Singapore Building Construction Agency has led the development of an integrated information system for the construction industry. As well as fulfilling a web-portal role, the system is encouraging the development and submission of virtual building models for building control (i.e. code compliance or building regulatory) checking. There are multiple agencies in Singapore with whom agreement had to be reached on interpreting the precise intent of the various codes. The system covers the regulations imposed by several authorities having jurisdiction over planning, fire prevention, servicing and other aspects. Its scope is intended to grow, leaving only the requirements or clauses allowing self-certification and waivers to be handled separately.

This same type of system is being adopted and adapted by the Norwegian state and local authorities to handle both planning (i.e. zoning) permission and building control. Trial implementations have been made for the UK Heath and Safety Executive to provide advice and for the Scottish Noise Transmission to assess a performance-based regulation. Trials using the Singapore system have also been prepared for some clauses of the New York City regulations.

17.2. Issues

However, some critical issues need to be resolved before the full potential benefits of such a system can be obtained. Most important of these are issues of acceptance and of scalability.

17.2.1. Acceptance

For such systems to be accepted, it is necessary for there to be a growing use of reliable and complete building information models (BIM). In the last few years, the key applications – previously known as CAD (computer-aided design) tools – have become significantly more widespread. They are being used as the preferred tools by the next generation of designers, who have used earlier versions at schools of architecture and colleges. Although there have been tentative attempts to share these models, there are now examples of whole projects being developed around a common object model. Many countries and client bodies are preparing BIM guides, which advise on the effective use of the application features to meet specific objectives. For instance, in the USA the National Institute of Building Sciences is leading a joint public/private initiative to develop a National BIM Standard. Outside of the application for code compliance, building owners, product manufacturers, insurance interests, the fire service, facility managers and others will also affect acceptance. For instance, the availability of product information in a format that provides for its use in virtual construction will be crucial. Another example is the reliability of a BIM for post-occupancy applications associated with facility maintenance or fire-fighting. If all of these interests see the value of BIM then they will support the development of information in a format that can be used in a BIM, and/or drive for the application and use of BIM in various building-related applications.

Another aspect of the acceptance issue is the credibility of the compliance testing applications. Both building control (code compliance) officers and designers need to develop the confidence that these applications are applying the regulations correctly.

17.2.2. Scalability

The second major issue is that of scalability. There have been many applications devoted to specific aspects of regulations, particularly energy and accessibility. However, for a system to genuinely reduce the transaction cost of preparing and evaluating a design approach or scheme, it must be comprehensive and correct. In federated jurisdictions, the federal, the state and the county are all competent to create and modify the body of codes. To apply the same approach to interpreting code atoms in the USA, as used in Singapore, presents some significant issues, since there may be distinct building codes in each of the over 40 000 code jurisdictions. (However, the majority of those entities adopt and apply the model codes of the International Code Council (ICC) and numerous reference standards developed in the private sector with public sector input.) Furthermore, many codes and requirements for compliance refer to a network of other documents and supporting publications, either as reference material or for actual requirements. Insurance and certification bodies are also generating requirements and best practice notes. In addition, the dynamic of change by governments in regulations or manufacturer information on products suggests that means need to be available for experts, practitioners or software developers to keep pace with this new information.

17.3. Coding model

The basic model for an automated solution to addressing the technical- and compliance-related aspects of building regulations has been to hand the assessment problem to software authors, who have then been expected to read a subsection of the regulations, create an information model for the problem, create a user interface and a solution algorithm. This is a broad list of skills, and requires the domain experts, the code officer and the building designers to validate the application as a 'black box'. The problem remains that the code – as developed by public and/or private sector interest and then passed, adopted or endorsed by the various legislatures or regulatory bodies – is paramount. The distance, essentially the number of interpretations, which can intervene between that primary source and the output of a conventional software solution can be so great that a credibility gap is created. Furthermore, such applications tend to focus on single-mode working: typically, data preparation, submission, evaluation, results. For many single-regulation applications, the number of inputs far exceeds the number of outputs. Whilst this is not an absolute criterion, it does indicate a suspicious imbalance.

Recently, in working with code-compliance agencies and European development partners as well as the International Code Council (ICC) in the USA, AEC3 Ltd have had the opportunity to revisit the problem of regulatory compliance checking, and are presenting an alternative paradigm, illustrated in Figure 17.1. Deep consideration was given to the way in which the code atoms were interpreted and how this was achieved. An approach was developed which turned the Singapore method upside down. Instead of code officials explaining to modelling specialists how they worked on a clause-by-clause basis, they were given a set of tools that enabled them to work directly with the written codes and specifically mark up their interpretations. The ICC is the developer of US model building codes that form the basis for the majority of federal, state and local building and fire regulations. This body has now demonstrated automated code-compliance checking for a range of clauses from its codes, using the paradigm described to deliver content directly to three different applications, including 'ePlanCheck' and a commercial product 'Solibri'. A report generated by a third,

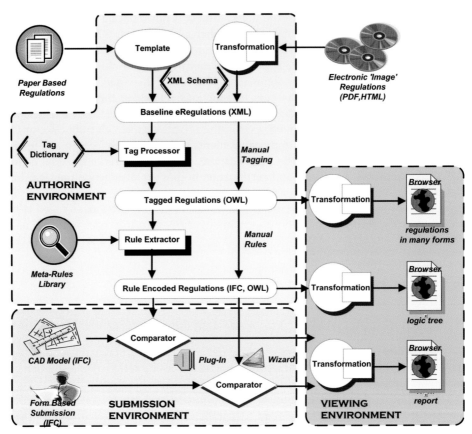

Figure 17.1 Marked-up regulatory documents are the key to innovative applications. (Figure is copyrighted to AEC3 and ICC, showing model provided to the US Coast Guard by AECInforSystems.)

web-based, application is shown in Figure 17.2. Work to complete all the ICC codes is anticipated to be completed in 2009 and federal, state and local amendments to those codes added shortly thereafter.

17.4. Analysis of regulations

Legislation and regulations typically present as apparently well-structured documents. Irrespective of the relative complexity of any particular document, it is possible and useful to identify the common constructs being described. It is proposed that regulations can be broken down into five fundamental concepts. These five concepts were selected to be most familiar to regulatory experts and designers, rather than to application analysts or BIM experts.

The most general of these has been named the 'check' and typically demarcates a section of the regulation that is distinct and independent of any other. 'Checks' are often, but not necessarily, closely related to the named or titled sections in the document. It is a characteristic of regulations that every 'check' must be in some way satisfied.

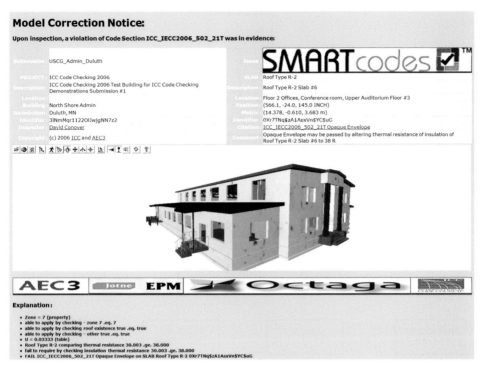

Figure 17.2   The system generates a detailed correction notice with the failed element highlighted in 3D. (Figure is copyrighted to AEC3 and ICC, with acknowledgement to the US Coast Guard and AECinfosystems for the use of the building model.)

A 'check' is not an indivisible (atomic) concept: it can be analysed down further into four subsidiary constructs. The most obvious and most easily identified are the 'requirements' as these are associated with the future imperatives 'shall' or 'must'. It is required that a check contains at least one 'requirement'. Secondly, there will be text that identifies the 'applicability' of the check. These are often compounded, for example 'external windows'. These phrases need not relate directly to the topic of the regulation or the topic of the overall check. For example, if a check applies in 'seismic zone X', this is a property of the building site, not of the structural integrity of a particular building material. In general, there will be one or more phrases defining the applicability. One special but distinct case is where a 'selection' of alternative subjects is offered, for example 'doors, windows and other openings'. Lastly, there may be one or more 'exceptions'. These are the opposite of 'applicabilities', and conversely work by exclusion.

To summarize, a regulation contains a number of 'checks', and each check contains a number of 'requirements', 'applicabilities', 'selections' and 'exceptions', as seen in Figure 17.3.

Each of these low-level constructs is identically attributed to have a topic, a property, a comparator and a target value. The topic and property will ideally be drawn from a restricted dictionary composed of terms defined within the regulation and normal practice. The value (with any unit) may be numeric, whereupon the comparators will

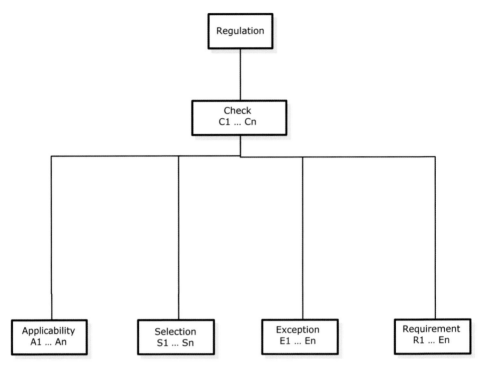

Figure 17.3   The mark-up introduces a simple hierarchy into the regulatory document. (Figure is copyrighted to AEC3 and ICC showing model provided to the US Coast Guard by AECInforSystems.)

include 'greater', 'lesser', 'equal' and their converses. If the value is descriptive, then only the 'equal' or 'not equal' comparators are relevant.

17.5. Passing and failing

'Passing' and 'failing' are two everyday concepts that need examination. In the context of regulatory compliance, an element can fail a check if there is sufficient information to say that it is selected and applicable, not excepted and does not meet the requirements. If a slightly broader definition for 'passing' a regulation is accepted, we can say that an element passes if either it is not applicable, or it is not selected, or it is excepted, or it is as required. In practice, refinements may be needed to these definitions: during design, having insufficient information may not be significant, and there may be mechanisms for self-certification and waivers within a specific compliance regime, there may be mechanisms for self-certification and waivers. Certificates may be issued subject to certain assumptions, such as satisfactory site-inspection, or the use of approved materials. A user may wish to know a number of things: does this element pass this check? Why does it fail? How can I get this element to pass? Conversely, a regulatory officer may question how could this element fail. The purpose of the deconstruction of the semantics of a regulation is to make the meaning of the regulation available to a variety of applications that respond to a broad range of use-cases.

17.6. Logical analysis

The deconstruction described above can be represented as:

Fail = is-Applicable
         and is-Selected
         and not is-Exception
         and not is-as-Required

Pass = not is-Applicable
         or not is-Selected
         or is-Exception
         or is-as-Required

    This analysis suggests differing approaches are possible depending on which use-case is being implemented and whether a pass or a fail is being sought. Applications can use these statements to evolve their heuristic methods to obtain the information needed. For example, automatic code compliance checking can explore each of the four types of clause, in any order, and may well be able to reach a conclusion before evaluating the most difficult. A web-based dialogue could identify all clauses by asking the user questions directed at securing building-related information and allow the user to respond to only those questions that can be confidently answered. Given a failure, a decision support system could back-track to find the easiest or even the cheapest way of obtaining a pass.

17.7. More complex examples

Actual checks may contain a number of requirements, applicabilities, selections and exceptions. It is important to identify how these combine. Requirements are typically cumulative: if there are several stated requirements, then it is to be expected that all must be satisfied. Similarly, applicabilities are cumulative and all must be met. However, if there are many exceptions or selections, then typically these are alternatives, and only one will be relevant.
    Using a shorter notation:

Requirement R = R1 and R2 and R3 and ... Rn
Applicability A = A1 and A2 and A3 and ... An
Selection S   = S1 or S2 or S3 or ... Sn
Exception E   = E1 or E2 or E3 or ... En

    Sometimes more complex linguistic structures are encountered, usually reflecting more complex logical intentions. Methods for representing these have been developed, using subtypes of the four concepts. However complex any actual examples prove, standard logical calculus allows these to be expanded and manipulated, as shown in Figure 17.4.

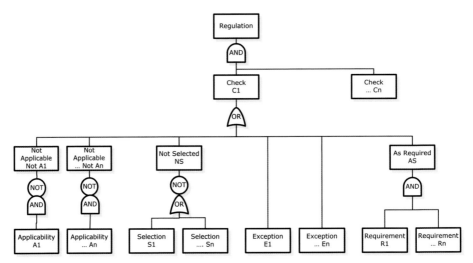

Figure 17.4   Applications use a logical expansion of the mark-up hierarchy. (Figure is copyrighted to AEC3 and ICC showing model provided to the US Coast Guard by AECInforSystems.)

17.8. Representation of existing regulations

It is now commonplace, and in some jurisdictions it is obligatory, to make available on the web all primary legislation and sometimes the supporting regulations as well. This can be achieved using document- or image-based technology with an appropriate viewer. Alternatively, it can be achieved by representing the legislation as HTML, XHTML or as a transformation of an XML mark-up such as 'docBook'. These document mark-ups are controlled by underlying schemas. Implementation of the new paradigm is based on extending the document schema to include the five extra entities required. Mark-up tags can be introduced which delimit the relevant text, but leave it otherwise undisturbed. These tags identify the distinct checks, and highlight each key phrase and identify its role. Initially, standard mark-up development tools (such as Altova XMLSPY or Microsoft® Visual Studio) are being used, but less sophisticated tools are just as relevant, and a schema-specific tool is envisaged for use by code domain experts. It is also possible to conceive of computer assisted or even partially automated applications based on recent advances in linguistic and grammatical recognition. The key point is that these tags have been designed to be well understood by domain experts, who, in general, have no difficulty in placing, filling and reviewing the tagging. An important point, regardless of how the regulations are presented in this format, is that those familiar with the genesis and application of the regulations must be an integral part of presenting the regulations in this format. Style sheets can re-present their work in a readable form, or as tables. This makes the results available for ongoing peer-review, discussion and quality control.

17.9. Ontologies

Ideally, the vocabulary used by the domain experts would be drawn from an existing ontology. At this early stage, the process is being reversed and key concepts and properties are being identified during the tagging process. Lists of synonyms and shades of meaning that depend on context are being catalogued.

Having both the logical structure of the regulations and the ontology behind them is not sufficient to enable automatic code-compliance checking. The final component is to create a recognizable association between the concepts and properties used in the regulations and their representation in a building model. The power of the new paradigm is that again we do not have to create and then seek approval for a single monolithic implementation: each concept can be reviewed individually and its representation within international standards agreed. Where the same information can be represented several ways, multiple definitions can be supported. Equally, web-based forms and dialogues can be used to collect and store the same information.

17.10. Standards

The methodology has been described without many references to the underlying standards that make it practical and economic. In summary, W3C standards, and in particular XML, XSLT and XLINK, allow the creation, presentation and interpretation of marked-up regulations. The general web infrastructure allows owners of documents to make them accessible dynamically in such a format without loss of control. A particularly important aspect of the paradigm is that checks can be derived as 'pseudo-rules' that can then, through the means of standard web-based techniques, be transformed into a manner that is directly usable by several different checking resources. For instance, the ICC is currently exploring mechanisms for user interfaces and storefronts that will allow users to 'load' a BIM and apply †SMARTcodes to facilitate automated code-compliance checking in the USA.

ISO/PAS 16739:2005 (IFC2x Platform) (http://www.iso.org/iso/iso_catalogue) allows all the major BIM applications to prepare descriptions of buildings in a form susceptible to automated checking based on IFCs. This means that the tools to actually generate a BIM that can be checked are already in daily use within the industry. It also provides a schema for the representation of the logical content of each regulatory framework, for use if the source regulations are not accessible.

Lastly, ISO 12006-3:2007 (Framework for object-oriented information) (http://www. iso.org/iso/iso_catalogue) provides an International Framework for Dictionaries (IFD) to manage the definition and sharing of common construction concepts independent of language and culture.

17.11. Conclusions

An approach has been developed which steers a path between highly specific applications for aspects of the regulations and monolithic systems that may prove difficult to maintain and check. It is based on developing and obtaining signing off two manageable sets of 'atoms': on the one hand, distinct mark-ups that are closely tied to the source phrases in the regulations; and, on the other hand, distinct implementations of well-defined concepts. Between these, any number of traditional and novel applications can be envisaged that consume an ever-widening library of regulations and advice.

It is to be hoped that by reducing the resources needed for code-compliance checking, more resources can be applied to generating options and assessing genuine fitness, to the benefit of the clients and the built environments.

Acknowledgements

The work described has been developed by AEC3 Ltd for the International Code Council of the USA, which has encouraged the publication of this overview. †SMARTcodes is a registered mark of ICC.

18 Virtual prototyping of financial flows as a form of management control

Richard J. Boland, Jr, Fred Collopy, Julia Grant and Lin Zhao

This chapter describes a new form of financial representation (the Business Animator) that provides a dynamic visualization of the flows of assets (money, materials, equipment and receivables), liabilities (borrowings, profits and losses), revenues and expenses for projects and individual firms. For projects, the Business Animator enables coordination and management control of the financial flows during planning, design and construction under varying technology infrastructures and practices. For the project managers, it provides a way to evaluate the financial strength and viability of potential project participants. In these two ways, it provides a key element in the financial management of large complex projects.

18.1. Introduction

Interpretation of financial information during the planning and execution of a large-scale design and construction project is a significant undertaking that relies on a full understanding of the firms involved and the project operations. It requires the ability to evaluate alternative sequences and timing of construction activity, as well as the ability to evaluate the financial viability of potential project members and to control financial flows during the project.

Little has changed in the graphical representations we use to present key financial data to company or project management since William Playfair published his *Commercial and Political Atlas* in 1786. Meanwhile, many areas of science and engineering have seen interactive computer graphics transform the way they explore and present data. Whether it is a weather pattern or a molecule, a tsunami or an as yet un-built building, computer graphics are being used both to explore and to explain the dynamics of complex systems.

A dynamic visual representation of economic activity offers promise for supporting the interpretation of financial information for a firm or a project, and in this chapter we describe the design of such a tool, called the 'Business Animator'. It is an interactive, animated model of the firm, designed to explore and visualize the financial implications of decisions and operations at various stages during the life of a firm or a project. Principles and findings from the accounting and information systems' literature were used to drive the design of the representation and the software used to control it. The Business Animator adds depth to traditional accounting and financial representations by conveying information about the momentum of a firm's or project's activities, and the rate of change at which various activities are occurring. The animation facilitates the identification of backlogs or breakdowns in operating processes, thus increasing the understanding of a firm's or a project's financial health.

18.2. Research on accounting representations

People have long developed accounting systems for recording their transactions with one another. Fra Luca Pacioli (1509) published a concise summary of the double-entry accounting system as it was used in Venice at the time, and his summary remains an accurate description of the basic accounting systems that are in wide use today. The financial statements that are produced by this system have been refined over the centuries and formalized by rules and regulations, yet the form in which financial statements represent a firm and its operations has seen little improvement over the centuries. We propose a dynamic, graphical method of representation that improves the understandability of financial statements, and provides a much needed innovation in the reporting of company and project-based financial information.

Basic mathematical techniques have been widely used through the centuries for both internal and external financial analysis. The most common mathematical technique is ratio analysis, which can provide useful information; but the interpretation of ratios, especially if reported in tables as numbers, is neither automatic nor instantly accessible. Analysts and researchers have frequently attempted to improve the interpretation of ratios by going beyond the standard tables with columns of numbers. Some of the representation techniques that have been explored include charts, graphs and even faces (in which facial features correspond to specific ratios, and increase or decrease in size or shape as the ratios increase or decrease). Another stream of literature has focused on what additional data might be added to standard financial statements to make them more meaningful and easier to understand. Ijiri developed the notion of triple-entry bookkeeping as a way to address a perceived weakness in the current system, namely, its inability to represent rates of change in financial variables and the forces behind those rates (Ijiri, 1982). We discuss his research more fully below.

Huff *et al.* (1981) applied this technique to financial data to determine whether it could detect potential business failures. Their results were consistent with Ijiri's proposal that his multi-dimensional approach would be more informative, but their sample size was inadequate to allow statistical testing. Ferris and Tennant (1982) proposed the use of the Chernoff (1973) face (with facial features changing as selected ratios changed) in order to help determine whether a company was approaching financial problems, which is an ongoing concern for internal audit and analytical review functions. Their results show that the facial graphics tool did provide users with early signals of possible problems. Scherr and Wilson (1985) compared three methods for estimating bond ratings: ratio tables, Chernoff faces and multiple discriminant analysis, and found that statistical discriminant analysis outperformed the other methods.

Other types of visualization tools have also been developed and tested. Dull and Tegarden (1999) compared three-dimensional to two-dimensional representations, finding the three-dimensional to be more accurate for company wealth predictions in a laboratory experiment. Lightner (2001) also used a laboratory setting and applied animation to graphical displays in order to measure how knowledge comprehension related to varying speeds of the animation. Benbasat *et al.* (1986) found that combining multiple forms (using both tabular and graphical presentations) was effective for improving decision-making, but the use of colour was helpful to only a subset of their laboratory subjects. Lim and Benbasat (2000) tested the perceived usefulness of combining text and video information. They found that the video presentation was more helpful in conveying less analysable information, while text and video presentations were equally useful in conveying more analysable information.

Another stream of literature has focused on what data might be missing in the traditional financial information being produced. Ijiri (1982, 1986, 1988) developed the notion of triple-entry bookkeeping as a way to address traditional financial systems' inability to represent rates of change in financial variables and the forces behind those rates. He proposed measuring dollars per time period, or 'momentum', to account for rates of changes in the various accounts, with the terms 'force' or 'impulse' used to identify the factors that affect rates, suggesting that the use of these sorts of measures would help managers to focus their attention more productively. Blommaert (2001) tested the usefulness of momentum and force information, finding that these data did facilitate decision-making by laboratory subjects. This triple-entry idea has not been adopted by professional accounting standards, but the concepts stand as a reminder that the current records and resulting statements provide no information about rates of change across specified periods, and that the information can be valuable.

In this chapter we integrate and extend these streams of research concerning the problems of representation. We propose a more complete picture of the firm than that which can be conveyed in a single face or static graph, while at the same time creating a more concise representation of the multiple dimensions than would be available in an array of charts and graphs. Our design does this by taking advantage of technological developments to exploit the power of animation. Our design is inspired by the notion that cognition is distributed and does not take place solely within our heads. Instead cognition is found in our interaction with the people, objects and representations in our environment.

18.3. The psychology of visual representation

A basic principle of distributed cognition theory is that a cognitive task includes both internal and external representations, which together contain the abstract structure of the task (Zhang and Norman, 1994). In order to perform a cognitive task, such as making a business judgement or decision, people need to process information distributed across the internal mind and the external environment. Whatever external representations we create will have to link to the internal representations that people produce. Computational offloading, re-representation and graphical constraining are three characteristics that can be used to explain the connection between internal and external representations (Scaife and Rogers, 1996). According to visual computing theorists and dual processing theory, a basic way to effectively facilitate the connection is to improve the visibility of the information embedded in the data (Friedhoff and Peercy, 2000; Stanovich and West, 2000). Indeed, a good external representation will be one that links naturally to the internal representations that people are capable of forming, thereby supporting their overall cognitive process.

18.4. Textual versus graphic representations

The first issue we face is whether textual or graphic representations should be favoured, given the task of understanding the financial and operational activities of an organization over time. Tables and graphs are both used to represent large data sets. Many studies have explored the features of these two representations and the conditions under which each is superior. A series of experiments summarized by Benbasat *et al.* concluded that graphical presentation was more useful in searches for optimal solutions, but that tables were more useful in tasks that required the determination of exact numbers (Benbasat *et al.*, 1986). In the study that is most closely related to the current project, Volmer compared graphical presentations of the financial ratios of a firm with numerical financial information, and concluded that graphical information not only saved time, but was also considered important in providing clear insights into the financial position of the firm, thus improving communication (Volmer, 1992). This literature provides a basis for optimism that a graphic representation will be useful given our objective of developing an understanding of the dynamics of a project or organization over time.

18.5. The power of animation

A second issue to consider is whether the graphic representations employed should be dynamic. There is a general belief that animations not only improve users' understanding, but also make interfaces easier and more enjoyable to use. This belief has been tested in many studies, although the results have not produced a consensus. In a review, Tversky *et al.* argued that empirical studies have not provided strong evidence that animated graphics outperform static graphics, but they are particularly promising in conveying real-time changes and temporal–spatial reorientations due to the cognitive congruence between the task and the representation (Tversky *et al.*, 2000). Other research suggests that animations can enhance comprehension, problem-solving, learning and presentation persuasiveness (Mayer and Sims, 1994; Large *et al.*, 1996; Morrison and Vogel, 1998). Movement and interactivity are two important attributes of animations. They stimulate the early preconscious stages of visual processing, and thereby improve the effectiveness of visualization for communicating and learning (Friedhoff and Peercy, 2000; Stanovich and West, 2000). Since our visualization is for use with data that have a strong temporal character, and in an environment where there will be repeated use and learning, we have reason to anticipate that animation will be helpful in improving the financial analysis of a firm or a project.

18.6. The cycle model

Visual models of a firm's operating activity have been proposed as a way of helping managers and accountants understand the critical components of a firm's operating cycle. Boland proposed the cycle model, which, in the most basic version, presents the operations of a firm as three interacting cycles: an input cycle; a transformation or value-adding cycle; and an output cycle (Boland, 1983). A manufacturing firm provides

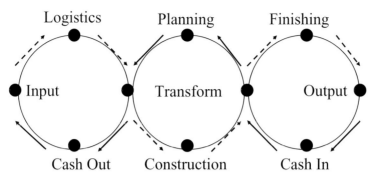

Figure 18.1   Material (dashed lines) and cash (solid lines) moving through input, transformation and output cycles.

the most accessible interpretation of these activities. The model assumes an existing business with the ability to purchase raw materials on credit. In the input cycle, a company acquires raw materials from its suppliers. In the transformation cycle, value is added to the raw materials through the fabrication process. The output cycle represents the installation of the fabricated materials in the construction project.

The cycle model contains representations of the stocks and flows that characterize the movement of materials and monies through the project or firm. In each of the three cycles there are several moments and arcs connecting the moments. These can be used to represent the ratio data that is often used to characterize a firm. Figure 18.1 shows the path that defines the movement of material in dashed lines. The size of nodes could be varied to represent the activity for that aspect of the enterprise: in this case how much material is on order. If we used a colour display, a node's brightness could also be varied to indicate important information about the activity. In this case varying levels of brightness could be used to represent how much is currently on backorder. This basic idea is used throughout the paths (Figure 18.1).

The next node along that path represents the delivery of materials and supplies. The connections between nodes can vary in speed, based on the time it takes for the system to make the transformations. If things are moving rapidly through a part of the system this will be reflected by the relative rate at which that link moves. The amount of material moving can be represented by the brightness of the link. From inbound logistics the material moves along the next arc into the middle cycle, which is where materials are assembled and constructed into a finished structure. In the transformation cycle, labour is applied to turn the materials into fabricated assemblies that are available for installation. Once transformed, the assemblies are put into a holding inventory to be delivered and installed at the construction site.

Figure 18.1 illustrates the movement of cash as solid lines. Cash moves in from right to left, starting at the right-most node. Based on completed work, the firm has receivables, which are collected as cash. As cash is received, the balances are accumulated in a planning and control process shown at the top of the middle cycle. As cash is disbursed, it flows to a cash-out node at the bottom of the left cycle, where it is used to pay for labour and materials received earlier.

This stylized representation of a firm's operating cycle, represents its day-to-day operations. For a firm to be able to operate, however, capital investments in property, plant and equipment are required, as well as the funding to make those investments.

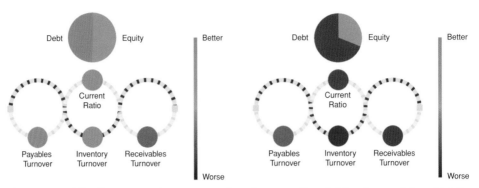

Figure 18.2   The diagrammatic representation of a strong firm (left) and a weak firm (right).

The simple model shown assumes an existing firm with adequate resources for its current level of operations. The possibility of changing those levels of resources requires appending two additional cycles to the model: an investing cycle and a financing cycle.

The overall activities of the firm represented in this model should seem familiar to financial statement users, since it conforms roughly to the structure of the cash flow statement and its labelled sections. This stylized representation illustrates most of the flows presented in the financial information currently produced by firms. However, a set of financial statements does not allow the user to measure characteristics of the flows occurring within the individual component cycles.

18.7. Adding dynamics to the cycle model

With the advent of more powerful and accessible computer animation, the cycle model can be adapted to provide a dynamic representation of the firm's activities. Business Animator utilizes control and display technologies developed for the real-time music industry (the Max/MSP programming environment from Cycling 74). Our goal is to provide a program through which managers and analysts, using sophisticated controllers (knobs, sliders, pedals, etc.), are able to develop an intuitive sense about the cycle model itself, while exploring and visualizing how firms at various stages of growth, sustenance and decay are affected by specific operating, financing and investing decisions. The animated version of the model incorporates a holistic vision of what characterizes the entire organization. The animation portrays temporality, allowing the theoretical construct momentum to be captured and depicted as the cycles change at differing rates (Figure 18.2).

The rules that govern transition and rates of change in a business are embedded within its financial outcomes, some of which are measurable by ratios that can be generated to help understand a period of operations. For example, the operating cycle's speed and efficiency are related to ratios such as the turnovers of inventories, accounts payable and accounts receivable. These turnovers reflect how quickly and efficiently inventory is managed, turned into finished projects and converted ultimately to cash to be paid to workers and suppliers and reinvested in the business.

The user can determine starting points or baseline acceptable conditions for a particular project, creating inputs commensurate with specified ratio relationships. The program then runs through the requested numbers of cycles, representing days, weeks, months, quarters or other periods of interest. The user can interact with the program using sliders and knobs to change the parameters of the simulations. The animation, using movement and colour, indicates whether the operation remains smooth or whether problems occur. Backups at nodes are indicated by both a slowdown in cycle movement and by gradually changing colours.

18.8. Implications and future research

The development and use of the tool described in this chapter has the potential to significantly improve understanding of the financial condition of a firm or a complex project viewed as a whole, from both external and internal perspectives. The cycle model of a firm's or a project's activity can be applied retrospectively to financial information to aid in the interpretation of performance ratios. It can quickly highlight areas in which the firm's or project's operations are not as efficient or as effective as possible. The model can also be applied prospectively as a planning tool to identify potential pitfalls in construction plans and schedules. This sort of simulation is greatly enhanced by the use of understandable knobs and sliders so that a user can quickly change inputs and observe the results. Growth plans and other strategic decisions can also be evaluated. The cycle model will never be better than the data provided to it, so the input parameters must be carefully considered and measured. Empirical validation will be required before we are able to assert the conditions under which the tool provides useful insights. Nevertheless, it has the potential to aid in understanding financial data for the management control of projects and firms in a way not presently available.

References

Benbasat, I., Dexter, A. S. and Todd, P. (1986) An experimental program investigating colour-enhanced and graphical information presentation – an integration of the findings. *Communications of the ACM* **29**: 1094–1105.

Blommaert, T. (2001) Additional disclosure: Triple-entry and momentum accounting. *The European Accounting Review, Doctoral section* **10**: 580–581.

Boland, R. (1983) *Organizations and Accounting Systems* [Unpublished].

Chernoff, H. (1973) Using faces to represent points in K-dimensional space graphically. *Journal of the American Statistical Association* **68** (June): 361–368.

Dull, R. B. and Tegerden, D. P. (1999) A comparison of three visual representations of complex multidimensional accounting information. *Journal of Information Systems* **13** (2): 117–131.

Ferris, K. R. and Tennant, K. L. (1982) New tools for analytical reviews. *The Internal Auditor* (Dec): 14–17.

Friedhoff, R. M. and Peercy, M. S. (2000) *Visual Computing*. New York, Scientific American Library.

Huff, D. L., Mahajan, V. and Black, W. C. (1981) Facial representation of multivariate data. *Journal of Marketing* **45** (Fall): 53–59.

Ijiri, Y. (1982) *Triple-Entry Bookkeeping and Income Momentum: Studies in Accounting Research.* Sarasota, FL, American Accounting Association.

Ijiri, Y. (1986) A framework for triple-entry bookkeeping. *The Accounting Review* **61** (4): 745–758.

Ijiri, Y. (1988) Momentum accounting and managerial goals on impulses. *Management Science* **34** (2): 160–166.

Large, A., Beheshti, J., Breuleux, A. and Renaud, A. (1996) Effect of animation in enhancing descriptive and procedural texts in a multimedia learning environment. *Journal of the American Society for Information Science* **47** (6): 437–448.

Lightner, N. J. (2001) Model testing of users' comprehension in graphical animation: the effect of speed and focus areas. *International Journal of Human–Computer Interaction* **13** (1): 53–73.

Lim, K. H. and Benbasat, I. (2000) The effect of multimedia on perceived equivocality and perceived usefulness of information systems. *MIS Quarterly* **24** (3): 449–471.

Mayer, R. E. and Sims, V. K. (1994) For whom is a picture worth 1000 words – Extensions of a dual-coding theory of multimedia learning. *Journal of Educational Psychology* **86** (3): 389–401.

Morrison, J. and Vogel, D. (1998) The impacts of presentation visuals on persuasion. *Information & Management* **33** (3): 125–135.

Pacioli, L. (1509) *De divina proportione* [On the divine proportion]. Venice, Luca Paganinem de Paganinus de Brescia (Antonio Capella).

Playfair, W. (1786) *The Commercial and Political Atlas: Representing, by Means of Stained Copper-Plate Charts, the Progress of the Commerce, Revenues, Expenditure and Debts of England during the Whole of the Eighteenth Century* [Republished, Cambridge University Press, 2005].

Scaife, M. and Rogers, Y. (1996) External cognition: How do graphical representations work? *International Journal of Human–Computer Studies* **45** (2): 185–213.

Scherr, F. C. and Wilson, J. E. (1985) The application of multiple discriminant analysis and Chernoff faces in estimating bond ratings. *The Mid-Atlantic Journal of Business* (Summer): 33–48.

Stanovich, K. E. and West, R. F. (2000) Individual differences in reasoning: Implications for the rationality debate? *Behavioural and Brain Science* **23** (5): 645–665.

Tversky, B., Morrison, J. B. and Betrancourt, M. (2000) Animation: can it facilitate? *International Journal of Human–Computer Studies* **57** (4): 247–262.

Volmer, F. G. (1992) Effect of graphical presentations on insights into a company's financial position: an innovative educational approach to communicating financial information in financial reporting. *Accounting Education* **1** (2): 151–170.

Zhang, J. and Norman, D. A. (1994) Representations in distributed cognitive tasks. *Cognitive Science* **18**: 1.

Part 6
Visions, roadmaps and future scenarios

19 VR – Roadmap: A vision for 2030 in the built environment

Nashwan Dawood, R. Marasini and John Dean

VR (virtual reality) is an emerging technology that will greatly benefit the construction industry and its supply chain in terms of capacity for experimentation, greatly improved communication, data visualization and the capture of ideas. This chapter presents the outcome of a research project that was aimed at developing a 'VR Roadmap: Vision for 2030 in the Built Environment'. The methodology used was to thoroughly review previous and current applications of VR in the construction- and manufacturing-based industries, and to conduct brainstorming sessions with experts in information technology (IT)/VR regarding future functionalities and the research and development (R&D) needed to develop VR tools and processes capable of supporting future built environment.

A total of 23 experts working in industry, academia and software development in the UK, EU and USA were invited to participate in brainstorming sessions during a two-day workshop held in Manchester in 2006.

The roadmap is focused on three main themes: the current state-of-the-art of VR in the built environment; technology and process specifications towards 2030; and R&D plans to deliver such specifications. Discussions centred on identifying enablers, barriers, opportunities and challenges that prevail in the industry, as well as those likely to be encountered towards 2030 with advancement of the technology and process changes. This chapter introduces and discusses the roadmap and its related methodology.

19.1. Introduction

For any industry to survive and be a world class player, it needs to innovate. Construction is no exception. The construction industry needs to innovate by searching for and experimenting with better products and processes that can ensure its survival. This need has been highlighted and emphasized in the Egan report (1998) and other government- and industry-sponsored initiatives.

The developments in VR and its use within construction have proved to be of great benefit to users who have embraced the technology. VR technologies can be applied to promote innovation by allowing the construction industry to better explore possible future products and services. This has been the subject of discussion and debate since the late 1990s. During this period a number of projects and reports have been commissioned by the EPSRC (Engineering and Physical Science Research Council) and the DTI (Department of Trade and Industry, 2000). Further developments in the field and innovative means of utilization are likely to continue.

From an industrial point of view, the introduction of VR will add value in the form of:

- Greater site safety – by its ability to view pinch-points of projects in terms of restricted access (spatial planning)
- Ability to test alternative construction methods and note the time implication of each
- Allow more accurate sequencing of operations via the ability to view element interfaces
- Present a new channel of communication between designers/suppliers/and contractors

Watts *et al.* (1998) stated that VR can help the innovative organization in the following respects:

- Low marginal cost → capacity to experiment
- Realism/accessibility → involving all
- Interactivity → capturing ideas

It is clear from the above that VR can play a major part in achieving innovation in the industry. Ample VR tools are available, developed from the results of research or/and commercial activities around the world, that support this concept. The built environment can learn from the successes of VR in various sectors of engineering, which have so far been tremendous (Wilson, 1996; Boyd, 1998).

VM (virtual modelling) has been shown to lead to the avoidance of costly mistakes, and to enable planners and managers envision the whole manufacturing process from design and assembly to product shipping. Factory simulation has helped to produce substantial savings on tooling, design, construction and installation. Compared to the use of conventional methods, VM has also been shown to dramatically reduce the amount of time it takes to analyse new design concepts and incorporate them into the production process. It has enabled decision-makers to make last minute changes and eliminate the need to build prototypes (Quayle, *et al.* 2005). The application of VR has made it much easier for factory workers to accomplish complex and error-prone tasks, and has also offered a safer environment for testing various manufacturing techniques.

Increasing globalization is driving the construction industry towards increased competitiveness and improved efficiency. The current key business drivers were seen to be the need for: efficiency improvements; cost reduction; increased collaboration; ease of communications; involvement of the client and end-users in the whole construction process; accurate progress measurement; and modelling of the as-built environment. Although VR could offer significant benefits to the industry, the construction industry is not adopting VR to its full capacity when compared to the aerospace, automotive and other manufacturing industries. Current applications of VR are used mainly for client walkthroughs, design reviews and the visualization of construction sequences. The technology is being pushed by academia and/or developers, but the fragmented and adversarial culture in the industry has to be changed in order to develop the industry and make it more technology-based and agile. However, simple and easy to use VR applications are needed in the built environment industry.

Currently, no R&D is focused on the use of VR in construction products and processes. However, it is hypothesized that, in the future, the built environment will be more intelligent (self-diagnostic, real-time communication for facilities management, etc.), flexible, secure, sustainable and with high value to stakeholders.

The remainder of the chapter discusses the methodology used and results of the 'VR Roadmap: Vision for 2030' initiative.

19.2. Development methodology of the roadmap

A thorough review of both academic and industry literature was conducted to identify the state-of-the-art of VR R&D. This included current roadmap initiatives like FIATECH, RoadCON and the nD initiative.

The following briefly shows the current state-of-the-art of R&D in the construction business processes:

- *Initiation and outline design*: R&D is being mainly conducted into 3D walkthrough visualization models, augmented reality applications and communication systems of outline design information, urban visual planning (Miles *et al.*, 2004)
- *Design development*: R&D in the area of visual design, 3D/nD modelling and sketching, product models and integration with VR and search-based engines for finding optimal design solutions (Lee *et al.*, 2004; Khatab *et al.*, 2005; Tizani *et al.*, 2005)
- *Contract and pre-construction*: VR is mostly used in marketing and information visualization, e.g. cost, quantities, 4D models, etc. (Rischmoller *et al.*, 2001; Dawood *et al.*, 2005, 2006)
- *Construction*: R&D is focused on visual construction, 4D modelling, construction sequencing, image analysis, simulation modelling, etc.
- *Maintenance*: Very little work is being performed in this area, with most of the work focused on asset information, visualization and simple augmented reality applications

Table 19.1 gives a list of the state-of-the-art developments in visualization and integrated technologies in the built environment with respect to their intended functions.

As mentioned at the beginning of this chapter, this review was followed by brainstorming sessions and presentations by VR experts during the two-day workshop. The workshop participants have a wealth of experience in the use of VR technology, including: the development of hardware and software technologies; VR modelling; socioeconomic analysis on the use of VR; teaching and training; and the use of VR in industrial projects.

The workshop included presentations that highlighted current practices, and three main workshops were used to brainstorm and develop the 2030 vision for the application of VR and also a vision for R&D towards 2030. Discussions focused on identifying the enablers, barriers, opportunities and challenges that prevail in the industry today and those likely to be encountered between now and 2030 with advancement of the technology and process changes. The following section discusses the pre-workshop data collection and its analysis.

Table 19.1 Current applications of VR in the built environment.

Functions	Details
Collaboration/interdisciplinary work	Compilation of VR models to build prototypes to represent interdisciplinary work Development of user interfaces for designers
Standards	IFC development for CAD CAD and VR data exchange Linkages with GIS, IAI, IFC mapping
Whole-life costing analysis	Visual costing What-if' analysis Operation analysis
Modelling and design	Digital architecture – design spatial analysis Early design stages – integration of VR model, with structural model, collusion-checking model
Planning	4D modelling process information linked to VR objects Urban planning/GIS/VR integration
Progress measurement and modelling as-built information	Image capture Laser scanning technology for the built environment, integration with presentation technologies, object recognition
Communication	Visualization/walk-through 3D/4D/nD visualizations Mobile and Grid technology to communicate visual data

19.3. Pre-workshop data collection and analysis

Experts were asked to complete a semi-structured questionnaire prior to the workshop event. The objective was to elicit their knowledge and expectations of a future VR roadmap. This was important to ascertain a successful outcome of the workshop.

The following lists the summary points of the participants' expectations of the workshop (this needs to be tied in with the major results of the network):

- Identify novel ideas, processes, trends and developments of VR
- Network to know what others are doing and identify opportunities for research collaborations
- Develop a clear strategy to promote the real benefits of VR as a cost-effective real application, not just an expensive toy
- Gain an appreciation of current research projects and opportunities for liaison with international practice
- Obtain feedback and vision to improve VR software
- Raise the topic of VR standards

- Agree a future vision and an policy on the requirements and functionality of VR
- Develop VR applications in the workflow of construction and civil engineering projects

The main key points identified by the experts prior to the workshop were:

- Key tool for project delivery
- Aid collaborative design, concurrent engineering, digital architecture and fabrication
- Simple for use in communicating design and construction processes – new generation tool
- Object classification
- Focus on how to deliver simple, cost-effective solutions and its easy accessibility for architects/designers/contractors/sub-contractors
- Increase in stereoscopic and immersive 3D usage; 3D displays (distinct from pseudo-3D displays)
- A paradigm shift of how we do things, from CAD to much more 3D visualization
- Stronger link to the industry and R&D to deliver what industry requires

The experts were of the view that:

- VR is just being used as a visualization tool and that the use of 2D computer-aided design (CAD) is prevailing in the industry
- There is a need for industry and academia to work very closely and deploy the technology to its full capacity
- The cultural, technical and other socioeconomic barriers must be tackled to bring efficiency and enhance the performance of the industry
- The development focus should consider education, training of the industry and stakeholders as well as the development of natural interfaces
- A paradigm shift is necessary

The next section discusses the operations of the workshop and its main conclusions.

19.4. The workshop

19.4.1. Setting the scene

To inform the formation of a new VR roadmap, the research team at the University of Teesside, UK, reviewed various reports, such as Construction 2020 (Hampson and Brandon, 2005), FIATECH (2004), the EU IT Roadmap (Zarli, 2003) and various nD workshops (Lee *et al.*, 2003, 2005) and intelligent infrastructure documents (Foresight, 2005). Issues regarding the role of VR as enabling technology towards 2030, barriers, challenges and requirements were identified and presented to the participants. The aim was to introduce and familiarize the participants with current IT roadmap initiatives and direct them towards achieving the aims and objectives of the workshop. This was followed by three brainstorming sessions to tackle the development of both the vision and the roadmap. Experienced facilitators for each session were appointed and

Table 19.2   Technology specifications, opportunities and challenges.

Technology specifications	Opportunities	Challenges
Tools to be used in whole-life cycle of the projects	Cost dropping	Change
	Generic objects/components/ library – office metaphor	• Legislation
Full control/interactive		• Globalization
		• Competition
Object modelling a 'new way of thinking'	Parametric modelling/object modelling	• Generation shift
		• Energy/sustainability, production + efficiency
Communicate 3D/objects to project team	Feed data in different ways	• Generic/proprietary standards, interoperability
Use 3D from the beginning	Realism vs. artistic (game) depends upon task/views	• Archive data (file not found)
Link and provide design process functions such as sustainability, energy, maintenance, etc.	Embedded intelligence 3D plane	• Education/new knowledge social gap
	Free channel of communication (motion, sound, etc.)	• Drive technology not just use it
	User interface 'Selling the dream' Structure data	• Dynamics of software industry
	Language/technology Visual scale	Skills, training/education

all sessions were videotaped for further analysis. The following sections discuss the various workshop sessions.

19.4.2. Brainstorming Session 1: Development of the VR vision for 2030

The objective of this session was to collaboratively develop a VR vision for 2030 that would form the foundation for the proposed roadmap. For this session, the participants were divided into two groups. The first group looked into the development of a technology specification for the roadmap and identified challenges and opportunities (see Table 19.2). The second group discussed the development of process specification for the roadmap and identified challenges and opportunities (see Table 19.3).

The participants came to the conclusion that VR will play a major role in the design, management and operation of the built environment, embracing the 'cradle-to-grave' concept. VR will be linked to other applications with a natural interface. Following the advancement in mobile technology, construction processes will be more intelligent and will utilize VR technology. The drivers for change in technology and research towards 2030 will be:

- Customer understanding: client lead required to force change
- Public participation
- End-user involvement; urban and building development in planning process
- Government regulations, legislation
- Sustainability
- VR software to simulate, assess and analyse
- Incentive: low cost technology
- Competition to provide efficient and quality products/services

Table 19.3   Process specifications.

Process specifications	Challenges	Opportunities

• Globalization/localization
• Energy conservation
  /sustainability
• Defragmentation – integration
• Computerization –
  optimization of technology
• Common standard
• Diversity/flexibility
• Automated (semi) building
  manufacture
• Specialization/complexity
• Whole-life cycle building model
• Information management
  (representation and use) to match
  technology and process changes
• Product model should support
  process model
	• Data interoperability
	• Cultural issues
	• Scale of construction projects
	• (height, etc.)
	• Sustainability
	• 'Smart' materials
	• Design structure integrity
	• User perception/clients'
	  requirement capture
	• Health and safety
	• Security
	• Shifting professional roles and
	  responsibilities
	• Retain creativity
	• Maintain energy efficiency
	• Include maintenance of facilities
	• Education of construction
	  workforce
	• Legal concerns
		• Mass customization
		• Reality/VR integration
		• Efficiency drivers –
		  time, cost performance
		• Increased safety
		• Increased buildability
		• Alternative materials
		• Professional integration
		• Shortened life cycle

- Requirements for knowledge economy
- Internet and mobile technologies: e-submission, VR planning
- Increased partnering and collaboration
- Energy conscious stakeholders (designers, builders, occupiers, owners)

19.4.3. Brainstorming Sessions 2 and 3: The VR Roadmap – vision for 2030

The participants envisaged that the result of the roadmap in 2030 will be the creation and use of highly intelligent electronic products and processes within a virtual environment for performing all business analyses of the built environment, and which considers all aspects of the whole development life cycle. This will allow built environment stakeholders (owners, designers, users, supply chain, etc.) to develop, design, test, rehearse, procure, build and maintain their facilities in a concurrent and integrated fashion with high certainty of value, usability and very low environmental impact.

The roadmap is highly influenced by and focused on the future built environment, business drivers and challenges. In delivering a 'VR Roadmap: A Vision for 2030', the experts were asked to define what they foresee as the built environment of the future and they agreed on the following:

> sustainable energy – renewable, intelligent controls, smart materials, common standards, advanced security, self diagnostic & healing buildings/facilities, intelligent interfaces with buildings/facilities, 100% certainty in time, very high cost performance & quality & H&S (health & safety), space efficiency and flexibility

This view was important in order to focus the mind on the tools, methods and functions needed to deliver the future built environment. In parallel to this, the experts developed futuristic business drivers and challenges.

Figure 19.1 shows a matrix of futuristic construction business processes (scenario-based project planning requirement capture; automated design; integrated procurement, intelligent and automatic construction jobsite; and maintenance) against current state-of-the-art, future VR specification business drivers, challenges and R&D up to 2030. The figure provides a summary of the conclusions of the roadmap workshop sessions.

VR prototyping was seen as being much simpler and efficient when it is assisted by the advent of standard object models of building components and structural members. As ambience is improved and global standards are developed, natural interfaces and immersive 3D usage will be better able to capture the requirements of both stakeholders and users.

There will be a high take-up by industry in the application of augmented reality (AR). The paradigm shift from the use of 2D CAD today to 3D visualization will make it possible to provide visual simulation of the whole construction process, therefore enabling the simulation and testing of several scenarios in a VR environment before investment decisions are made. Moves towards full collaboration and potentially a polarization of the industry will take place. Integration of the supply chain into the whole construction process will take place, partially as a result of the adoption of modular construction methodology but also as a result of increased standardization. Seamless supply chain streams to expert construction operations will speed up the construction process, and improved efficiency/productivity will evolve from the capability to rehearse and simulate construction activities. Moves towards the integration of VR into university teaching courses were seen as being important in the further development of the knowledge and use of the technology in industry.

More intelligence will be incorporated into buildings, with improved control applied to their maintenance. The interior fitting-out process will be modelled with the full involvement of the end-user/client. Buildings will be constructed in a sustainable manner with 'smart' materials, and recognition given to the deconstruction process and the carbon footprint of the materials and components used during construction.

The view of the experts was that there would be a move towards the increased utilization of stereoscopic and immersive 3D usage as opposed to pseudo-3D displays. R&D activity into the extended use of VR and simulation in the town and road planning aspects of the built environment was seen as being capable of contributing major improvement in these areas. Finally, industry and academia will need to work more closely to develop simple methods for the use of VR tools required by the construction industry.

Academia has not yet convinced the industry of the benefits of VR. Little measurement of the undoubted benefits accruing from the use of VR platforms has been undertaken. Industry will become more aware of the proven value benefits inherent in the use of VR during the execution and the whole-life cycle phases of a construction project; for instance, in planning and measuring progress and in improved efficiency and collaboration. There are opportunities to be had from the use of ambient interfaces, improved standardization, image capture, end-user involvement in design, laser scanning, object recognition and other emerging technologies that can be seamlessly integrated with VR technology. The workshop participants saw the breaking down of the traditional construction approach as the greatest barrier to further development of the technology, with major opportunities to increase collaboration, improve communication and efficiency from the adoption of the technology.

	Initiation Scenario based project planning requirement capture	Design Development Automated Design	Contract Integrated procurement	Construction Intelligent and automatic construction jobsite	Maintenance Intelligent self maintaining operational facilities
Current State of the Art	Walkthrough 3D CAD Augmented Reality	3D CAD/Product Modelling	Method Statements/Tender Documents Marketing	Walkthroughs, PDAs Visualisation of construction sequences Image capture	Asset management tools Augmented/mixed reality applications
Future VR Technology Specifications	End users as designers Generic object blocks VR as a laboratory	3D Object Modelling Parametric modelling Energy efficiency Sustainable design tool VR linked to structural design VR models of interior and utilities	E-Automated Portals (On-line viewing of models, construction sequences)	Ambient Access Embedded Intelligence and interaction	Self Diagnostic systems Smart materials
Business Drivers	Realistic Client requirements Professional Integration Globalisation	Wireless technologies Flexible working patterns Multidisciplinary-collaborating teams Powerful computing power Ageing population	Internet technologies Mobile/Wireless/GPS/Grid technology Globalisation/localisation	Sustainability Construction Population educated in ICT Complex projects Scale of construction Integration of Product and processes 100% certainty in time, cost, performance and quality	Sustainability Technology Smart Materials Computer Power Educated Users
Challenges	Retaining Creativity Low cost Technology Personal Preferences User friendliness of the technology	Sustainability Social Context Intellectual Property Rights Security of Information Design friendly technology Common standards	Security IPR Educated and trained users Cost of the technology	Data protection, Security, Human and Natural Resources Educated and trained users Low cost and affordable technology	Aging population Renewal energy Sustainability Social Inclusion
Research & Development initiatives for the next 25 years	Natural Interfaces Haptic Interfaces Business case Thematic and Topological visualisation Knowledge Management Spatial reasoning	Haptic Interactions, Augmented and Mixed realities Standardisation Intelligent Objects/ 3D object modelling Spatial reasoning, resource optimisation Knowledge Management	Ambient Access Collaborating (virtual) teams/ Technology Value Engineering (i.e. Smart materials)	Ambient Access Standardisation Natural Interfaces Construction process simulation	Socio-economic use of VR Whole life cycle modelling Real-time integration and presentation Intelligent data models Simulation

Figure 19.1 'VR vision 2030': Research, development and application of VR in the built environment.

There was significant debate as to whether or not VR modelling was being pushed by academia or pulled by industry. The consensus was that it is still being pushed by academia, though there are some positive signs of change. VR technology is currently used mainly for client walkthroughs. There is an essential need to change the industry culture from a fragmented structure, with adversarial relationships still existing in some areas. The current key business drivers were seen to be the need for efficiency improvements, cost reduction, increased collaboration, ease of communication, involvement of the client and end-users in the whole construction process, accurate progress measurement and modelling of the as-built environment.

19.5. Research and development activities from now to 2030

The R&D activities identified to achieve the VR vision 2030 focus on the development of: VR hardware and software systems; common standards to ensure data interoperability; skills required for changing business requirements; and VR to include whole-life cycle analysis of the built environment. These R&D activities will bridge the gap between the current state and that of the 2030 vision, and will contribute to forward thinking processes. Figure 19.2 shows a summary of the R&D needed to deliver this vision. It shows the required R&D against each business process and function need.

The following summarizes the proposed R&D (Figures 19.3 and 19.4 give an elaborate summary of the roadmap):

- Product and process visualization
- Development of standard/flexible VR object models that are easy to access/use, web-enabled
- Exploration of XVL, X3D and other web-enabled VR
- Development of intelligent/visual search/agent-based tools to enable the outline and detailed design process
- Development of predictive and interactive tools for design and construction planning
- Development of gesture and tangible interfaces to communicate with the VR models to retain creativity and accurate requirement capture from users and stakeholders
- Development of tools for whole-life cycle (WLC) analysis of the built environment and infrastructure (cost, H&S, time, environmental impact, material modelling, security)
- Development of innovative product data manager
- Development of wireless- and Grid-enabled technology
- Development of visual site construction process simulation (space analysis, logistics, 4D, resource interaction, GPS (global positioning system) control)
- Development of real-time and visual control of facilities operations through 'smart' materials
- Development of cost/value analysis of visual/intelligent tools and methods
- Application of AR throughout the construction process
- Development of education and training materials
- Development of accurate image and video analysis for process control (progress/materials control, schedule monitor, etc.)

Process / Functions	Initiation Scenario based project planning requirement capture	Design Development Automated Design	Contract Integrated procurement	Construction Intelligent and automatic construction jobsite	Maintenance Intelligent self maintaining operational facilities
Visualization of Products	VR planning/walkthrough Common Standards Automated Requirement Capture	3D object modelling Common standards Design structure integrity 4D/nD modelling	Model-based Visualisations Integrity of the model Development of automated procurement systems	Real time image capture	As-built information capture and representation
Communications	Web-based model exchanges Mobile technologies Interaction between stakeholders	Links with structural analysis software Web-based design development Security	Building models with bottom-up approach (Inter-company) Component level details Wireless/GRID technologies	Construction sequence, Real time interactions Digital site	Real time status and maintenance information prompts Self reporting systems
Concurrency	Data Exchange Common standards	Decision support systems Information Access to all stakeholders	Data Exchange Common standards	Data Exchange Common standards	Data Exchange Common standards
Collaboration	Design reuse 3D model communication Link of 3D objects with Extended Enterprise functions	3D model communication Link of 3D objects with Extended Enterprise functions Collaborative structural analysis	3D model communication Link of 3D objects with Extended Enterprise functions	3D model communication Link of 3D objects with Extended Enterprise functions	3D model communication Link of 3D objects with Extended Enterprise functions
Tangible Natural Interface	User and design friendliness Application Programming Interfaces	User and design friendliness	User and design friendliness	User and design friendliness	User and design friendliness
Intelligent/Knowledgebased	Knowledge reuse Generic objects with intelligence	Knowledge Encapsulation and reuse, Generic objects with intelligence	Encapsulation, Knowledge reuse	Encapsulation, Knowledge reuse	Encapsulation, Knowledge reuse Linking acaptive systems to VR visualisation
Integration with smart materials and RFID	Interfacing knowledgebase of materials, specifications/properties	Interfacing knowledgebase of materials, specifications/properties	Interfacing knowledgebase of materials, specifications/properties	Real time data collection and visualisation Linking material managements/ stock management, Off-site manufacturing	Linking real time data from sensors and their visualisation Knowledgebase development Decision support systems
AR Application	Object models Parametric objects Business cases	Object models Parametric objects Business cases	Object models Parametric objects Business cases	Object models Parametric objects Business cases	Object models Parametric objects Business cases
Novel Interface	User friendliness User requirement capture Retaining creativity	User friendliness Speed and complexity of VR models Retaining creativity	User friendliness User requirement capture	User friendliness Links to extended enterprise functions	User friendliness User requirement capture Retaining creativit
Life Cycle Analysis	Intelligent Objects Decision support systems Cost and value analysis	Decision support systems Integration of design process functions (sustainability, energy, maintenance)	Decision support systems	Decision support systems	Decision support systems for maintenance of facilities
Safety Analysis	Rules encapsulation Automatic safety checks	Rules encapsulation Automatic safety checks	Rules encapsulation Automatic safety checks Rehearsed method statements	Rules encapsulation Automatic safety checks Method statement adherence	Rules encapsulation Automatic safety checks
Environment Impact Analysis	Rules encapsulation Availability of carbon-foot print information for decision making Urban regeneration	Rules encapsulation Use of smart materials database to produce internal energy requirements	Rules encapsulation Securing VR models, Procurement of minimum impact materials	Intelligent objects Rules encapsulation	Rules encapsulation Environment change control systems via VR database
Visual Process Simulation	Intelligent objects What-if analysis, value Engineering Design Concept Analysis (Outline design)	Intelligent objects Structural analysis such as FE linked to VR models Alternative design concepts	Intelligent objects, risk analysis Scenario-based evaluations, value Engineering	Objects with behaviours/Intelligence Simulation models for training , Optimisation of resource use (materials, energy)	Intelligent objects What-if analysis Simulation databases

Figure 19.2   R&D towards 2030.

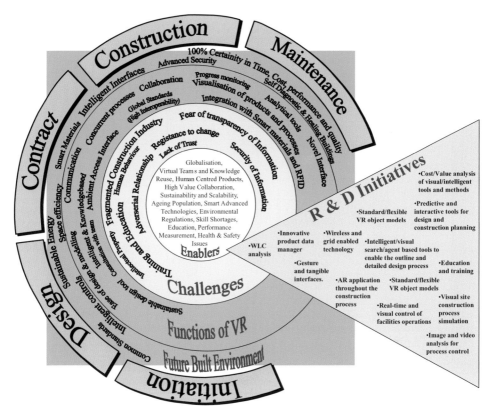

Figure 19.3   Abstract of VR-Roadmap, vision for 2030.

The industry structure will advance in 2030; many current barriers will not exist but new barriers will emerge. The experts envisaged that the built environment and its business processes towards 2030 will be different and will comprise the following characteristics and functions:

- *Semi-automated construction*: The design of buildings and infrastructure will be semi-automated, technology will be used to integrate supply chains and will routinely and seamlessly communicate all the product and process information. The 3D and VR models will encompass all parametric designs, and the design will be facilitated through the use of intelligent object libraries. The construction process will also be more or less automatic.
- *Integration*: Design and construction will run concurrently; decisions will be made prior to execution through integrated VR technologies, i.e. early in process of the project development. There will be coherent integration of product and process models.
- *Flexibility*: The industry will be more flexible to accommodate changes in the design or processes quickly with the support of advanced IT systems. The VR systems will be flexible to accommodate late changes and will embrace manufacturing philosophy, thereby providing better modelling.

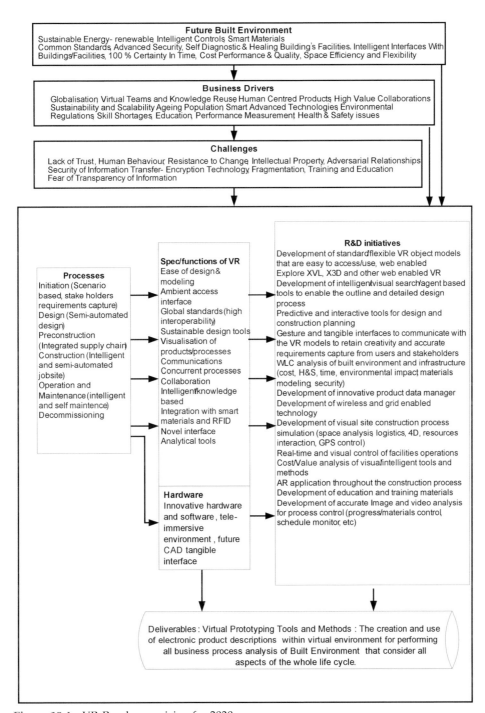

Future Built Environment
Sustainable Energy- renewable, Intelligent Controls Smart Materials Common Standards, Advanced Security, Self Diagnostic & Healing Building's Facilities. Intelligent Interfaces With Buildings/Facilities, 100 % Certainty In Time, Cost Performance & Quality, Space Efficiency and Flexibility

Business Drivers
Globalisation, Virtual Teams and Knowledge Reuse Human Centred Products, High Value Collaborations Sustainability and Scalability Ageing Population Smart Advanced Technologies, Environmental Regulations, Skill Shortages, Education, Performance Measurement Health & Safety issues

Challenges
Lack of Trust, Human Behaviour, Resistance to Change, Intellectual Property, Adversarial Relationships, Security of Information Transfer- Encryption Technology, Fragmentation, Training and Education Fear of Transparency of Information

Processes
Initiation (Scenario based, stake holders requirements capture)
Design (Semi-automated design)
Preconstruction (Integrated supply chain)
Construction (Intelligent and semi-automated jobsite)
Operation and Maintenance (intelligent and self maintence)
Decommissioning

Spec/functions of VR
Ease of design & modeling
Ambient access interface
Global standards (high interoperability)
Sustainable design tools
Visualisation of products/processes
Communications
Concurrent processes
Collaboration
Intelligent/knowledge based
Integration with smart materials and RFID
Novel interface
Analytical tools

Hardware
Innovative hardware and software, tele-immersive environment, future CAD tangible interface

R&D initiatives
Development of standard/flexible VR object models that are easy to access/use, web enabled
Explore XVL, X3D and other web enabled VR
Development of intelligent/visual search/agent based tools to enable the outline and detailed design process
Predictive and interactive tools for design and construction planning
Gesture and tangible interfaces to communicate with the VR models to retain creativity and accurate requirements capture from users and stakeholders
WLC analysis of built environment and infrastructure (cost, H&S, time, environmental impact materials modeling, security)
Development of innovative product data manager
Development of wireless and grid enabled technology
Development of visual site construction process simulation (space analysis, logistics, 4D, resources interaction, GPS control)
Real-time and visual control of facilities operations
Cost/Value analysis of visual/intelligent tools and methods
AR application throughout the construction process
Development of education and training materials
Development of accurate Image and video analysis for process control (progress/materials control, schedule monitor, etc)

Deliverables : Virtual Prototyping Tools and Methods : The creation and use of electronic product descriptions within virtual environment for performing all business process analysis of Built Environment that consider all aspects of the whole life cycle.

Figure 19.4   VR-Roadmap, vision for 2030.

- *Automated requirement capture*: Due to technological advancements, a common design standard and seamless integration of data, users will be able to input their requirements, thereby providing clients with a realistic briefing.
- *Single stop*: Products will be supplied by a single multidisciplinary company.
- *Integration*: Sustainable, virtual, distributed concurrent collaborative design will be achieved through networking, with most projects utilizing e-planning and that takes into account environmental considerations, aesthetic appearance and urban environment.
- *Educated and skilled workforce*: The industrial workforce will be educated and competent in using the technology. VR will be used as a training tool for engineers and the construction workforce.
- *Standardization*: A common standard for designing and communicating designs and associated applications will emerge. It is likely that CIS/2, IFC (Industry Foundation Classes) or a similar standard will be commonplace. Generic standard building blocks will be available for design, and applications will support ease of modelling to concentrate on design creativity.
- *Hardware interfaces*: Haptic interfaces, multi-modal interfaces, augmented reality, mixed reality and other advanced VR technologies will find their application in the industry for interaction with VR models.
- *Design*: VR will be used to examine design performance. The development of VR models will use natural sketches and informal and parametric VR. The design will be created using 3D intelligent objects. The design process will include whole-life cycle costing, value engineering analysis and health and safety issues.
- *Mobile technology*: Mobile technology will be advanced to a greater extent, thereby supporting an enhanced capacity for seamlessly transferring and communicating large VR models. Wireless and Grid technologies will be the enablers for the interaction, processing and visualization of information using holistic models.
- *Intelligent infrastructures*: VR will be a common interface for visualizing and interacting with the information provided by the intelligent elements of the building/infrastructures, for example sensors, transmitters and microprocessors. The intelligent elements will be embedded into structures such as walls, furniture, etc. VR will be used in the selection and design of 'smart' materials, thereby incorporating sustainability and other environmental considerations in a virtual laboratory.

Figures 19.3 and 19.4 show a more elaborate presentation of the proposed roadmap. Figure 19.3 shows the built environment business process against enablers, challenges, functions and future built environment. Figure 19.4 gives a more elaborate summary of the whole business processes.

19.6. Drivers for change

VR will play a major role in the design, management and operation of the built environment embracing the 'cradle-to-grave' concept. VR will be linked to other applications through a natural interface. Following advances in mobile technology, all the buildings constructed will be intelligent and will utilize VR technology. Eventually, the construction industry will follow the route of the manufacturing and aerospace industries where

VR modelling and simulation technology is used extensively. The drivers for change in technology and research towards 2030 will be:

- *Customer understanding*: Client lead is required to force change. Using visualization, clients will have a better idea of the project and they will demand a higher uptake and use of VR in projects as they are the people paying for facility. Clients and major firms will demand advanced VR technologies to evaluate design, construction and operational performances.
- *Public participation*: Public participation will be increased in the development of projects. VR will provide a tool for planning permission, consultation, end-user involvement and their links.
- *End-user involvement*: End-users can be involved in urban and building development during the planning process.
- *Government regulations*: Government regulations are getting tighter and include aspects of safety, environmental considerations, energy reuse and sustainability issues. The legislation on the use of technology will be a strong driver for change.
- *Sustainability*: Sustainability requirements – e.g. minimization of negative impacts on the natural environment, minimization of waste and the use of renewable resources – will force the industry to undertake several analyses before documentation, design and use of the facilities, and VR will be a common technology used in the construction industry. A procurement system will use sustainability as the main criterion, as opposed to the current, low cost bid criteria.
- *Simulation capability*: VR will be linked to other IT systems for environmental and life cycle analysis to inform decision makers in the design and construction of the built environment, and to simulate, assess and analyse different scenarios using different material components and different construction methodologies.
- *Incentive*: The development of low cost technology with its interoperability will increase the use of VR.
- *Competition*: Due to growing competition between the construction companies to provide efficient and quality products/services, VR will be used by the companies to be upfront and maximize savings by 'testing before construction'.
- *Requirements for knowledge economy*: There is a need to develop trust among participants in the built environment so that knowledge is shared using advanced IT for their mutual benefit.
- *Internet and mobile technologies*: Internet and mobile technologies are developing at a fast pace and their enhanced capability will make it easier to use VR models of any complexity. Therefore, e-submission of tenders and use of VR in planning will become routine. Mobile devices such as hand-held computers, personal digital assistance (PDA) and other equipment will be utilized on construction sites for data collection, project monitoring and quality control, thereby minimizing waste.
- *Partnering, collaboration*: Partnering and collaboration between companies requires the sharing of resources between all those involved in a project, and VR will be utilized better to exchange information. Development of international alliances between researchers and users of VR working in collaboration will lead to mutual benefits, as construction workforce skills will be developed and, therefore, the interactions between the stakeholders will be more effective.

19.7. Delivering the vision

The realization of the Vision 2030 requires the adoption of several steps by the industry and R&D. The key elements for achievement are:

- *Networking*: Members of academia, industry and technology should work together to identify the real needs of the industry and to develop customized applications and best practice examples of using VR.
- *Education*: Construction industry practitioners should be provided with education and training opportunities on the technology. VR should be an inherent part of the built environment courses, and technology transfer programmes should be developed and delivered to the industry. Best practice guides should be developed and shared across the industry.
- *Product and process model integration*: VR should be used in modelling processes and products and their integration to make informed decisions, from initiation to decommissioning of the project. The integration of products and process models is key to the increased use of VR.
- *Standardization*: Initiatives to develop standards need to be prioritized so that a common standard will be adopted to model products and process information and included with the object model. Initiatives such as IFC development is an example.

19.8. Conclusions

The objective of this chapter was to report on the development of a 'VR Roadmap: A Vision for 2030 for the Built Environment'. The methodology and process for delivering the VR roadmap was discussed.

The future thrust for R&D will be to make the technology simple to use and accessible/integrated to key new generation tools. It is hoped that these tools will aid collaborative design, concurrent engineering and provide digital design in the form of a single internet-based interface that retrieves related and shared data in the form of images and drawings, etc. and visualizes a project's whole-life cycle.

The key challenge in VR development is to retain individuality and artistic design/creativity, while retaining the capability and flexibility to capture and model user/client requirements. Finally, members of industry, academia and technology need to work more closely together in the future to develop simple to use VR tools to cater for the requirements of the construction industry.

References

Boyd, L. (1998) Digital factories. *Computer Graphics World* **May**: 45–52.
Dawood, N., Scott, D., Sriprasert, E. and Mallasi, Z. (2005) The virtual construction site (VIRCON) tools: An industrial evaluation. *ITcon* **10** [Special Issue: *From 3D to nD modelling*]: 43–54 [Available at http://www.itcon.org/2005/5].
Dawood, N. and Sriprasert, E. (2006) Construction scheduling using multi-constraints and genetic algorithms approach. *Construction Management and Economics* **24** (Jan): 19–30.

DTI (Department of Trade and Industry) (2000) *UK Business Potential for Virtual Reality, Executive Summary, Information Society Initiative.* London, HMSO.

Egan, J. (1998) *Rethinking Construction Report.* London, Construction Task Force, Department of the Environment, Transportation and the Regions.

FIATECH (2004) *Capital Project Technology Roadmap Report.* FIATECH.

Foresight (2005) *Intelligent Infrastructure Futures.* London, Office of Science and Technology.

Hampson, K. and Brandon, P. (2005) *Construction 2020, a Vision for Australia's Property and Construction Industry, CRC Construction Innovation Report.* CRC.

Khatab, A., Dawood, N. and Hobbs, B. (2005) Development of an integrated virtual reality decision support system (IVR-DSS) for outline design for R.C.B.S [Reinforced Concrete Building Structure]. *Construction Application of Virtual Reality Conference, CONVR 2005,* 12–13 September, Durham.

Lee, A., Wu, S., Marshall-Ponting, A. J., Aouad, G., Cooper, R., Tah, J., Abbott, C. and Barrett, P. S. (1994) nD modelling RoadMap: A vision for nD-enabled construction. Salford University Report.

Lee, A., Marshall-Ponting, A. J., Aouad, G., Wu, S., Koh, I., Fu, C., Copper, R., Betts, M., Kagioglou, M. and Fischer, M. (2003) *Developing a Vision for nD-Enabled Construction, Construction IT report.* University of Salford.

Lee, A., Wu, S., Marshall-Ponting, A. J., Aouad, G., Copper, R., Tah, J. H. M., Abbott, C. and Barrett, P. S. (2005) *nD Modelling Roadmap, a Vision for nD-enabled Construction Report.* University of Salford.

Miles, J., Cen, M., Taylor, M., Bouchlaghem, N., Anumba, C. and Shang, H. (2004) Linking sketching and constraint checking in early conceptual design. *10th International Conference on Computing in Civil and Building Engineering,* 2–4 June, Weimar, Germany, pp. 150–161.

Quayle, S. D., Taylor, S. M., Rennie, A. E. W. and Rawcliffe, N. (2005) From concept to manufacturing: effective use of CAD and FEA without compromising design intent. *The 6th National Conference on Rapid Prototyping, Rapid Tooling and Rapid Manufacturing,* June, Buckinghamshire Chilterns University College, High Wycombe, Bucks.

Rischmoller, L., Fischer, M., Fox, R. and Alarcon, L. (2001) 4D planning and scheduling (4D-ps): grounding construction IT research in industry practice. *Work Commission (W78) Conference, 2001: IT in Construction in Africa 2001* (ed. G. Coetzee and F. Boshoff), 30 May–1 June 2001, Mpumalunga, South Africa.

Tizani, W., Smith, R. and Ruikar, D. (2005) Virtual prototyping for engineering design. *Construction Application of Virtual Reality Conference, CONVR 2005,* 12–13 September, Durham, pp. 12–13.

Watts, T. G. A., Swann, P. and Pandit, N. (1998) Virtual reality and innovation potentials. *Business Strategy Review* **9** (3): 45–54.

Wilson, J. (1996) *Virtual Reality for Industrial Applications: Opportunities and Limitations.* Nottingham, Nottingham University Press.

Zarli, A. (2003) *RoadCon, Building a Roadmap for Research into ICT in Construction Report.* France, CSTB.

20 Future collaborative workspaces for the construction industry

Terrence Fernando

This chapter summarizes visions, scenarios and research challenges for creating future collaborative workspaces for the construction sector. The content of the chapter is based on the FutureWorkspaces roadmap project (www.avprc.ac.uk/fws) and MOSAIC project (www.mosaic-network.org), funded by the European Commission under the Framework V programme.

20.1. Introduction

A typical construction project will comprise 20 or more organizations, formed into a temporary project team. Such teams are likely to be a unique combination of partners for each major project. These organizations are geographically separated and have a pressing need to follow efficient project processes, set up integrated communication infrastructures and develop shared models of the project and the buildings they are constructing. The organizations involved in a construction project need to work as a virtual enterprise, using collaborative technologies to share information and to implement concurrent engineering principles with the view to reducing cost, time and improving their quality of work. However, despite the interest and efforts applied by leading companies, advanced virtual technologies used by the construction industry are still in their infancy. While there are clear drivers for encouraging the use of advanced virtual technology and collaborative workspaces in the construction sector, the speed of change is hampered by the following barriers:

- Organizational readiness for deploying advanced virtual technology
- Lack of understanding of business benefits
- Resistance to change
- High cost in developing and deploying appropriate collaborative workspace solutions
- Lack of business models for supporting collaboration between stakeholders
- Lack of understanding of collaboration models
- Lack of security and control for supporting seamless access to business information and tools
- Lack of integrated platforms which can provide seamless access to information, tools and people with appropriate security and QoS (quality of service)

- Lack of integration with existing enterprise solutions
- Non-intuitive nature of virtual interfaces and collaborative workspaces for performing real industrial tasks
- Difficulty in configuring and using current virtual environments for supporting design reviews of complex projects
- Difficulty in integrating current information sources with virtual environments without losing the integrity of sensitive data
- Lack of standards for creating interoperable software platforms for supporting the entire product life cycle stages with adaptable interfaces
- Lack of training for workers to use advanced technologies

Although there are many such barriers to deploying advanced virtual technologies, the following drivers can be considered as the ones that will drive the deployment of virtual technologies in the construction sector:

- Need to improve business processes to compete in the global market
- Increasing need for collaboration between workers in the same organization and between stakeholders from different organizations
- Company image for attracting potential clients and skilled workforce
- Importance of service-centric products
- Environmental/sustainability objectives
- Customer demand for smarter products
- Need for offering stimulating new work environments for workers providing improved quality of life
- Demand for implementing democratic processes in urban planning
- Need for considering safety and security issues (terror attacks, floods, etc.)
- Maturity and affordability of advanced IT products, such as graphics workstations and communication technology
- New generation of young workforce who are more IT literate than the older generation

Giving due considerations to drivers and barriers, this chapter presents envisioned future workspaces for construction with some scenarios. A roadmap is then presented to implement the envisioned future collaborative workspaces with key research challenges.

20.2. Characteristics of envisioned future workspaces for construction

In the future, there will be a growing demand for process integration between internal and external organizations involving all the project partners and suppliers. Thus information and communication technologies' (ICT) infrastructures will have to support transparent, reliable and flexible execution of cross-organizational construction processes through advanced distributed workflow systems with open and standardized information and process technology. Integrated technologies will allow the creation of virtual workspaces to support such distributed workflow systems based on advanced technologies such as the Grid, wireless communication and ambient interfaces. Construction networks will use distributed workspaces that are specific to their engineering domains.

These workspaces will support co-located or geographically distributed participants to work together to achieve a common objective. The ICT tools and the infrastructure will dynamically be configured to support this team of individuals to achieve their objective. The configuration of individuals, their tasks, roles and organization, the information they need and the tools and equipment they use will be used to create context-based collaborative workspaces.

Ultimately, engineers will be able to work seamlessly in their workspace environment with documents, scientific models and virtual prototypes, both alone and collaboratively with distant colleagues as if they were in the same room. Virtual and hybrid prototypes will be available as a means for engineers to design new products. They will access specialized services via intelligent and secure network infrastructures that can detect, predict and satisfy user demand at any time and any place through location- and device-independent applications, which are able to seamlessly migrate across network technologies. The computing power necessary for executing compute-intensive simulation tasks in real-time will be available through Grid technologies, transparently to the user.

In the future, flexibility and mobility will replace many fixed and scheduled ways of working. We will see wireless communication as a secure and realistic means of data transfer, which will facilitate the evolution of the organizational structure into one realistically allowing distributed and mobile workers. The 'distributed and mobile workers' group will expand beyond its current capability, with fixed workers interchanging freely with this group when the environment demands. Data storage must reflect the evolving structures of real and virtual spaces around the project team, allowing flexible allocation of resources. The concept of working in a virtual team will become more of a reality when flexible and mobile working with increased communication capabilities allows collaboration with many companies at once, communicating across multiple geographical locations with richer, more interactive virtual meetings. The workplace must accommodate these communication tools with allocated space and integrated technology.

The reduced cost of technological equipment will enable companies to implement interactive physical spaces, housing large embedded displays, networked furniture, wireless devices for tracking people and remote access to supercomputers, etc. The integration of technology with physical space will make the present computer systems and interfaces less visible or transparent in the future environment. Workstations, meeting rooms, touchdown points and videoconferencing suites will be fitted with high-speed broadband links with high levels of security. Narrow viewing angle monitors will be used alongside this secure network connection to facilitate confidential work in a mixed-business environment. Content in the environment will be created by user profiles and actions or their position or activity in the environment.

Human factor issues need to be taken into consideration throughout all related aspects of work environments. Understanding the characteristics, needs, limitations and likely behaviours of end-users and other stakeholders, and of the organizations in which they work, and the societies in which they live, will be vital for successful future workspaces. The collaborative engineering workspace of the future will be user-centric.

All devices, environmental conditions and input/output modalities will be adjusted to each individual. 'Smart' environments, 'smart' objects and 'smart' virtual humans will allow multi-modal interaction with human users. The real meeting places of the future will be continually changing spaces, with virtual environments encompassing many of the fixed advantages of working today. Project and social groupings will be re-created online, with dynamic groupings and chance meetings being virtually simulated. This will allow chance communication to remain, even though the people you would once have interacted with are no longer in the same office or building. Profiling of each user can be achieved by tagging and tracking in an intelligent environment, allowing personalization throughout the workspace. This personalization could include automatic log-on facilities, virtual work and social grouping, ambience control and health- and mood-monitoring. Both passive and dynamic inputs would be in place to achieve this personalization, to give each user a level of control in their environment.

20.3. Workspace types for virtual prototyping

As explained in the early sections, engineering companies need to operate as a virtual enterprise that includes a wide variety of different team structures engaged in a blend of planned and on-demand interactions, with contributions from knowledge workers operating both from mobile locations and fixed locations. In addition, the staffing on project teams is often constantly changing. These circumstances fuel the need for better integration of working environments for fixed and mobile knowledge work, and involve the development of systems that increase the flexibility and level of interactivity, as well as increasing the feeling of presence, trust and control in different types of teams.

Current tools and methods do not provide adequate support for many types of dispersed teams involved in knowledge work. For instance, neither videoconferencing systems nor desktop conferencing systems are able to provide the necessary support for current engineering design processes, where sketching and graphic communication is of primary importance. The support offered to complex, dynamic team structures by current video- and data-conferencing systems, is particularly poor. Therefore, there is a need to improve ICT infrastructure (low-cost, ubiquitous/wireless high-speed network) and underlying software systems to support workers to collaborate effectively.

New collaborative services and tools in the next generation collaborative working environments should address these issues. Large scale displays and tracking technologies can be used to improve the ubiquitous collaborative performance of highly dynamic teams within fixed workspaces. We propose four types of workspaces for supporting the design phase of a construction project in order to empower individuals, teams and communities: individual workspaces; co-located workspaces; distributed workspaces; and community-based workspaces. A brief description of these workspaces is presented below:

Individual workspaces:
One important challenge here is to explore how flexible and easy to use individual workspaces could be developed using virtual technologies. Members of the product

development team should be able to access the relevant product information, computing power, simulation data from anywhere, anytime, using the mobile technology. Such mobile workspaces should provide interfaces with various types of displays to empower users to carry out tasks depending on their needs and location (workplace, home, public place). For example, one might want to explore his/her design data in an immersive environment at his/her workplace and therefore may need mobile interfaces for freely exploring the product data in a 3D environment. One might be working from home and hence need access to a smaller display environment. Furthermore, another user might be temporarily working at a public place and hence need access to data in a secure fashion using devices such as micro-optical glasses.

Co-located workspaces:

Design teams, comprising various disciplines, meet regularly to discuss the evolving design. One key research theme in the area of a new work environment is to explore how such co-located meeting spaces can be constructed using large displays with access to various simulation and visualization tools. Such environments have the potential to empower individual team members to communicate his/her design ideas to the rest of the team, hence improving the quality of design review meetings. One of the key challenges in creating such an environment is the construction of ambient interfaces so that the members can interact with the simulation environment without heavy wire-based hardware interfaces. Mobile technology has the potential to create remote interfaces through optical tracking and gesture recognition techniques.

Distributed workspaces:

While it is important for the team to meet face-to-face to make certain decisions, most of the product development work is done in a distributed fashion. Therefore, the creation of flexible distributed workspaces, which can mimic the human interaction capabilities within a distributed workspace, is a challenging issue. Such meetings could be spontaneous or scheduled to discuss various design issues. However, the gathering of all the stakeholders for a scheduled meeting or contacting someone for an unscheduled meeting will require a flexible IT infrastructure to bring such parties to a single workspace from various remote places.

Community-based workspaces:

We hypothesize that, in the future, mobile workers will form their own expert communities which can provide complimentary or extra help to execute various projects or to solve multi-disciplinary problems. Through mobile technologies, workers will be able to get appropriate expert help anywhere, anytime. By creating a pool of reputed international experts, they will be able to identify the experts who can help them on demand to solve a given problem on time.

However, the creation of those workspaces, within an integrated product development environment, to support seamless interaction among workers is a challenging problem requiring a sophisticated underlying IT infrastructure. As shown in Figure 20.1, the distributed virtual product development should provide access to product data management systems, simulation tools and various workspace types in which to work as an individual and as a team. The members of the product team should be able to access the product information and simulation anytime, anywhere, to engage in their product development tasks.

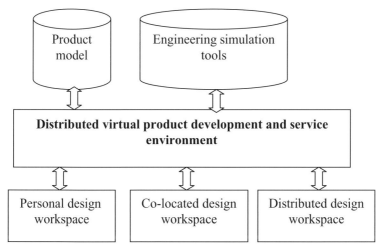

Figure 20.1   Workspace types for supporting virtual prototyping.

20.4. Examples of scenarios

The following two scenarios are used in this chapter to illustrate the envisioned future workspaces, with ambient interfaces and collaborative infrastructure with appropriate design tools.

Scenario 20.1: Distributed design work environment

Purpose: To illustrate the use of virtual technologies during the design phase of a product to improve flexible collaboration between teams.

Actors: Tom – self-employed FEA (Finite Element Analysis) consultant; Giovanna – design manager for a bridge

Script:
- Giovanna and her team of five designers, who are co-located at the recently opened collaboration centre, are preparing for a milestone meeting with their main customer.
- Instead of flying the entire team down, the milestone meeting is organized as a distributed session, by connecting the co-workspace installation at the collaboration centre and the matching facilities at the customer's headquarters in China with a large, interactive display for visualization and virtual tools for collaboration.
- During the preparations, one of the test engineers notices a mismatch between the results of two tests used to discover complex load conditions. The collaboration centre has been developed to support both planned and on-demand interaction between collaboration centre participants and others, and is thus well equipped to support situations like this.
- After discussing the possible consequences, the team decides to check if it is possible to run a new set of simulations to determine if it is necessary to change the

design or not. While her five designers are busy preparing the presentation and different physical and digital models they are going to show and discuss with their customer in 4 hours' time, Giovanna enters the 'External Team Resources' section of their digital project workspace by interacting directly with a large, interactive display. The optical tracking set-up installed around the display allows her to use her hand to directly interact with the virtual screen. She notices that Tom's status is 'personal device/online', and calls him by using her finger to press his virtual picture on the display.

- 'Hello, Giovanna', Tom says. She greets him back and asks if he has a couple of minutes. He responds positively, and Giovanna explains the situation to him, and asks him if it is possible to quickly check if the new load conditions will require any design changes to prevent failure. 'I'll see what I can do', Tom says, and leaves the session. Giovanna returns to her other tasks, comforted in the knowledge that Tom is taking care of the business.

- Some 25 minutes later, the communication system announces 'Incoming call from Tom, asking for video mode with shared viewing of simulation results'. Giovanna responds 'accept, engage in full collaboration'.

- The system adapts to her request and optimizes cameras and displays to support shared interaction with complex graphics. Tom appears in full size on the main communication screen, wearing his see-through 3D stereo glasses. The assuring look on his face is a great relief. Tom says 'Hello folks, I have good news for you. As you can see on the shared display, the new load conditions do not require a redesign. Under normal operating conditions, the design should still exceed the design requirement of trouble-free operation for the specified period by the client. You have the output files in the shared space; do you need anything else before your meeting today?' Giovanna and the others all thank him for helping them out, wish him a great, well-earned weekend and close the video session to continue preparing for their meeting.

- Giovanna looks at her watch, $3\frac{1}{2}$ hours to go. They might even have time for lunch and a coffee before the meeting.

Scenario 20.2: Co-located workspace

Purpose: To illustrate the use of high-tech co-located workspaces to support design reviews.

Actors: Hans – electrical engineer from the STAD; Luke – project manager from COWIT; Obiwan – air-conditioning engineer; Tom – main contractor; Leïa – architect

Script:
- In order to solve a potential space clash between the electrical subsystem and the air-conditioning subsystem, Hans requests a meeting with several project partners to find a suitable solution. Luke, the project manager, schedules a meeting at the design review workspace room at his office in Denmark. This room has been designed by COWIT to support both co-located design reviews and distributed design reviews.

- Designed as a multi-screen environment, it allows participants to visualize complex information on several display types such as Powerwall, plasma screens, etc. It offers participants the choice to visualize their data on separate screens or to superimpose them on a single stereoscopic screen. Various mobile interaction devices, such as optically tracked pens and hand gestures, flexible thin PDAs (personal digital assistants) and micro-headsets, are available in this workspace to allow the participants to interact with the data presented on the screens.

- Through mobile technologies and e-tags, the workspace automatically recognizes the people and their devices and offers automatic connections to the shared virtual workspace and their resources. All the documents, schemas and annotations produced during a session are saved on a virtual area for future reference.

- At the specified time, Luke, Hans, Tom and Obiwan go to the design review workspace for the meeting. They have been informed by Leïa that she will join them later since she has another meeting in the same premises. As they enter the co-located workspace room, their personnel devices are automatically connected to the shared virtual workspace, giving access to their data and simulation tools to support the design review. The access rights, depending on their role, also get established in the shared workspace environment for each participant.

- The work session begins. Luke explains the purpose of the meeting and invites Hans to present his concerns regarding the potential clash problems in the design between the electrical subsystem and the air-conditioning system. He brings out the 3D design on the Powerwall and navigates through the model to demonstrate the problem areas. The team discusses the problem and decides to make changes to the wiring paths and the ducting for the air-conditioning. Obiwan makes the suggested changes by re-routing the ducting and passes the new design to Hans. Hans runs a new clash detection on the model and asks the workspace to deliver the results in under 5 minutes. The underlying software of the design review workspace automatically identifies a super-computer that can deliver the results within the specified time and submits the task to the super-computer. In under 5 minutes, Hans receives the results and the new analysis shows that the new paths are now satisfactory.

- Meanwhile Leïa has found some 'idle time' during her meeting and has been monitoring the progress of the design review meeting through her mobile workspace, by reading the automatic meeting reports generated by the design review workspace. Oral indexing of the discussions helps Leïa to easily find and listen to all Luke's interventions. She is now fully aware of the ongoing discussions and Hans' concerns. After her meeting, Leïa arrives at the design review workspace and joins the discussion. She has automatically been detected and connected by the design review workspace with appropriate access rights and tools. Luke asks Leïa if she is happy with the proposed changes. She confirms that she is happy with the proposed changes after navigating through the design.

- Luke authorizes the new design and saves the solution in their PDM system. Luke concludes the meeting and all the participants leave the room. The design review workspace generates the final meeting report and saves it in the PDM system for future reference.

20.5. Roadmap for implementing future workspaces

This section summarizes the roadmap developed by the Future Workspaces project consortium to support the creation and deployment of the future workspaces envisioned in the previous sections. It has been defined over short-term (<2 years), mid-term (<5 years) and long-term (>10 years) periods. Over the short-term period, it has placed the emphasis on *integration* of the relevant research results for industrial implementation. Thus, the short-term target is mainly to consolidate the current technologies to develop the foundation for future collaborative workspaces and to reach a high usability level. The mid-term target is to build on the software integration efforts to improve *ubiquity* and mobility of the collaborative workspaces and to establish its deployment among all the stakeholders of an construction project. The long-term target is to achieve a higher degree of *intelligence* in self-learning and self-organizing collaborative workspaces to offer appropriate computing power, bandwidth and context-aware information to allow the workers to work more productively.

20.5.1. Short-term goals for collaborative workspaces

There are sufficient research results available right now that can improve the engineering workspaces of the present day, especially in their facilitation of collaboration between the various partners in an enterprise. Therefore, what is required in the short term is to bring these research results together to create industrial solutions in workable forms. Hence, the emphasis for the next two years is on the *integration* of the relevant research results for industrial implementation. Thus, this target is mainly a consolidation of current technologies to develop the foundation for future collaborative workspaces to reach a high usability level. It is reasonable to set the following short-term research targets, based on the current maturity level of the technology and the readiness of the industry to deploy collaborative workspaces:

- Aim to deploy collaborative workspace technologies as a standard tool for supporting collaboration between geographically separated teams who belong to the same organization.
- Establish a software framework that brings together deployable technological components and distribution services which support the easy creation of distributed collaborative workspaces.
- Deploy the advances in the communication technologies to provide transparent usage of networking services, with adequate quality of services for real-time interaction and a basic level of security for supporting inter-organizational collaboration.
- Establish a user interface framework which integrates various technologies such as audio, video, visual, haptic, advanced tracking, sensors, avatars, interaction metaphors which allows the workers to create easy-to-use personalized interfaces.
- Implement a 'workspace authoring toolset' to allow the easy construction of various workspaces by invoking various interface technologies, engineering simulation modules, engineering data flows and appropriate resources.
- Provide basic models, theories, systems and knowledge to better understand and design for human factors in complex distributed sociotechnical systems.

- Develop new and expanded business models and human-centred collaboration models, matching the progressive development of collaborative workspaces in detail. This is essential to the understanding of, and thus ultimately to the delivery of, business benefits.

20.5.2. Mid-term goals for collaborative workspaces

The mid-term target is to build on the software integration efforts to improve *ubiquity*, usability, mobility, context-aware and security features of the collaborative workspaces, and to establish its deployment among all the stakeholders of an engineering project. This collaborative workspace will support collaboration between multi-site workers, including mobile workers, at any time wherever they may be. This will require interfaces with inbuilt context-awareness and transparent access to resources that are adaptable to the individual user and his/her terminal's capabilities, with greatly reduced intrusiveness of the devices upon the users, thus embodying unrestricted mobility in these collaborative workspaces, with ubiquity as an essential underlying characteristic. This target is based on the vision of an evolution of the engineering organization taking advantage of expected ICT progress. The co-location of product integrated teams is still considered necessary for programme development, but the number of people involved in these teams is decreased significantly due to the remote involvement of the key project partners and supply-chain partners. Only a small team from each subcontractor will be fully involved in the product integrated team while a large number of subcontractor people will remain on their home sites and/or will be able to contribute to the decision-making process from a mobile or remote location. To support this distributed team activity, it will be necessary to set the following research targets:

- Aim to deploy collaborative workspace technologies as a standard tool for supporting collaboration between geographically separated teams who work for different organizations.
- Achieve an integrated network platform infrastructure that brings together various networking technologies.
- Employ the proliferation of IP-enabled wireless devices in a variety of environments (home, car, public transportation, office, shopping mall, conference centre, airports, etc.). This raises the challenge of supporting secure *ad-hoc* connectivity and reconfiguration in a multiplicity of environments.
- Establish network-independent and scalable services to access audiovisual materials, 3D and mixed-reality objects, distributed speech recognition over a heterogeneous infrastructure having variable bandwidth and quality of service availability with a high trust level.
- Support context awareness within collaborative workspaces. The ambient environment will be based on a large number of various sensors that track users' location, activities and behaviour. The interpretation of these data will allow the recognition of the users' context, allowing a contextualized presentation of engineering knowledge to support individual and team work. Fundamental work will be required to better understand the nature of the context awareness. The contextualized interpretation of the project team members will also be used to create a visual illusion of 'virtual project communities', encouraging social interaction and team work.

- Enhance the usability of the multi-modal interface technologies, and implement a range of ambient workspaces by integrating interface technologies (displays, sensors, wireless communication, speech recognition, etc.) with background objects such as walls, tables, chairs, etc.
- Understand true collaborative network models, methods to assess performance, attitudes and effects on individuals of all new methods of working and human factor guidelines for developing collaborative workspaces.
- Utilize validated models, theories and methods to reflect complex sociotechnical system needs, and the use of cognitive/physical artefacts in a social setting.
- Gain a better understanding of issues such as knowledge management, motivation to share knowledge, measurement and management of trust, effects of pace of change on performance, health and attitudes.

20.5.3. Long-term goals for collaborative workspaces

The long-term target is to achieve a higher degree of *intelligence* in collaborative workspaces which are self-learning and self-organizing, to offer appropriate computing power, bandwidth and context-aware information to allow the workers to work more productively. These workspaces will be served by intelligent agents in the background to enable the workers to be more productive. These software agents will act on behalf of the user for actions such as finding and filtering information, ensuring procedures are followed, detecting changes or anomalies, automating workspace tasks, establishing *ad-hoc* workflows, retaining process and decision-making knowledge. They will provide a fundamental building block for a new generation of workspace innovations especially well suited for ambient intelligence environments. The underlying technologies, such as interfaces, network bandwidth, computing power and security, will be dynamically configured to provide both the appropriate setting for the collaborative workspaces to cater for user preferences and quality of services for various collaborative sessions. This challenging target is based on the vision of a completely new engineering organization centred on intelligent workspaces. In this organization, the product integrated team becomes fully distributed: the network of 'business domain' teams becomes accessible and can provide knowledge at any time for any life cycle activity. A supervision team controls true engineering activities performed by a network of engineering competence centres (the engineering activities themselves become truly ubiquitous!). This organization is expected to reach a high level of agility (high reactivity to business opportunities), adaptation (ability to adjust to key business parameters) and self-awareness (dealing with risks in a dynamic manner), bringing the best answers to foreseen business constraints. The following research targets are set to achieve such intelligence:

- Aim to achieve complete deployment of collaborative workspaces, supported by trusted intelligent agents, among all the partners in a value chain, wherever located and in all stages of the product life cycle.
- Establish self-organizing capabilities (ability to find and incorporate resources dynamically in the network, ability to restructure the organization to face workload peaks).

- Establish self-configuration capabilities (the workspace adapts itself to the context and the team members depending on their role) with quality of service determined by willingness to pay.
- Implement transparent management of contractual issues (including IPR (intellectual property rights)) and security issues.
- Employ intelligent user interfaces based on biocentric control and systems that understand relevant human communication, including mental and emotional state as expressed through the observed person's body language and explicit spoken language through sensors, as well as generating output that is naturally understood.
- Implement adaptive training systems, supported by intelligent agents, embedded into the intelligent workspaces.
- Establish true distributed joint cognitive systems, engineering thinking and decision-making carried out by networks of people and computer working as a synergistic whole dynamic allocation function, whilst distributed in time and space.

20.6. Summary

This chapter has presented visions and scenarios for creating future collaborative workspaces for the construction sector. A roadmap with key research challenges over three landing places (short term, medium term and long term) were presented which will promote the creation and deployment of future collaborative workspaces within the construction sector.

Acknowledgements

This chapter is based on the Future Workspaces roadmap project (FP5-IST-2001-38346) and the MOSAIC (FP6-IST-2003-2 004341) project, funded by the European Commission during the Framework V programme. The author would like to acknowledge the European Commission for financial support and the following organizations for contributing to the final roadmap. A large number of people have contributed to the ideas and the roadmap presented in this chapter. The author would like to acknowledge the contributions of Prof. John Wilson (University of Nottingham, UK), Mr Gary Dalton (B, UK), Dr Manfred Dangelmaier (Fraunhofer IAO, Germany), Mr Pierre-Henri Cross (Cerfacs, France), Mr Yves Baudier (EADS, France), Mr Jens Skjaerbaek (COWI, Denmark), Dr Giuseppe Varalda (CRF, Italy), D. Hans Schaffers (Telematica Institute, The Netherlands), Mrs Liz Carver (BAE, UK), Sarah Bowden (Arup, UK), Marc Pallot (Eso-CENet, France), Alain Zarlie (CSTB, France), Kjetil Kristensen (NTNU, Norway) and Pekka Huovila (VTT, Finland).

21 The future organization: Sustainable competitiveness through virtual prototyping

Mustafa Alshawi

This chapter addresses organizational 'soft issues' that hinder or otherwise facilitate the successful implementation of virtual prototyping into design and construction work practices. It will examine the impact of the main four elements (people, process, work environment and IT (information technology) capability) on the creation of a work environment, within which virtual prototyping can be the main driver and enabler for sustainable business competitiveness. The need and requirements for future 'e-readiness' tools to measure the capability of organizations to successfully absorb virtual prototyping into their business will be addressed in the context of the 'future organizations'. This chapter also presents a brief futuristic scenario that portrays the alignment of advanced technology with the development of matching IT capabilities in organizations.

21.1. Background and introduction

There have been significant advances in the technological development of virtual prototyping (VP) both in hardware and software, including virtual reality environments, simulations, integrated databases (building information model), nD modelling, CAD (computer-aided design), etc. Elements of such technologies have clearly influenced work practices. In general, it is widely accepted that IT is becoming a key element of any organizational infrastructure. Indeed, many like to think that the level of an organization's reliance on IT in the twenty-first century is similar to the reliance on electricity in the previous century, where it was not expected that an organization would function without electricity. For example, networks, internet, e-mail and office automation are seen as standard applications for 'reasonable' size organizations. For small businesses, stand-alone applications such as e-mail, presentations and report writing are seen to be essential components for running any business. However, the picture is more complex where an IT infrastructure plays a key role in supporting core business functions within and among larger organizations. In this context, IT is being increasingly used to support business strategies as an enabler to leverage its potential to gain a competitive advantage and therefore new markets/clients.

However, there is ample evidence that the implementation of large information systems (IS)/IT has failed to bring about a competitive advantage to organizations in spite of the large investments over the past decade. A large percentage of IS/IT systems have failed to achieve their intended business objectives. Previous studies in the area of 'IS/IT failure' have shown that 80–90% of IT investments did not meet their performance objectives. Such projects were abandoned, significantly redirected, or, even worse, they were 'kept alive' in spite of their failure. The cost of funding such projects and the missed opportunities of not benefiting from their intended capabilities constituted a tremendous loss for organizations. This dissolution in the strategic benefits of IS/IT is currently forcing many not to invest in IT for any competitive advantage but for the reasons of bringing efficiency and effectiveness to business processes.

The main attributes contributing to the high percentage of systems' failure are rarely purely technical in origin. They are more related to the organizational 'soft issues' which underpin the capability of the organization to successfully absorb IS/IT into its work practices. IT is still, in many cases, considered by management to be a cost cutting tool (owned and managed by their IT departments). A 'technology push' alone, even though to some extent it is still dominant in many industries like construction and engineering, will not harness the full business potential of IS/IT and thus will be unable to lead to competitive advantage. Although the implementation of a few advanced IT applications might bring about a 'first comer' advantage to an organization, this will not last long as it can be easily copied by competitors. It is the innovation in process improvement and management, along with IT as an enabler, that is the only mechanism to ensure sustainable competitive advantage. This requires an organization to be in a state of readiness which will give it the capability to positively absorb IS/IT-enabled innovation and business improvement into its work practices.

21.2. Factors affecting IT capability: Organizational readiness

The competencies that an organization needs to develop in order to acquire the capability to strategically benefit from VP, prior to investment, fall into four main elements: people; process; work environment; and IT infrastructure. The first two elements are the key to change and improvement, while the other two are enablers without which the first two elements cannot be sustained. The 'acceptable' level of VP that can be successfully utilized in an organization, i.e. ensuring its business benefits are realized, depends on assessing a range of critical issues needed to ensure a balance between its readiness (mainly factors required to adapt to the proposed change) against the level and complexity of the proposed technology (which often hinders or limits success). This balance often includes many issues such as: capital expenditure; resource availability; the organization's maturity and readiness, culture and vision; and available IS/IT skills.

For an organization to achieve the required level of capability to address VP-based innovation, it has to:

- *Create an innovative work environment*. This should focus on developing and sustaining a highly skilled and flexible workforce, which will have the skills and the competence to introduce continuous improvement through better and more streamlined business processes enabled by advanced IT. In this context, organizational learning and knowledge management become a necessity for sustaining business improvements and the competitive advantage out of their VP investments.

- *Achieve effective alignment between business strategies and IS/IT strategies.* The focus should be on improving the organization's efficiency by directly integrating VP with the corporate, strategic and operational needs. This ensures VP resources are 'in line' with business imperatives.

In order to develop a VP capability, an organization needs to rethink its processes, structure and work environment. This necessitates the development of a 'forward looking' management tool that will enable managers to:

- Measure their current capabilities in the relevant areas, i.e. those that affect the development of the required VP capabilities.
- Predict the required level of change and resources to develop the target capabilities, i.e. identifying the organization's 'readiness gap' for developing and adopting specific VP capabilities.

This chapter first refers to previous predictive studies in industry and attempts to highlight their successes and failures. It then focuses on the organizational soft issues that hinder the successful implementation of VP in industry, and goes on to portray future scenarios where technology and organizational IT capability are integrated.

21.3. Previous technology forecast: Building IT 2005

In 1990, the Construction IT Forum carried out one of a few studies 'Building IT 2005' to predict the level of IT usage in the construction industry (CICA, 1992). The study failed to predict the arrival of global communication networks, although the internet was well established in universities. Then, five years later, the Forum reviewed its successes and failures and tried another 10-year forecast in a report launched at *Interbuild* in 1995 (Alshawi, 1995). A number of other technology forecasts for construction were launched at about the same time, including Technology Foresight studies by the UK Government Office of Science and Technology and 'Construct IT – bridging the gap' carried out by Arthur Andersen for the then DoE (Department of Environment) Construction Sponsorship Directorate (UK).

Howard (2005) examined the outcome of the 'Building IT 2005' study and evaluated its outcome in the context of the present situation. The conclusions of the Building IT 2005 study were aimed at four groups: clients; government; the construction industry; and the information technology industry. Table 21.1 shows a few quotations drawn out of the recommendations of the 'Building IT 2005' study (Howard, 2005) and evaluated with the present situation.

Typically, developments in IT (such as mobile communication, building information model (BIM), intelligent agents, graphics, etc.) have developed faster than expected, while those dependent upon human and organizational change are taking longer. However, one catalyst for the fundamental reappraisal of the construction process, particularly in the *Rethinking Construction* report, was the 'feeling' that technology was not being fully exploited due to the way the industry was organized. In this context, the recent success of UK construction can be attributed, in part, to the organizations that were set up to integrate different groups, to the sharing of information more effectively

Table 21.1   Building IT 2005 conclusions (Howard, 2005) versus the present situation.

Quotes from Building IT	Present situation
Clients	
IT has already contributed to the 30% cost saving target but faster, more efficient data exchange via partnering could add to this	It has still not been established what gains in productivity are due to IT, but partnering is being used and project extranets can increase the speed and efficiency of communications
Establish metrics for monitoring costs and benefits of IT investments and benchmarking clubs to raise firms to levels of best practice	The key performance indicators were established by constructing excellence covering most aspects of construction, but good data on the benefits of IT is still hard to find
Construction information should be linked via network gateways and that on Internet classified using UC/ci via search engines	Information such as product data is now widely available on the web, but is in diverse formats
The technology and ergonomics of work away from the office need to be studied with the possibilities of the virtual project team	Communication technology is available, but has not yet created the virtual project team
Construction industry	
Standard product descriptions and object libraries need to be defined by users to help model buildings for use in any CAD system	There are protocols for defining objects for the single building model, but the proprietary methods of CAD suppliers make limited use of standards, e.g. Industry Foundation Classes
The ergonomics of new communications technology which is portable requires study of how it will be used to capture data on site	'Picture phone' technology has superseded all expectation; wearable computers on suit or helmet are still awaited
The growing quantity of information available requires development of intelligent agents for filtering this to suit each user	Software is now available to filter out unwanted information and prioritize users' needs; but, network security and trust have become a greater problem than anticipated
Information technology industry	
VR to be used, not only for visualisation, but also for environmental and other forms of analysis of the performance of buildings	Virtual reality is available, but is mainly used to sell projects; research continues into modelling human behaviour and analysing performance
Cheap diskless computers could be widely deployed and dependent upon networks for current software and data	The 'thin' client offered economies and more central control of users, but the rise of graphics put excessive loads on central servers
To encourage wider usage, restrictions on the UK costs of broadband should be relaxed for conformance with the rest of Europe	Broadband connections to the internet are now offered across much of the UK and prices have started to come down through competition between suppliers

on the global network now available to all firms, large and small, and to changes in the procurement process.

21.4. Difficulties in integrating business and IS/IT

In the UK, the government, industry and clients are all seeking to bring about change in the construction industry to improve quality, competitiveness and profitability, and to increase value to clients. The implementation is being undertaken through initiatives such as the Construction Task Force (Egan, 1998), the Government Construction Clients Panel (GCCP), the activities of the Construction Industry Board (CIB), the Construction Clients Forum (CCF) and other CIB umbrella organizations. These initiatives are established to secure a culture of cooperation, teamwork and continuous improvement in the performance of the industry. Where the emphasis has traditionally been on the need to manage the interface between the project and the client's organization, it is now shifting towards the need to manage the flow of activities across the supply chain, concentrating on those activities that actually add value. Activities that facilitate the management of project information need to be identified and streamlined. This is not an easy task in construction, simply because of the project-oriented nature of the industry.

The construction sector operates in a wide range of sizes and types of organizations. It deals with an enormous range of projects in terms of size and complexity. The sector also draws on skills and resources of a highly varied nature, from highly skilled designers to non-skilled site workers, pulled together from different organizations and with different experience and backgrounds, working in teams to deliver one project. The design stage involves a highly professional design team comprising members such as architects, project managers, structural engineers and mechanical and electrical engineers. This is a highly information-intensive stage, where more than 80% of the project information is generated. However, much of the assembly process and the materials' supply part of the construction process creates much less information and is highly vulnerable to changes in, and exchange of, information. The complexity of the different parties and the lack of communication between them is one of the major problems that face the construction industry.

In order to understand the cause of the problem, a simple process-focused analysis of the current situation is required. Figure 21.1 represents the hierarchical structure of a typical organization. Each organization has strategic or core business functions which identify its speciality and the type of activities it undertakes. Each core business function is accomplished through a number of high-level functions at either a department or project level. These high-level functions are of two types: business functions and supporting functions. The former has the responsibility of achieving the core business objectives by directly executing activities to that end. The latter is necessary to facilitate the achievement of the former, including administrative and marketing functions. Each high-level function is accomplished by a large number of low-level processes, which are mainly concerned with low-level data capturing and manipulation. The low-level processes of each high-level function are normally isolated from the other low-level processes of a different function, but they both are considered to have a 'client–customer' relationship. The variation in the execution of these processes increases with their location in the hierarchy, i.e. the greater the variation at operational level.

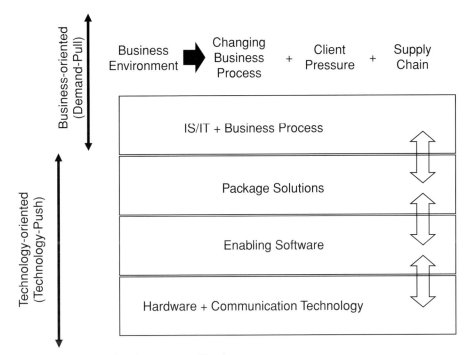

Figure 21.1 Hierarchical structure of business processes.

To illustrate the above, consider a design office with a vision of providing clients with a 'one-stop shop' for the design phase. The core business functions for such an organization could be: architecture; structural; and mechanical and electrical design. Each of these core business functions is conducted through a representative department which runs its own specific processes, and is very likely to have its own future plans and targets to achieve in isolation from other departments. Although there is a strong functional relationship between the three departments, as the outcome of one department is the input for the others, the level of collaboration between them or the integration of their business processes is very low. Typically, the relationship that exists between them is more of a 'client–customer' type. The processes of one department are very likely to have been created based on the inherited experience and backgrounds of its employees, with little or no integration with other departments. Therefore, communication between the various departments is normally limited to the exchange of 'formal' documents.

The picture becomes more complicated when IS/IT is considered. In the construction and engineering industry, investment in IS/IT over the years has resulted in isolated applications, in various departments, taking place at various times. Such decisions are normally taken to automate a certain function in a department with the aim of relieving an operational bottleneck. Such a scenario leaves an organization with different IS/IT systems, which are incompatible and extremely difficult to communicate within one department let alone throughout the organization. However, with advances in the internet, organizations have started to use the internet infrastructure as a communication platform between the various IT applications (see Figure 21.1). Although this type of communication is beneficial to organizations in improving the speed of communication, it adds little value, if any, to improving business performance. For example a 'building drawing' sent from the architecture department to the structural engineering department through an e-mail attachment will add no value to how designers in both departments can work together; it will add no value, for instance, to finding a solution to problems like the coordination between architecture and structure drawings.

A construction project which involves a large number of partners with differing core business functions and IS/IT applications makes improvement to the flow of project information across the supply chain extremely difficult. In addition, as partners work together for a limited period, i.e. over the project's duration, long-term business improvements across the supply chain through IS/IT (beyond the exchange of documents over the internet) are extremely difficult to achieve.

21.5. Business dynamics and technology

In spite of the above, many construction organizations have started to use IT, not only for performance improvement and cost reduction but also to utilize the potential of IT to open up new markets and/or gain an advantage over their competitors. With a better understanding of IS/IT at the executive level, IS/IT strategies are being aligned with business strategies with the aim of deploying advanced systems in support of innovative business processes that best achieve business objectives. This alignment process requires a careful and balanced approach between the level and complexity of the enabling technology and the required level of (expected) process change within the organization. Achieving this balance is a difficult process, requiring highly skilled professionals who fully appreciate the strategic needs of the business and the benefits and functionalities that advances in IT could bring about to achieve the business strategy.

21.6. Relation between business process and technology: The five-layer model

To explain the integrative nature of construction business and technology and to highlight the key organizational elements which underpin this concept, a five-layer model is presented (Figure 21.2). This conceptual model clearly demonstrates the relationship between the dynamic nature of business and the supporting IT infrastructure.

- *Layer 1 – Business environment*:
  In a project-oriented industry, like construction and engineering where projects are at their core, business processes are highly dynamic and can vary from one project to other, even in the same organization. The roles and responsibilities of project partners are defined by the type of procurement method that legislates the relationship between members of the supply chain. For example, the processes of designing and constructing a project in a 'partnership' agreement are different from those outlined in a traditional contract. This variation in business processes, within and among partner organizations, will affect not only the internal performance of partners but also the efficiency of the communication process between the partners. This complexity reflects the nature of the construction and engineering industry, which makes it difficult to align IS/IT with business strategies to achieve a competitive

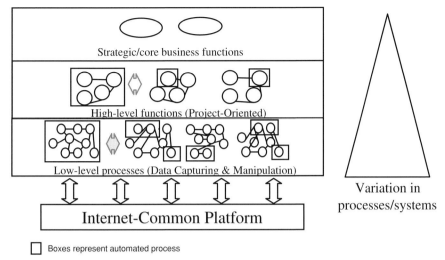

Strategic/core business functions

High-level functions (Project-Oriented)

Low-level processes (Data Capturing & Manipulation)

Internet-Common Platform

Variation in processes/systems

☐ Boxes represent automated process

Figure 21.2   Integration between business and IS/IT.

advantage. This is reflected by the top layer in Figure 21.2, i.e. changing business requirements.

- *Layer 2 – IS/IT and business processes*:

  IS/IT systems are deployed in organizations either to support their business functions or to bring about innovation and competitive advantage. This deployment will lead to the integration of IS/IT into their work practices, which can be accommodated by:

  – introducing new business processes which are enabled by advanced IS/IT, i.e. can only be realized by the deployment of IS/IT; or
  – aligning the existing business processes with new functionalities of the IS/IT system.

  The second layer in Figure 21.2 reflects the strategic business decision on the role of IS/IT in achieving the organization's business strategy, and whether the organization has the capability to successfully integrate the proposed IS/IT into their business processes. Thus, this is an important layer which portrays not only the capability of the organization to successfully deploy IS/IT but also the capability of IS/IT to bring about its improvement and competitive advantage.

The first two layers are business-oriented and highly affected by the 'demand–pull' mechanism. The effective utilization of IS/IT requires highly skilled business managers who understand both the internal needs of the organization and the external demands on the business. In addition, they should have the capability not only to make decisions on the selection of the appropriate strategic IS/IT projects but also to assess the required changes in business processes as a result of the deployment of the new IS/IT. Such decisions are critical to achieving success and maximizing the benefits of IS/IT to the organization.

- *Layer 3 – Package solutions*:

  The third layer represents the software packages that support decisions made at the second layer. Once a decision is made to invest in improving a particular business

process (strategic, core, or non-core), the next stage is to implement the most appropriate software solution in support of this process. This would involve decisions such as whether the software solution should be:
– internally developed and implemented
– a commercially available package, or
– outsourced to a third party
In all cases, such decisions are technology-based and require highly skilled IS/IT people with a good understanding of IS/IT implementation strategies. This layer is the main link between business decision-makers and technology professionals, i.e. the link between the second and fourth layers. It is also prone to the 'demand-pull' mechanism, as any selected IS/IT solution should meet the functions and specifications identified in the upper layer. It is important to mention that at this stage the 'technology-push' mechanism also plays a key role, as technology might identify new opportunities to further improve or even re-engineer the identified business processes through its advanced capability.

- *Layer 4 – Enabling software*:
  Once the software solution/package is selected, whether bespoke or commercially available, a full development and implementation plan should be established. This would include the design, development and implementation of the various components of the IS/IT solution, such as databases, application software modules and communication protocols. Decisions of this nature require experienced IT professionals with project management skills, who can optimize the use of the organization's resources, such as the existing functionalities and capability of the IT infrastructure. This could also include the ability of the system to adapt to the new requirements.

  The 'enabling software' layer represents this process and shows that decisions of this nature are strongly affected by 'technology-push'. Technology is normally evaluated and its impact on achieving the project's objectives assessed. This layer creates a strong link between the software development and implementation, and the hardware and communication technology.

- *Layer 5 – Hardware and communication technology*:
  The last layer in Figure 21.2 is the 'enabling hardware' layer. This layer represents the design and implementation of the hardware infrastructure which is capable of delivering the proposed solution effectively and efficiently. This requires highly skilled professionals in hardware and communication technology, such as types of network, internet communication and hardware specifications and performance. This layer is very much subject to 'technology-push', which considers the best technological solutions that are reliable, flexible and cost-effective.

This five-layer model highlights four main elements that could influence the effective selection, development and implementation of IS/IT. These elements are: people and skills; business processes; IT infrastructure; and work environment.

21.7. Building IT capability

The previous section has clearly demonstrated that 'technology-push' alone, even though to some extent still dominating in many industries like construction, will not lead to competitive advantage. Although the implementation of advanced VP

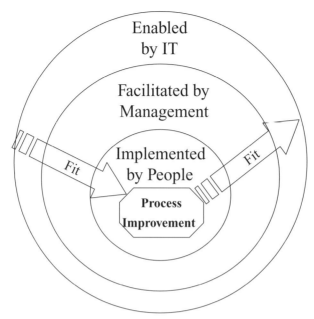

Figure 21.3   The four organisational elements for IT success.

applications might bring about 'first-comer' advantage to an organization, this will not last very long as it can be easily copied by competitors. Innovation in process improvement and management, along with IT as an enabler, is the only mechanism to ensure sustainable competitive advantage. This requires an organization to be in a state of readiness, which will give it the capability to respond positively to innovation in business improvement and advancement in IT.

'Organizational capability' is defined as the ability to initiate, absorb, develop and implement new improvement ideas in support of an organization's objectives. Kangas (1999) and Moingeon *et al* (1998) referred to organizational capability as the strategic applications of competencies. The development and deployment of specific organizational competencies is the process of building organizational capability. In the context of VP and organizational capability to ensure sustainable competitive advantage, competencies refer to many factors, such as highly flexible skills, awareness of change, flexible management structure, process improvement schemes, clear business goals as well as advanced and flexible IT infrastructure.

The competencies that an organization needs to develop in order to acquire the capability to benefit strategically from VP falls under four main elements: people; process; work environment; and IT infrastructure. These elements are highly interrelated, i.e. developing competencies in one element must be accompanied by improvement in the others. Figure 21.3 shows the relationship between the four elements. Process improvement is shown as the core competency that an organization needs to develop to achieve the VP capability sought. This element needs people with the necessary skills and power to implement process improvements. That, of course, cannot be undertaken without the management's consent and the creation of an environment that facilitates the proposed changes through activities such as motivation, empowerment and management of change. A high level of integration between the three elements can be enabled by flexible and advanced IT infrastructure.

The first two elements (people and process) are the key to change and improvements, while the other two elements (work environment and IT infrastructure) are enablers, without which the first two elements cannot be successfully implemented. Through good and effective management and an advanced IT infrastructure, an attractive and innovative work environment can be created where:

- People are motivated, empowered and made aware of the expected change. Thereby, they will be ready to innovate, absorb new ideas, and develop and implement them effectively.
- Business goals and improvement targets are clearly communicated to employees, with strong support from top management.

The time required for an organization to build up a VP capability is highly dependent on the level of maturity of the organization in each of these elements.

21.8. Measuring organizational readiness

The above analysis demonstrates clearly that a successful implementation of an innovative system such as VP in organizations is highly related to their ability to absorb and integrate the proposed systems into their current practices. Investments in such projects should be in support of business objectives. The success of these projects can not simply be measured by the successful installation of the system or by the level of users' satisfaction, because even a fully operational system with satisfactory functionalities can be alienated in organizations and not used by employees. In order for such systems to achieve their intended objectives, they have to be fully integrated and absorbed into the organization's current work practices and be accepted and supported by its employees and management.

The two critical elements that can significantly influence the level of IS/IT project integration into organizations' work environments are:

(1) *Process alignment*: Ability to align the organizations' processes with the proposed functionalities
(2) *People*: Ability of professionals to accept and adapt to the new system quickly and efficiently

These two elements are enabled by: (a) the level of awareness of management and employees of the benefits (operational and strategic) that VP can bring to businesses; and (b) their experience and know-how on how IT, and VP in particular, can be utilized to achieve such benefits. The issues are underpinned by two capabilities:

(1) The presence of a work environment that supports, motivates and empowers employees to innovate and seek improvement changes in line with the business objectives.
(2) The availability of advanced IT infrastructure that allows employees to develop an appropriate level of experience and understanding of the capability of the hardware and software to improve the performance of their current business processes, and to achieve their department's or corporate goals.

Figure 21.4   Domains of readiness.

In order for an organization to achieve that level of capability and thrive, it has to carefully develop and implement plans for the creation of a work environment for innovation and continuous improvement. The focus should be on developing and sustaining a highly skilled and flexible workforce, one that will have the skills and the competencies to continuously introduce improvements through better and more streamlined business processes enabled by advanced IT.

However, managers need guidance to effectively address the organizational soft issues prior to IS/IT investments. Such requirements could lay the foundation for developing a balanced measurement model, with the aim of identifying the current and the target status of organizational readiness to successfully absorb new information systems into their work practices. Readiness models (Alshawi, 2007) should have appropriate structures to embrace the four key elements: process; people; IT infrastructure; and work environment (see Figure 21.4). Levels of readiness (both the current and target) should be clear and indicative of the organization's situation in terms of measurable attributes and in maturity-like levels.

The rationale behind the readiness models are that for a new system to be successfully implemented in an organization, the four key soft elements need to be at a level of readiness which is adequate for that particular system. For example, it is extremely unlikely that an organization based on a heavy hierarchal management structure will successfully utilize internet-based management systems which are designed to ease the communication and the decision-making process between the different stakeholders. Also, an organization that tries to overcome the large backlog and heavy maintenance load of its IS/IT is unlikely to be able to develop substantial strategic information systems.

21.9.   Future implementation of VP

Lessons learnt from previous technology forecast studies indicate that technology development needs to be carefully considered together with the development of matching capabilities in organizations. It is important that benefits from innovative technology, such as VP, are fully taken advantage of through the ability to adapt to new business practices. In this context, successful implementation of VP, i.e. maximizing business benefits through the implementation of this technology, will have an impact on many

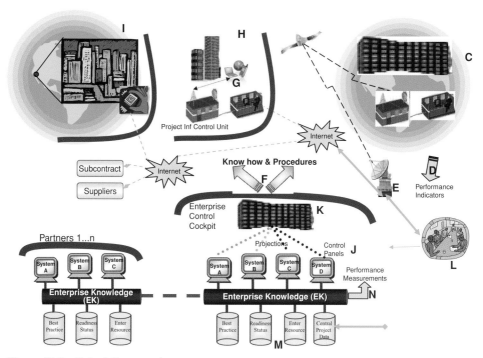

Figure 21.5 Futuristic scenario.

factors, such as the current organization of the industry, IT processes, roles of professions and procurement methods. Given the current level of industry awareness of this
issue (both construction and IT industries), it is expected that more effective solutions
and approaches will be developed in the future to enable not only individual organizations to improve their IT capabilities and performance but the entire supply chain to
work together in order to maximize overall performance and product quality.

Figure 21.5 portrays a futuristic scenario which brings together technology and the
work environment in a holistic picture. It is expected that VP will be the core of construction projects in future. Through its product model (which will be based on the
implementation of industry standards), project information will be effectively shared
and exchanged. The VP environment will not be considered as a technological tool as
such, but it will be an integral part of agreed business practices which all project partners will implement. The VP technology will not only help in the exchange of project
information and documents at the interface of the various projects phases, but it will
enable common business processes between the various professions to take full advantage of the available functionalities provided by the virtual prototyping environment.
This will be achieved through better partnering agreements and an agreed level of IT
readiness in all of the main soft elements, i.e. process, people, work environment and
IT infrastructure.

The heart of future construction projects will be the 'Enterprise Control Cockpit'.
This will be a graphical intensive environment which will be supported by the various elements of the virtual prototyping technologies, i.e. virtual reality, simulation,
databases, etc. The core element of the virtual prototyping technology will be the

'Enterprise Knowledge Database', which will provide the business intelligence to the VP elements by interpreting and analysing the organization and project information such as best practices, readiness gap(s), enterprise resources and project information. The result of this process will be:

(1) *Performance management*: A full analysis of the current and expected performance of the project (design and construction)
(2) *Graphic representation*: Representing all the project's phases, including simulation
(3) *Resource management*: Managing enterprise resources effectively, including internal capabilities to use the technology
(4) *Communication management*: Managing communication and project performance and resources among the supply chain partners

The 'Cockpit' will enable professionals (within and among partner organizations) to interact with the VP environment to facilitate running alternative scenarios, test 'what-if' situations, etc. Due to the ability to communicate with construction sites through the internet, project partners can assess site productivity and performance through the graphical simulation provided by the virtual prototyping environment. Partners can either be in the same physical place or be linked through virtual meeting spaces.

21.10. Conclusions

Developments in technology (such as mobile communication, BIM (building informa-tion model), intelligent agents, graphics, etc.) have developed faster than expected, while those dependent upon human and organizational change are taking longer. 'Technology-push' alone, even though to some extent it is still dominant in many in-dustries like construction, will not harness the full business potential of advanced technologies such as VP, and thus it is unable to lead to competitive advantage. Inno-vation in process improvement and management, along with IT as an enabler, is the only mechanism to ensure sustainable competitive advantage. This requires an organi-zation to be in a state of readiness, which will give it the capability to positively absorb VP-enabled innovation and business improvement into its work practices.

This chapter has highlighted the critical elements that can influence the level of success of VP in construction. These elements are business processes, people, work environment and IT infrastructure. The chapter has also outlined a futuristic scenario which brings together technology and work environment in a holistic picture.

References

Alshawi, M. (1995) *Virtual Reality in Construction, Building IT 2005*. London, CICA (Con-struction Industry Computer Association). [CD-ROM publication]
Alshawi, M. (2007) *Rethinking IT in Construction and Engineering: Organisational Readiness*. London, Taylor and Francis.
CICA (Construction Industry Computer Association) (1992) *Building IT 2000*, Sutherland Lyall. Manchester, CICA. [http://www.cica.org.uk]

Egan, J. (1998) *Rethinking Construction: The Report of the Construction Task Force to the Deputy Prime Minister.* Norwich, Department of the Environment, Transport and the Regions.

Howard, R. (2005) Will the future ever come? In: *Real Estate & Construction Annual Reviews.* Southfield, MI, Construction Communication.

Kangas, K. (1999) Competency and capabilities based competition and the role of information technology: the case of trading by a Finland-based firm to Russia. *Journal of Information Technology Cases and Applications* **1** (2): 4–22.

Moingeon, B., Ramanantsoa, B., Métais, E. and Orton, J. D. (1998) Another look at strategy–structure relationships: the resource-based view. *European Management Journal* **16** (3): 298–304.

Index